Environmental Chemistry for a Sustainable World

Volume 56

Other Publications by the Editors

Books
Environmental Chemistry
http://www.springer.com/978-3-540-22860-8

Organic Contaminants in Riverine and Groundwater Systems
http://www.springer.com/978-3-540-31169-0

Sustainable Agriculture
Volume 1: http://www.springer.com/978-90-481-2665-1
Volume 2: http://www.springer.com/978-94-007-0393-3

Book series
Environmental Chemistry for a Sustainable World
http://www.springer.com/series/11480

Sustainable Agriculture Reviews
http://www.springer.com/series/8380

Journals
Environmental Chemistry Letters
http://www.springer.com/10311

More information about this series at http://www.springer.com/series/11480

Sophie Fourmentin • Margarida Costa Gomes
Eric Lichtfouse
Editors

Deep Eutectic Solvents for Medicine, Gas Solubilization and Extraction of Natural Substances

Springer

Editors
Sophie Fourmentin
Unité de Chimie Environnementale et Interactions sur le Vivant (UCEIV, UR 4492), SFR Condorcet FR CNRS 3417
Université du Littoral Côte d'Opale
Dunkerque, France

Margarida Costa Gomes
CNRS Chemistry Laboratory
École Normale Supérieure de Lyon
Lyon, France

Eric Lichtfouse
CNRS, IRD, INRAE, Coll
France, CEREGE
Aix-Marseille University
Aix-en-Provence, France

ISSN 2213-7114 ISSN 2213-7122 (electronic)
Environmental Chemistry for a Sustainable World
ISBN 978-3-030-53068-6 ISBN 978-3-030-53069-3 (eBook)
https://doi.org/10.1007/978-3-030-53069-3

This Springer imprint is published by the registered company Springer Nature Switzerland AG
The registered company address is: Gewerbestrasse 11, 6330 Cham, Switzerland

Preface

Similia similibus solvuntur – Like dissolves like. Anonymous

This aphorism is used to explain that polar solvents dissolve polar solutes, whereas non-polar solvents dissolve non-polar solutes. It appeared in the Corpus Pharmaceutico-Chymico-Medicum Universale in 1711.

Deep Eutectic Solvents (DES) are liquid mixtures at ambient conditions, for which the eutectic point temperature is lower than that of the ideal mixture. Initially considered as a sub-class of ionic liquids, eutectic mixtures are formed by low cost, often biodegradable Lewis or Bronsted acids and bases. As a consequence, a large number of possible deep eutectic solvents can be designed and synthesized for green chemistry. DES have thus recently attracted academic and industrial interest for various applications such as metal processing, biomass treatment and pharmaceuticals. Since the expression Deep Eutectic Solvent was coined in Prof. Abbott's paper in 2003, the number of related publications has increased exponentially, reaching about one thousand in 2019. This book gathers contributions by the most active research groups that use eutectic mixtures for separation, extraction and medical applications. The reader will discover ground-breaking results in different disciplines.

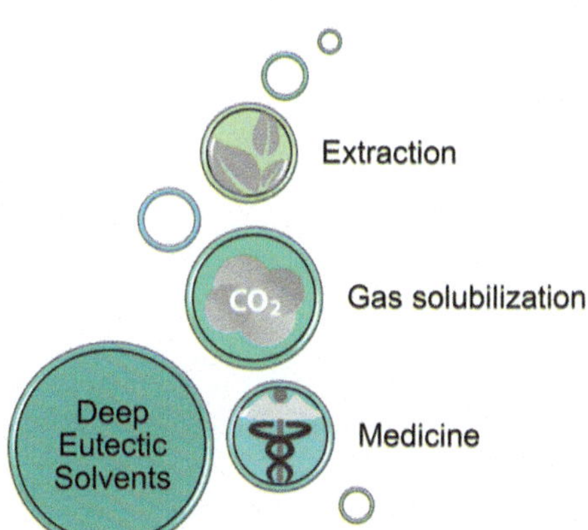

The first chapter by El Achkar et al. presents an overview of DES and their physicochemical properties. Chapter 2 by Nguyen et al. reviews pharmaceutical applications and toxicity of DES for living organisms and the environment. A new system prepared with an active pharmaceutical ingredient, named therapeutic deep eutectic systems (THEDES), is described by Filipa Santos and Ana Rita C. Duarte in Chap. 3. Understanding of how DES dissolve various solutes is of major importance for further use, as explained in Chap. 4 by Moura et al. who review the solubility of gases in DES. Chapter 5 by Byrne et al. discloses new hydrophobic DES, formed by mixing fatty acids with ammonium or phosphonium salts, thus conferring hydrophobicity without fluorinated species. These solvents appear promising for gas capture and liquid-liquid extraction. Chapter 6 by Nakhle et al. presents extraction methods that use deep eutectic solvents. Then, extraction of polyphenols by DES and a review of recent DES applications is presented by Percevault et al. in Chap. 7.

The editors extend their thanks to all the authors who contributed to this book for their efforts in producing timely and high-quality chapters. The creation of this book would not have been possible without the assistance of several friends deserving acknowledgment. They have helped by choosing contributors, reviewing chapters and in many other ways. Finally, we would like to thank the staff at Springer Nature for their highly professional editing of the book.

Dunkerque, France Sophie Fourmentin

Lyon, France Margarida Costa Gomes

Aix-en-Provence, France Eric Lichtfouse

Contents

About the Editors

Sophie Fourmentin works at Université du Littoral Côte d'Opale, Dunkerque, France. She conducts research on the interface between supramolecular chemistry and environmental chemistry. Her group first published a paper on deep eutectic solvent with supramolecular properties in 2019. Prof. Fourmentin supervised and/or co-supervised 12 PhD students, 20 Master students and 2 postdoctoral fellows. She has now 122 publications listed in Scopus, with a total of 2441 citations and an h-factor of 31. Prof. Fourmentin also holds a patent and coordinated three books. She is the President of the French Cyclodextrin Society.

Margarida Costa Gomes is a Physical Chemist and Chemical Engineer working at the French National Centre for Scientific Research in Lyon, France. Her current research interests in the field of molecular thermodynamics of fluids and solutions aim to contribute to greener and more sustainable chemical processes by using environmentally friendly solvents like ionic liquids or eutectic mixtures. She was awarded the CNRS Bronze Medal and was an invited Researcher at the Institute of Chemical and Biological Technology, Portugal, and she is a visiting scholar at the Massachusetts Institute of Technology, USA, where she maintains a position as Research Affiliate. Prof. Costa Gomes has supervised or co-supervised 26 PhD theses and 19 postdoctoral researchers and has published more than 140 papers with a WoS h-index of 42.

Eric Lichtfouse is a Biogeochemist working on climate, pollution and carbon sequestration at Aix-Marseille University, France, and Xi'an Jiaotong University, China. He has invented carbon-13 dating and has discovered temporal pools of individual substances in soils. Prof. Lichtfouse is teaching scientific writing and communication and has published the book *Scientific Writing for Impact Factors*. He is Founder and Chief Editor of the journal *Environmental Chemistry Letters*. Prof. Lichtfouse got the Analytical Chemistry Prize from the French Chemical Society, the Grand Prize of the Universities of Nancy and Metz, and a Journal Citation Award by the Essential Indicators. He is World XTerra Vice-Champion.

Contributors

Luc Augis Université Paris-Saclay, CNRS UMR 8612, Institut Galien Paris-Saclay, Châtenay-Malabry, France

Gillian Barratt Université Paris-Saclay, CNRS UMR 8612, Institut Galien Paris-Saclay, Châtenay-Malabry, France

Emily L. Byrne The QUILL Research Centre, School of Chemistry and Chemical Engineering, Queen's University Belfast, Belfast, UK

Margarida Costa Gomes Chemistry Laboratory ENS Lyon, CNRS & Lyon University, Lyon, France

Ana Rita C. Duarte LAQV, REQUIMTE, Departamento de Química da Faculdade de Ciências e Tecnologia, Universidade Nova de Lisboa, Caparica, Portugal

Tracy El Achkar Bioactive Molecules Research Laboratory, Faculty of Sciences, Lebanese University, Beirut, Lebanon

Unité de Chimie Environnementale et Interactions sur le Vivant (UCEIV, UR 4492), SFR Condorcet FR CNRS 3417, Université du Littoral Côte d'Opale, Dunkerque, France

Sophie Fourmentin Unité de Chimie Environnementale et Interactions sur le Vivant (UCEIV, UR 4492), SFR Condorcet FR CNRS 3417, Université du Littoral Côte d'Opale, Dunkerque, France

Fabienne Gauffre Univ Rennes, ISCR, CNRS UMR 6226, Rennes, France

Mark Gilmore The QUILL Research Centre, School of Chemistry and Chemical Engineering, Queen's University Belfast, Belfast, UK

Hélène Greige-Gerges Bioactive Molecules Research Laboratory, Faculty of Sciences, Lebanese University, Beirut, Lebanon

John D. Holbrey The QUILL Research Centre, School of Chemistry and Chemical Engineering, Queen's University Belfast, Belfast, UK

Miriana Kfoury Bioactive Molecules Research Laboratory, Faculty of Sciences, Lebanese University, Jdaidet el-Matn, Lebanon

Laura Kollau Chemistry Laboratory ENS Lyon, CNRS & Lyon University, Lyon, France

Corinne Lagrost Univ Rennes, ISCR, CNRS UMR 6226, Rennes, France

David Landy Unité de Chimie Environnementale et Interactions sur le Vivant (UCEIV, UR 4492), SFR Condorcet FR CNRS 3417, Université du Littoral Côte d'Opale (ULCO), Dunkerque, France

François-Xavier Legrand Université Paris-Saclay, CNRS UMR 8612, Institut Galien Paris-Saclay, Châtenay-Malabry, France

Emmanuelle Limanton Univ Rennes, ISCR, CNRS UMR 6226, Rennes, France

Isabelle Mallard Unité de Chimie Environnementale et Interactions sur le Vivant (UCEIV, UR 4492), SFR Condorcet FR CNRS 3417, Université du Littoral Côte d'Opale (ULCO), Dunkerque, France

Leila Moura The QUILL Research Centre, School of Chemistry and Chemical Engineering, Queen's University Belfast, Belfast, UK

Lamia Nakhle Bioactive Molecules Research Laboratory, Faculty of Sciences, Lebanese University, Jdaidet el-Matn, Lebanon

Unité de Chimie Environnementale et Interactions sur le Vivant (UCEIV, UR 4492), SFR Condorcet FR CNRS 3417, Université du Littoral Côte d'Opale (ULCO), Dunkerque, France

Canh-Hung Nguyen Université Paris-Saclay, CNRS UMR 8612, Institut Galien Paris-Saclay, Châtenay-Malabry, France

Department of Pharmaceutics, Hanoi University of Pharmacy, Hanoi, Vietnam

Ludovic Paquin Univ Rennes, ISCR, CNRS UMR 6226, Rennes, France

Lucie Percevault Univ Rennes, ISCR, CNRS UMR 6226, Rennes, France

Filipa Santos LAQV, REQUIMTE, Departamento de Química da Faculdade de Ciências e Tecnologia, Universidade Nova de Lisboa, Caparica, Portugal

Małgorzata Swadźba-Kwaśny The QUILL Research Centre, School of Chemistry and Chemical Engineering, Queen's University Belfast, Belfast, UK

Chapter 1
Understanding the Basics and Properties of Deep Eutectic Solvents

Tracy El Achkar, Hélène Greige-Gerges, and Sophie Fourmentin

Contents

T. El Achkar (✉)
Bioactive Molecules Research Laboratory, Faculty of Sciences, Lebanese University, Beirut, Lebanon

Unité de Chimie Environnementale et Interactions sur le Vivant (UCEIV, UR 4492), SFR Condorcet FR CNRS 3417, Université du Littoral Côte d'Opale, Dunkerque, France
e-mail: tracy.el-achkar@etu.univ-littoral.fr

H. Greige-Gerges
Bioactive Molecules Research Laboratory, Faculty of Sciences, Lebanese University, Beirut, Lebanon
e-mail: hgreige@ul.edu.lb

S. Fourmentin
Unité de Chimie Environnementale et Interactions sur le Vivant (UCEIV, UR 4492), SFR Condorcet FR CNRS 3417, Université du Littoral Côte d'Opale, Dunkerque, France
e-mail: lamotte@univ-littoral.fr

S. Fourmentin et al. (eds.), *Deep Eutectic Solvents for Medicine, Gas Solubilization and Extraction of Natural Substances*, Environmental Chemistry for a Sustainable World 56, https://doi.org/10.1007/978-3-030-53069-3_1

Abstract The disclosure of deep eutectic solvents represents a landmark in green chemistry. These systems were reported only 17 years ago and yet resulted in a plethora of publications covering various research areas and diverse fields of application. Their economic and straightforward preparation using highly accessible and natural compounds together with their extremely great tunability were enough to consider them as alternatives to the conventional organic solvents. Besides, deep eutectic solvents hold interesting properties suggesting a promising contribution in multiple domains.

Herein, the classification of deep eutectic solvents and the reported preparation methods to date are presented. Further, their physicochemical properties, namely, their phase behavior, density, viscosity, ionic conductivity, surface tension, and polarity, are summarized. The stated properties are affected by various parameters such as the choice of the forming compounds, their molar ratio, the temperature, and the water content. Finally, in the last section, we entirely present the impact of water on both the physicochemical properties and the characteristic supramolecular network of deep eutectic solvents.

Keywords Deep eutectic solvents · Eutectic point · Phase behavior · Natural deep eutectic solvents · Hydrogen bond acceptor · Hydrogen bond donor · Choline chloride · Physicochemical properties · Water · Supramolecular network

Abbreviations

BA	Butyric acid
CA	Caprylic acid
ChCl	Choline chloride
DEG	Diethylene glycol
DES	Deep eutectic solvents
DSC	Differential scanning calorimetry
EG	Ethylene glycol
G	Glycerol
HBA	Hydrogen bond acceptor
HBD	Hydrogen bond donor
LA	Lactic acid
Lev	Levulinic acid
MA	Malonic acid
MD	Molecular dynamics
NADES	Natural deep eutectic solvents
NMR	Nuclear magnetic resonance
PEG	Polyethylene glycol
RAMEB	Randomly methylated β-cyclodextrin
TBABr	Tetrabutylammonium bromide
TBACl	Tetrabutylammonium chloride

TEG	Triethylene glycol
THEDES	Therapeutic deep eutectic solvents
U	Urea
VA	Valeric acid

1.1 Introduction

The discovery of the deep eutectic solvents (DES) was a major breakthrough in the world of green chemistry. Deep eutectic solvents are frequently defined as binary or ternary mixtures of compounds that are able to associate mainly via hydrogen bonds. Combining these compounds at a certain molar ratio results in a eutectic mixture (Zhang et al. 2012). The word "eutectic" comes from the Ancient Greek εὔτηκτος or *eútēktos*, which means easily melted, and a eutectic point represents the chemical composition and temperature at which a mixture of two solids becomes fully molten at the lowest melting temperature, relative to that of either compounds. However, defining a deep eutectic solvent is still a controversial subject, and there are various reported definitions that do not really distinguish deep eutectic solvents from other mixtures, since all the mixtures of immiscible solid compounds present a eutectic point and considering that numerous compounds are able to form hydrogen bonds when put together (Coutinho and Pinho 2017). Given that the presence of a eutectic point or hydrogen bonding between components is not a sufficient condition to define a "deep eutectic solvent" and in order to clarify what a deep eutectic solvent is and what makes it special compared to other mixtures, Martins et al. recently defined deep eutectic solvent as "a mixture of two or more pure compounds for which the eutectic point temperature is below that of an ideal liquid mixture, presenting significant negative deviations from ideality ($\Delta T_2 > 0$)," where ΔT_2 stands for the temperature depression which is the difference between the ideal and the real eutectic point (Martins et al. 2019). The same authors stated that it is important that the temperature depression results in a liquid mixture at operating temperature, regardless of the mixture composition. The fact that there is no fixed composition offers an even greater tunability for these systems.

Although deep eutectic solvents were extensively studied, especially in the past decade, there is still a lack of understanding the principle behind deep eutectic solvent's formation and properties. It all started almost 20 years ago, when Abbott et al. were looking for liquids that can overcome the moisture sensitivity and high cost of some common ionic liquids (Abbott et al. 2001). In this study, numerous mixtures based on different quaternary ammonium salts and metal salts were tested, and it turned out that choline chloride (ChCl) mixed with zinc chloride in a 1:2 molar ratio presents the lowest freezing point (23–25 °C). Thereafter, the same authors investigated eutectic mixtures of quaternary ammonium salts and hydrogen bond donors (HBD) and named them "deep eutectic solvents" (Abbott et al. 2003). The lowest freezing point (12 °C) was obtained with 1:2 ChCl:urea. This significant depression of the freezing point, compared to that of ChCl (302 °C) or urea (U) (133 °C), is due to hydrogen bonding between urea molecules and chloride ion as proved by nuclear magnetic resonance (NMR) spectroscopy. What is interesting about these solvents is that they are not only liquid at ambient temperature but also tunable and highly solubilizing. After that, other deep eutectic solvents based on ChCl and carboxylic acids were characterized and were also shown to have important solubilizing ability toward some metal oxides (Abbott et al. 2004a). Other liquids were also obtained when mixing ChCl with a hydrated metal salt like chromium (III) chloride hexahydrate (Abbott et al. 2004b). Later on, an additional class of ambient temperature solvents based on metal salts and hydrogen bond donors such as amides (urea and acetamide) and diols (ethylene glycol and 1,6-hexanediol) were reported, but it turned out that only a restricted number of metal salts and hydrogen bond donors can lead to their formation (Abbott et al. 2007a).

Few years later, Choi et al. coined the term "natural deep eutectic solvents" (NADES) (Choi et al. 2011). This category covers the deep eutectic solvents that are made of primary metabolites such as organic acids, amino acids, sugars, polyols, and choline derivatives (Choi et al. 2011; Dai et al. 2013). Besides, water can also be part of natural deep eutectic solvents' composition. They were introduced as a way to explain the omnipresence of metabolites in high concentrations in cells. Since different combinations of these candidates led to the formation of liquids which also succeeded in the solubilization of some natural compounds, natural deep eutectic solvents were proposed as a new cellular phase, together with water and lipids. These mixtures might be engaged in the biosynthesis, storage, and transport of some poorly water-soluble compounds as well as some other processes like dehydration, drought resistance, and cryoprotection. Further, their consideration is highly encouraged owing to the advantages that they provide from an environmental and economic point of view.

1.2 Classification

In order to differentiate between the possible eutectics, deep eutectic solvents were classified into four types based on the general formula $Cat^+ X^- zY$, where Cat^+ is generally an ammonium, phosphonium, or sulfonium, while X is a Lewis base (usually a halide anion). Y represents a Lewis or Brønsted acid and z is the number of Y molecules that interact with the corresponding anion (Fig. 1.1) (Abbott et al. 2007a; Smith et al. 2014).

Type III eutectics are the most studied in the literature and are usually based on ChCl and various hydrogen bond donors. ChCl has been extensively adopted since it is relatively cheap, nontoxic, and biodegradable, considering it is approved as a natural additive for several animal species ("Scientific Opinion on Safety and Efficacy of Choline Chloride as a Feed Additive for All Animal Species," 2011). In fact, the first type III deep eutectic solvent was primarily based on ChCl. Since then, a plethora of compounds have been successfully used in deep eutectic solvents' formation. The hydrogen bond acceptors (HBA) mainly include quaternary ammonium or phosphonium salts, whereas the most common hydrogen bond donors are

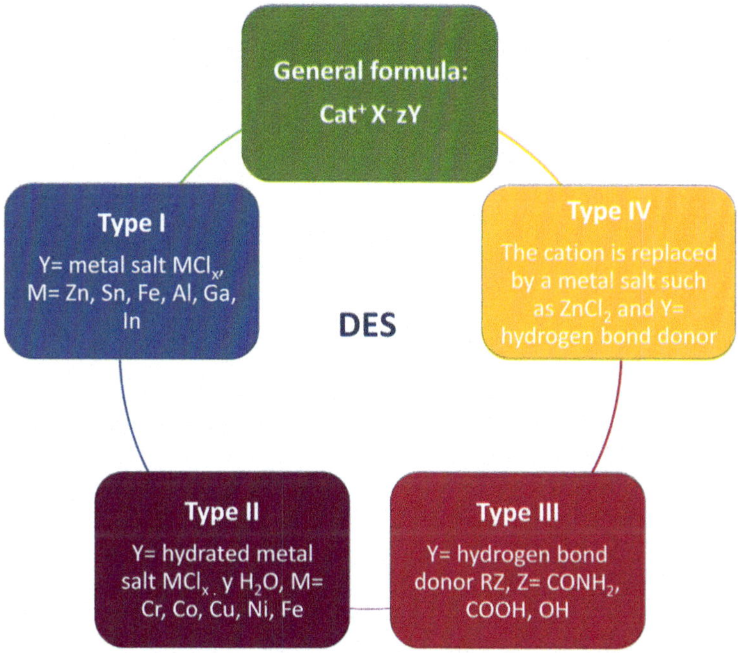

Fig. 1.1 The four types of deep eutectic solvents based on the general formula $Cat^+ X^- zY$, where Cat^+ is generally an ammonium, phosphonium, or sulfonium, while X is a Lewis base (usually a halide anion). Y represents a Lewis or Brønsted acid and z is the number of Y molecules

amides, alcohols, and carboxylic acids. In addition, compounds like sugars, sugar alcohols, and amino acids are also considered for natural deep eutectic solvents' preparation (Dai et al. 2013). More recently, hydrophobic deep eutectic solvents were introduced, and they are based on the use of hydrophobic compounds such as tetrabutylammonium bromide (TBABr), menthol, thymol, and fatty acids as hydrogen bond acceptors together with long alkyl chain alcohols and carboxylic acids as hydrogen bond donors (Florindo et al. 2019; Osch et al., 2015). Furthermore, deep eutectic solvents can be made of active pharmaceutical ingredients like ibuprofen, lidocaine, and phenylacetic acid. In that event, the solvents are named therapeutic deep eutectic solvents (THEDES) (Duarte et al. 2017; Paiva et al. 2014). Some of the frequently used hydrogen bond acceptor and hydrogen bond donor counterparts described in the literature are illustrated in Fig. 1.2.

On the other hand, although natural deep eutectic solvents can sometimes be considered as type III deep eutectic solvents, it is not always the case. That said, natural deep eutectic solvents were recently classified into five main groups (Dai et al. 2013; González et al. 2018):

- Ionic liquids, made of an acid and a base
- Neutral, made of only sugars or sugars and other polyalcohols
- Neutral with acids, made of sugar/polyalcohols and organic acids

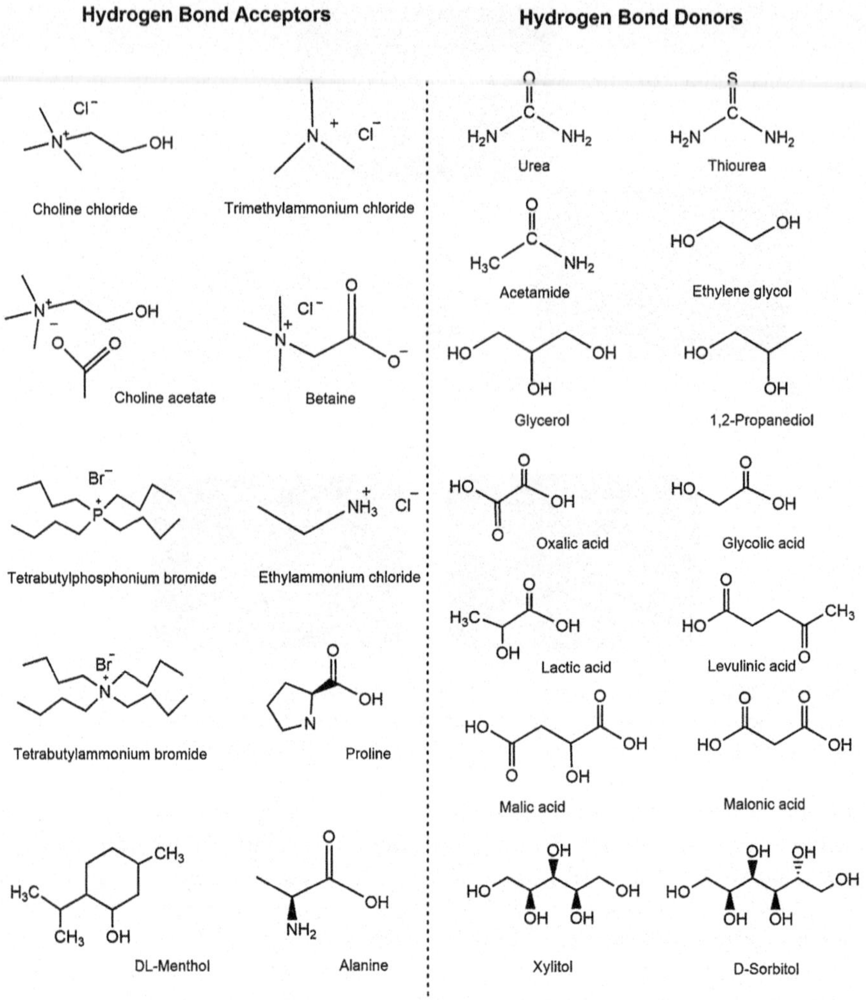

Fig. 1.2 Commonly used hydrogen bond acceptor and hydrogen bond donor compounds in deep eutectic solvents' preparation

- Neutral with bases, made of sugar/polyalcohols and organic bases
- Amino acid-containing natural deep eutectic solvents, made of amino acids and sugars/organic acids

Nevertheless, the reported deep eutectic solvents do not certainly fall into one of the above-mentioned classes given their versatility and the myriad of the considered starting compounds. As a result, Coutinho and coworkers proposed type V deep eutectic solvents composed of nonionic species. In their study, they proved that mixing thymol with menthol led to severe negative deviations to ideality due to a strong interaction between the components. The latter is attributed to resonance effects

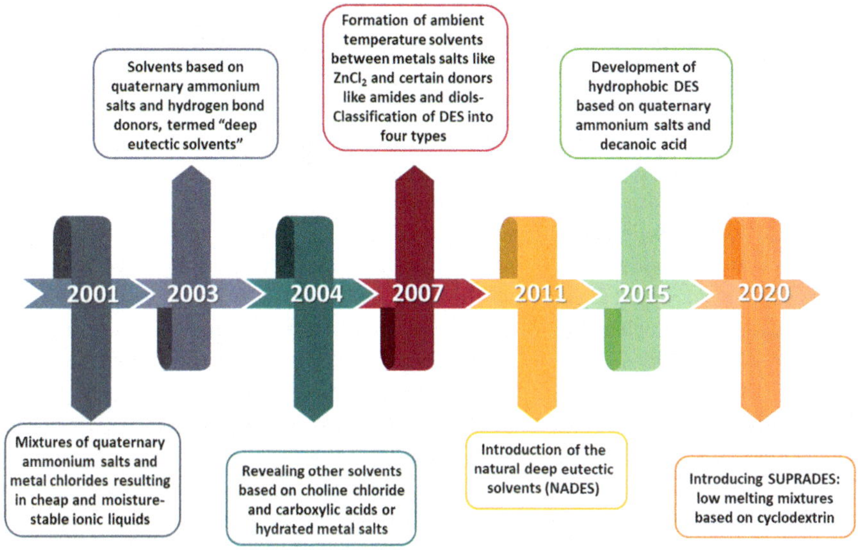

Fig. 1.3 Major events marking the development of deep eutectic solvents throughout the years

related to the structure of phenolic compounds which acted as strong hydrogen bond donors (Abranches et al. 2019). On another note, two recent studies reported the use of cyclodextrins, which are nontoxic cyclic oligosaccharides, as hydrogen bond acceptors resulting in the formation of liquid supramolecular mixtures at room temperature (El Achkar et al. 2020a; El Achkar et al. 2020b). The important events marking the development of deep eutectic solvents so far are presented in Fig. 1.3 (Abbott et al. 2001, 2003; Abbott et al. 2004b; Abbott et al. 2004a; Abbott et al. 2007a; Choi et al. 2011; El Achkar et al. 2020a; van Osch et al. 2015).

1.3 Methods of Preparation

As mentioned above, deep eutectic solvents are obtained by mixing two or more compounds capable of associating through hydrogen bonds, thus forming a eutectic mixture at a well-defined molar ratio. Researchers generally use one of the two main methods to prepare deep eutectic solvents: the heating method and the grinding method. The heating method consists on mixing and heating the compounds, under constant stirring, until a homogeneous liquid is formed (Abbott et al. 2004a). The heating temperature usually ranges between 50 and 100 °C. However, a high temperature may potentially lead to a degradation of the deep eutectic solvent due to an esterification reaction regardless of the preparation method, as demonstrated by Rodriguez et al. for solvents based on ChCl and carboxylic acids (Rodriguez Rodriguez et al. 2019). The grinding method is based on mixing the compounds at room temperature and crushing them in a mortar with a pestle, until a clear liquid is formed (Florindo et al. 2014). Another method based on the freeze-drying of the aqueous solutions of the components of deep eutectic solvents was also revealed by Gutierrez et al. (Gutiérrez et al. 2009). Indeed, separate aqueous solutions of ChCl and urea (or thiourea) were mixed to form an aqueous solution of 1:2 ChCl:U (or ChCl:thiourea), having 5 wt% solute contents. The obtained solutions were then frozen and freeze-dried, resulting in the formation of clear and viscous liquids. However, water was detected in the freeze-dried mixture because it can interact with deep eutectic solvent's components and be part of the deep eutectic solvent's network (Choi et al. 2011; Dai et al. 2013). That said, different deep eutectic solvents are obtained when using different methods of preparation. An evaporation method was also reported by Dai et al., consisting on dissolving the components of deep eutectic solvents in water, followed by an evaporation at 50 °C. The resulting liquid is then placed in a desiccator in the presence of silica gel (Dai et al. 2013). Considering the optimization of time and energy consumption, a greener microwave-assisted approach was proposed for the preparation of natural deep eutectic solvents within seconds (Gomez et al. 2018). Lastly, an ultrasound-assisted synthesis of natural deep eutectic solvents was recently introduced (Santana et al. 2019).

1.4 Physicochemical Properties

The physicochemical properties of deep eutectic solvents are one of the main reasons behind the rising researchers' interest in these solvents. Besides having a low volatility, nonflammability, low vapor pressure, and chemical and thermal stability, deep eutectic solvents are chemically tunable, meaning they can be designed for specific applications given the wide variety of the possible deep eutectic solvents' forming compounds. All these properties encouraged the scientists to explore deep eutectic solvents and apply them as a good alternative to conventional solvents. Herein, the main physicochemical properties of deep eutectic solvents, namely, their phase behavior, density, viscosity, ionic conductivity, surface tension, and polarity, are presented and discussed.

1.4.1 Phase Behavior

As mentioned above, deep eutectic solvents are not pure compounds but mixtures of two or more pure compounds. This system is represented by a solid-liquid phase diagram, which shows the melting temperature in function of the mixture composition. Therefore, if we consider a binary mixture of compounds A and B, the eutectic point represents the composition and the minimum melting temperature at which the melting curves of both compounds meet (Fig. 1.4).

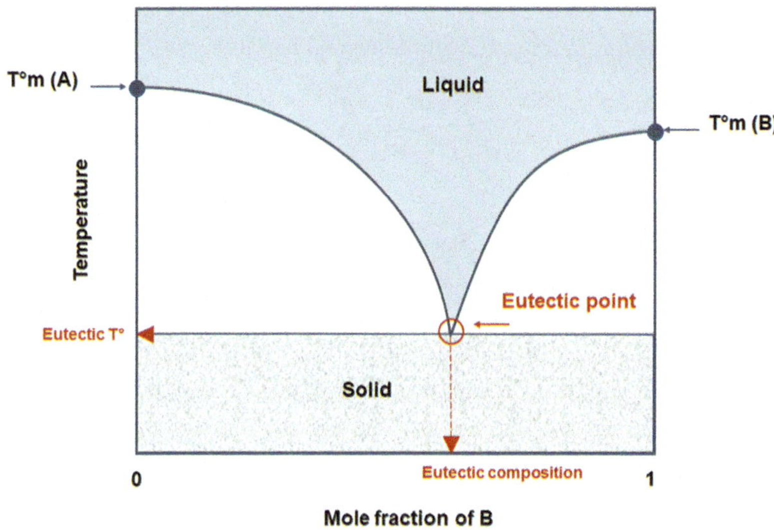

Fig. 1.4 General solid-liquid phase diagram of a binary mixture. Tm(A) and Tm(B) represent the melting temperatures of compounds A and B, respectively

According to Martins et al., the deep eutectic solvent appellation should only cover mixtures with a melting point lower than the ideal eutectic temperature; otherwise, deep eutectic solvents would not be called "deep" and could not be differentiated from other mixtures (Martins et al. 2019). In addition, they stated that a deep eutectic solvent must be liquid at operating temperature even if this requires a different composition than the eutectic one. Consequently, having a phase diagram is essential and knowing the melting properties of the pure compounds is necessary to determine the ideal solubility curve. Nevertheless, very little is reported about the thermodynamic behavior of the deep eutectic solvents to date. The freezing points of most of the deep eutectic solvents usually range between – 69 and 149 °C (Zhang et al. 2012). A number of deep eutectic solvents with a melting point lower than 60 °C were summarized by García et al. (García et al. 2015). The choice of the hydrogen bond donor (Abbott et al. 2004a; Abbott et al. 2003), the nature of the organic salt and its anion (Abbott et al. 2003), and the organic salt/hydrogen bond donor molar ratio (Shahbaz et al. 2011b) can all affect the freezing point of the mixture. The method of preparation can also influence the value of the freezing point, but not the eutectic composition which must remain unchanged no matter the method used (Abbott et al. 2006). On the other hand, no correlation was found between the freezing point of a deep eutectic solvent and the melting points of its pure components (Abbott et al. 2004a; Zhang et al. 2012). The hydrogen bond donor did however affect the freezing point depression ΔT (Abbott et al. 2004a; E. L. Smith et al. 2014). For instance, Abbott et al. found that the lower the hydrogen bond donor's molecular weight, the greater is the depression of the freezing point (Abbott et al. 2004a). But unlike Abbott and coworkers who considered the

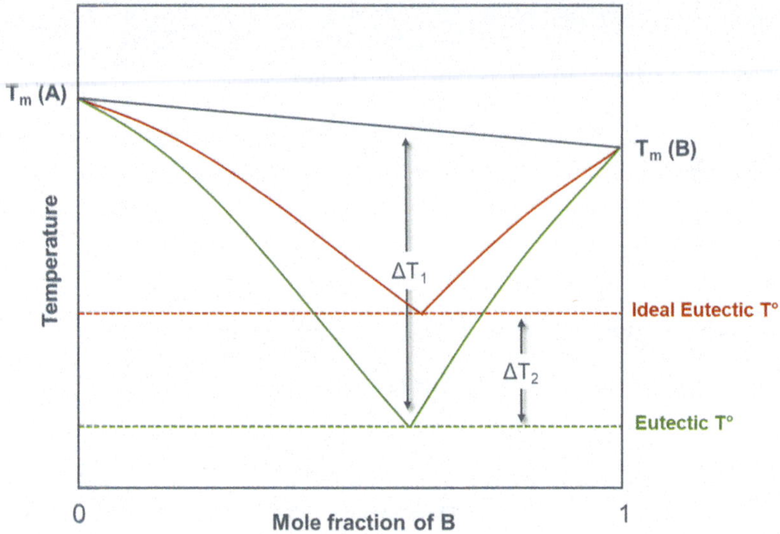

Fig. 1.5 Solid-liquid phase diagram representing a simple ideal eutectic mixture (red line) and a deep eutectic mixture (green line). ΔT stands for the considered freezing point depression. (Adapted from Martins et al. 2019)

temperature depression as the difference between the linear combination of the melting points of the pure components and the real eutectic point (ΔT_1), Martins et al. thought it would be more appropriate to define the temperature depression as the difference between the ideal and the real eutectic point (ΔT_2); otherwise, any mixture of compounds would be referred to as a deep eutectic solvent (Fig. 1.5) (Martins et al. 2019).

Nevertheless, several other reported mixtures presented only a glass transition and no melting point was detected (Dai et al. 2013; Florindo et al. 2014; Francisco et al. 2012; Savi et al. 2019a; Savi et al. 2019b).

1.4.2 *Density*

Density is one of the fundamental physical properties of liquids. Most of the reported deep eutectic solvents present higher densities than water with values ranging between 1.0 and 1.3 g.cm^{-3} at 25 °C, while deep eutectic solvents based on metal salts have densities in the 1.3–1.6 g.cm^{-3} range (Tang and Row 2013). Contrarily, lower densities than water are obtained for hydrophobic deep eutectics (Florindo et al. 2019). The deep eutectic solvent's density shows a temperature-dependent behavior, and it decreases linearly with the increasing temperature (Cui et al. 2017; Florindo et al. 2014; Ibrahim et al. 2019; Shahbaz et al. 2012a). Moreover, the density depends on the choice of the hydrogen bond donor (Abbott et al. 2007b; Cui et al. 2017; Florindo et al. 2014; García et al. 2015) and the molar ratio (Abbott et al. 2011). According to Shahbaz et al., a higher hydrogen bond donor mole fraction lowers the density of a deep eutectic solvent whenever the hydrogen bond donor's density is lower than that of the corresponding deep eutectic solvent and vice versa (Shahbaz et al. 2011a; Shahbaz et al. 2012a). In addition, when the hydrogen bond donor contains hydroxyl groups, the density of the ChCl-based deep eutectic solvent increases with the number of hydroxyl groups but decreases with the addition of aromatic groups. Also, when the deep eutectic solvent is made of an acid, its density decreases when the chain length is increased (Florindo et al. 2014; García et al. 2015; Mitar et al. 2019). The effect of the hydrogen bond donor type on the density of some ChCl-based deep eutectic solvents obtained by different studies is represented in Fig. 1.6. These results were in accordance with Yusof et al. who also proved that tetrabutylammonium bromide:alcohol deep eutectic solvents present a higher density when a hydrogen bond donor with a higher number of hydroxyl groups is adopted. The same group also noticed a decrease in density as the hydrogen bond donor's chain length increased (Yusof et al. 2014).

The deep eutectic solvent's density is weakly affected by the alkyl chain length of the ammonium salt (Z. Chen et al. 2017). However, looking at results from different studies reviewed by García et al., one can clearly see that the organic salt and its anion affect the density of deep eutectic solvents. Indeed, phosphonium salts and bromide salts result in denser deep eutectic solvents than ammonium salts and

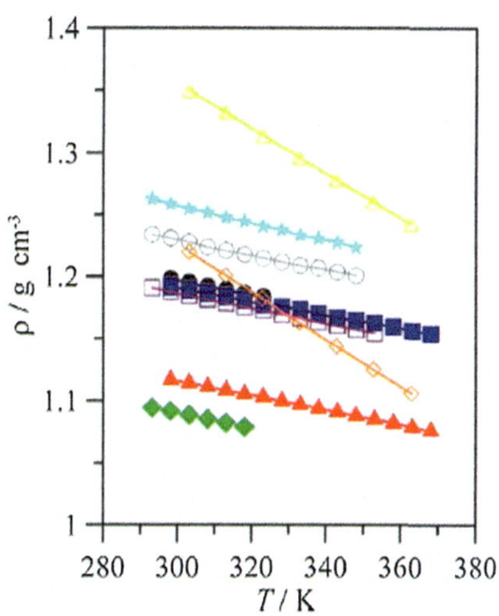

Fig. 1.6 Effect of the hydrogen bond donor on the densities of some choline chloride-based deep eutectic solvents (yellow, 1:3 choline chloride:ethanolamine; light blue, 1:1 choline chloride:oxalic acid; grey, 1:1 choline chloride:malonic acid; black, 1:2 choline chloride:urea; dark blue, 1:2 choline chloride:glycerol; purple, 1:1 choline chloride:glutaric acid; orange, 1:3 choline chloride:2,2,2-trifluoroacetamide; red, 1:2 choline chloride:ethylene glycol; green, 1:3 choline chloride:phenol). (Reprinted with permission from (García et al. 2015). Copyright (2015) American Chemical Society)

chloride salts, respectively (García et al. 2015). On another note, Florindo et al. proved that there is no significant difference in density values whether the heating method or the grinding method was used for the preparation of deep eutectic solvents (Florindo et al. 2014). Yet, differences of up to 4% were detected between the available literature sources when it comes to the density of the most studied 1:2 ChCl:U deep eutectic solvent (García et al. 2015). A series of studies aiming to efficiently predict the density of deep eutectic solvents were conducted by Mjalli et al. via several theoretical approaches (Mjalli 2016; Mjalli et al. 2015; Shahbaz et al. 2011a, 2013; Shahbaz et al. 2012b). The mass connectivity index-based correlation, taking into account the molecular structures of deep eutectic solvents' forming compounds, allowed the prediction of the density of different type III deep eutectic solvents as a function of temperature with a very high efficiency (Mjalli 2016).

1.4.3 Viscosity

The viscosity is another important and extensively studied property of deep eutectic solvents. Most of the reported deep eutectic solvents to date are highly viscous at room temperature ($\eta > 100$ mPa.s) which is mainly ascribed to the extensive hydrogen bond network taking place between deep eutectic solvents' components. In addition, they present a very broad viscosity range. In fact, ChCl:EG (1:2) is known to have a very low viscosity (37 mPa.s at 25 °C), while sugar-based deep eutectic solvents present extremely large viscosities (12,730 mPa.s for 1:1 ChCl:sorbitol at 30 °C and 34,400 mPa.s for 1:1 ChCl:glucose at 50 °C), and even higher viscosities were recorded for metal salt-based deep eutectic solvents (85,000 mPa.s for 1:2 ChCl:zinc chloride at 25 °C) (Zhang et al. 2012). Yet, very low viscosities were recorded for hydrophobic deep eutectic solvents based on DL-menthol (7.61 mPa.s at 25 °C for 1:3 DL-menthol:octanoic acid) (Nunes et al. 2019; Ribeiro et al. 2015). The viscosity of a eutectic mixture is clearly affected by the nature of its components (Abbott et al. 2007a; D'Agostino et al. 2011), their molar ratio (Abbott et al. 2011), the temperature (Abbott et al. 2004a; Abbott et al. 2003, 2006; Dai et al. 2015; Kareem et al. 2010), and the water content (D'Agostino et al. 2015; Dai et al. 2015; Du et al. 2016; Florindo et al. 2014; Shah and Mjalli 2014). The effect of water will be discussed in detail in the upcoming sections. Moreover, the viscosity not only depends on the intermolecular forces between the hydrogen bond donor and the ion but also on the steric effects which can be quantified by the hole theory. The latter considers the existence of holes or voids in the fluid which affects the fluid's viscosity and ionic conductivity (Abbott et al. 2006). The distribution of holes of radius r is influenced by the hydrogen bond donor and the salt. It also seems that deep eutectic solvents containing large holes are less viscous because they allow a certain ionic motion (García et al. 2015). On a separate note, it is worthy to mention that large differences were noticed when comparing the viscosity data obtained by different researchers for the same deep eutectic solvent (e.g., 152 mPa.s vs 527.28 mPa.s for 1:2 ChCl:U at 30 °C and 202 mPa.s vs 2142 mPa.s for 1:1 ChCl:oxalic acid at 40 °C) (García et al. 2015). These major differences can be attributed not only to the preparation method as stated by Florindo et al. (Florindo et al. 2014) but also to the experimental method and the impurities (García et al. 2015).

1.4.4 Ionic Conductivity

Since the viscosity is the main controller of the conductivity, most of the deep eutectic solvents tend to have poor ionic conductivities ($\kappa < 2$ mS cm^{-1} at room temperature). Therefore, increasing the temperature results in a decrease in the viscosity and an increase in the conductivity (Lapeña et al. 2019; Zhang et al. 2012). This property is also affected by the hydrogen bond acceptor/hydrogen bond donor molar ratio (Abbott et al. 2004a), the nature of both the organic salt and the hydrogen bond donor as well as the salt's anion (García et al. 2015), and of course the water addition (Dai et al. 2015).

1.4.5 Surface Tension

The studies related to the surface tension of deep eutectic solvents are quite limited compared to the studies dealing with other physicochemical properties. Yet, it is an essential property since it is highly dependent on the intensity of the intermolecular forces taking place between the hydrogen bond donor and the corresponding salt. That said, highly viscous liquids present high surface tensions. The values relative to the reported deep eutectic solvents generally vary between 35 and 75 mN m^{-1} at 25 °C (García et al. 2015; Ibrahim et al. 2019). Once again remarkable high values were recorded for sugar-based deep eutectic solvents such as ChCl:D-glucose (Hayyan et al. 2013) and ChCl:D-fructose (Hayyan et al. 2012), reflecting their extensive hydrogen bond network. Besides, the surface tension is influenced by the salt mole fraction and the cation type since an additional hydroxyl group or a longer alkyl chain in the quaternary ammonium salt leads to higher surface tensions. Also, the surface tension linearly decreases with increasing temperature (García et al. 2015; Lapeña et al. 2019; Nunes et al. 2019).

1.4.6 Polarity

Polarity is a key property since it reflects the overall solvation capability of solvents. Despite its significance, the polarity of the deep eutectic solvents was not considerably studied and was not addressed until recently. This property is often estimated via the solvatochromic parameters which consider the hypsochromic (blue) shift or bathochromic (red) shift of UV-vis bands for the negatively solvatochromic dyes (e.g., Reichardt's betaine dye) or the positively solvatochromic dyes (e.g., Nile red), respectively, as a function of the solvent's polarity (Reichardt 1994). The most frequently used scales are the polarity scales of Dimroth and Reichardt (E_T and E_TN) (Reichardt 1994) and the multiparameter scale of Kamlet and Taft (the hydrogen bond donating ability α, the hydrogen bond accepting ability β, and dipolarity/polarizability π*) (Kamlet et al. 1977; Kamlet and Taft 1976). The common probes adopted for the establishment of Dimroth and Reichardt's scale include Reichardt's betaine dyes and Nile red, while molecules like 4-nitroaniline and *N,N*-diethyl-4-nitroaniline are used to determine the parameters following the Kamlet and Taft multiparameter scale. However, it is worthy to mention that the polarity scales are not universal and are probe dependent, which means that we cannot compare polarity parameters obtained by different solvatochromic probes (Valvi et al. 2017). A general overview on the studies conducted on deep eutectic solvents' polarity is presented in Table 1.1 following a chronological order.

1.5 Effect of Water

Given the omnipresence of water and the hygroscopic character of some deep eutectic solvents and their forming compounds, the water uptake by the eutectic solvents is inevitable (Du et al. 2016; Florindo et al. 2014). While traces of water in deep eutectic solvents are usually considered as impurities, a plethora of papers intentionally added water to their solvents in order to fine-tune their properties so they can respond to the requirements of some desired applications and water allowed, in many cases, to improve the performance of deep eutectic solvents. On the other hand, the presence of water not only affects the physicochemical properties but may also jeopardize the integrity of deep eutectic solvents (El Achkar et al. 2019), which explains the inconsistency in the literature given that deep eutectic solvents are prepared in different operating conditions. Therefore, studying the effect of water on the eutectic systems is of utmost importance. This section highlights the impact of water on the physicochemical properties of deep eutectic solvents and the characteristics of their supramolecular organizations.

1.5.1 Effect on Deep Eutectic Solvents' Physicochemical Properties

Herein, the effect of water on the main physicochemical properties (melting point, density, viscosity, conductivity, surface tension, and polarity) will be discussed according to the reported studies so far. Some investigated the effect of low water

Table 1.1 Overview of the reported studies related to the polarity of deep eutectic solvents

Deep eutectic solvent(s)	Solvatochromic probe(s)	Main results	Reference
ChCl:G (1:1, 1:1.5, 1:2, and 1:3)	Reichardt's dye 30; 4-nitroaniline; *N,N*-dimethyl-4-nitroaniline	The studied DES at different molar ratios make up polar fluids with the $E_T(30)$ values increasing with the increase in ChCl in a nearly linear trend	Abbott et al. (2011)
Numerous NADES based on ChCl, sugars, alcohols, organic acids, amino acids, and water	Nile red	The organic acid-based NADES are the most polar, followed by amino acid- and sugar-based NADES, while the sugar- and polyalcohol-based ones seem to be the least polar	Dai et al. (2013)
ChCl:U; ChCl:G; ChCl:EG; ChCl:MA (1:2)	Several solvatochromic probes such as betaine dye 33 and Nile red	All the DES are considered highly polar. The structure of the HBD clearly affects the solvent's polarity which is highest with alcohol-based DES followed by those having urea and malonic acid	Pandey et al. (2013)
13 binary or ternary ChCl-based DES using urea, glycerol, ethylene glycol, thiourea, or formamide as HBD	Reichardt's dye 30; 4-nitroaniline; *N,N*-diethyl-4-nitroaniline	The polarity of the studied DES highly depends on the polarity of the HBD. A correlation was also found between the solvatochromic parameters and the influence of DES on the activity or stability of lipase	Kim et al. (2016)
4 DES using different N-oxides as HBA and phenylacetic acid as HBD	Nile red	Four N-oxides were adopted as HBA (three amphiphilic and one non-amphiphilic) along with phenylacetic acid as HBD. The non-amphiphilic HBA gives rise to a more polar DES than the ones based on amphiphilic HBA	Germani et al. (2017)
19 DES based on ammonium salts as HBA and carboxylic acids as HBD	4-Nitroaniline; *N,N*-diethyl-4-nitroaniline	Increasing the alkyl chain length of both HBA and HBD results in a decrease in the hydrogen bond acidity and an increase in the hydrogen bond basicity, while the dipolarity/polarizability is mainly affected by the HBD given that it decreases when the HBD's alkyl chain length is increased	Teles et al. (2017)

(continued)

Table 1.1 (continued)

Deep eutectic solvent(s)	Solvatochromic probe(s)	Main results	Reference
ChCl:U; ChCl:G; ChCl:EG (1:2)	Several molecular probes	The studied DES present an overall higher polarity than most of the organic solvents. The HBD type might extremely affect the polarity response of the probe. Some polarity parameters displayed different temperature-dependent trends when comparing the three studied DES, and the polarity did not always decrease with the increasing temperature, thus revealing a possible "nonclassical" nature of DES compared to other solvents	Valvi et al. (2017)
7 DES based on ChCl; 4 DES based on TBACl; 4 DES based on DL-menthol	Reichardt's betaine dye; 4-nitroaniline; *N,N*-diethyl-4-nitroaniline	The HBA seems to play a major role in the polarity of the DES, especially the hydrophobic ones that are based on TBACl or DL-menthol as HBA, while for the hydrophilic ChCl-based DES, a diacid HBD like malonic acid generates a more polar DES than a monoacid-based one	Florindo et al. (2018)
12 DES based on thymol and (L)-menthol as HBA and 6 different monocarboxylic acids as HBD	4-Nitroaniline; *N,N*-diethyl-4-nitroaniline; pyridine-*N*-oxide	Thymol-based DES present higher polarity and hydrogen bond acidity character than (L)-menthol-based ones. On the other hand, a more pronounced hydrogen bond basicity character is seen with (L)-menthol with a slight positive correlation with the alkyl chain length of the acid	Martins et al. (2018)
17 ChCl-based DES using polyalcohols (ethylene glycol, glycerol, propanediols, and butanediols) as HBD at molar ratios 1:1, 1:2, and 1:3	Nile red	Polarity is affected by the molecular structure of the HBD. Polarity is also an important property to consider while using DES as an extraction solvent	Mulia et al. (2019)

(continued)

Table 1.1 (continued)

Deep eutectic solvent(s)	Solvatochromic probe(s)	Main results	Reference
9 DES based on PEG as HBD and carboxylic acids or amides or ammonium salts as HBA	Nile red	The HBA controls the DES polarity since acid-based DES are more polar than the amide- and ammonium-based DES. A negative correlation was also detected between the pH values and the polarity of these PEGylated DES	Chen et al. (2019a)
9 ternary DES based on ChCl, 1-ethyl-3-methylimidazolium chloride, or tributylmethylammonium chloride as HBA and a binary mixture of 3-hydroxycarboxylic acids as HBD at a 1:1 or 1:2 HBA:HBD molar ratio	Nile red	Similar polarities were obtained for all the studied DES. The DES based on 3-hydroxycarboxylic acids were slightly more polar than the ones based on their aliphatic carboxylic acid analogues. No difference was detected between the two adopted molar ratios	Haraźna et al. (2019)
ChCl:BA; ChCl:VA; ChCl:CA at both 1:2 and 1:3 molar ratios	Nile red; 4-nitroaniline; *N,N*-diethyl-4-nitroaniline	Increasing the alkyl chain length of the HBD results in a decrease in DES polarity. Yet, no significant effect was observed when changing the HBA:HBD molar ratio	Dwamena and Raynie (2020)
ChCl:Lev and ChCl:LA (1:2)	Nile red	ChCl:LA is more polar than ChCl:Lev, which is mainly ascribed to the difference in the HBD's polarity, while the two DES were more polar compared with some alkyl cholinium lactate and alkyl cholinium levulinate ionic liquids	Fahri et al. (2020)

content that can naturally be present in the deep eutectic solvent, and others considered a full range of water content. After being in contact with the atmosphere for 3 weeks, ChCl:U deep eutectic solvent absorbed up to 20 wt% water. That said, Meng et al. tested the effect of water (up to 10 wt%), which can be naturally absorbed by the deep eutectic solvents, on the melting point of ChCl:U. The latter was determined via three different and complementary techniques: a thermostated bath, optical microscopy, and differential scanning calorimetry (DSC). A linear decrease of the melting point was observed as a function of the water content. The melting point of the mixture dropped from 30 °C for the dry deep eutectic solvent to 15 °C in the presence of 5 wt% of water. This tremendous water effect can explain the dissimilarities obtained by different studies for the same deep eutectic solvent (Meng et al. 2016). These results were somehow in accordance with the findings of Smith et al.

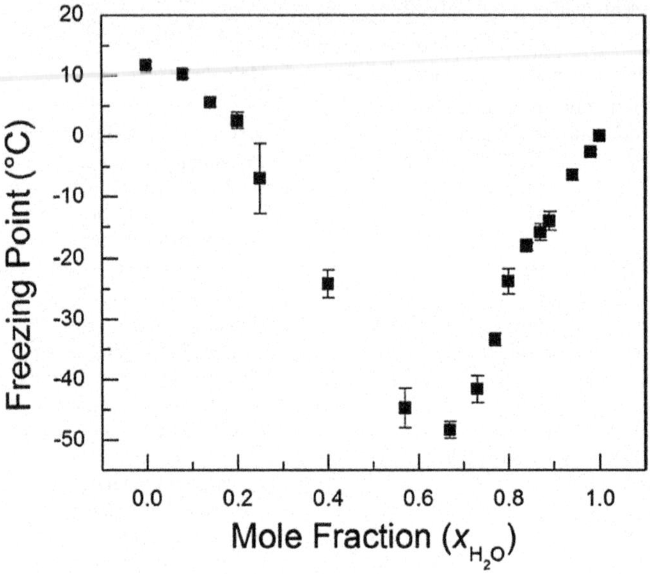

Fig. 1.7 Variation of the freezing point of 1:2 choline chloride:urea deep eutectic solvent with the added mole fraction of water. (Reprinted with permission from (Smith et al. 2019). Copyright (2019) American Chemical Society)

who also followed the variation of the melting point of the same deep eutectic solvent but with a full range of water content. Though a similar linear trend was obtained up until 10 wt% of water by the two studies, further increase in the water content yields a minimum melting point of −48 ± 2 °C at 0.67 mole fraction of water. Above this point, the melting temperature linearly increased as shown in Fig. 1.7. Owing to the behavior of the studied mixture, the authors proposed that 1:2:6 ChCl:U:water makes a ternary deep eutectic solvent (P. J. Smith et al. 2019). Nevertheless, this behavior of ChCl:U was not observed by the study of Shah et al. in which the melting point only decreased as a function of water content studied in full range (Shah and Mjalli 2014). Contrarily, the addition of up to 10 wt% water slightly increased the melting point of 1:1 ChCl:boric acid which was explained by a possible reaction between water and boric acid (Häkkinen et al. 2019).

On the other hand, all the studies have agreed that unlike the density, both viscosity and conductivity are highly sensitive to the presence of water in deep eutectic solvents. Agieienko et al. noticed a slight decrease of 0.14% in the density of ChCl:U at around 0.008 mass fraction of water, while 0.005 water mass fraction decreased its viscosity by around 22%. The authors stated that different water contents along with the chosen experimental method and associated instrument calibration may be the reasons behind the poor agreement between the reported viscosity values of ChCl:U (Agieienko and Buchner 2019). Du et al. showed that both viscosity and conductivity of ChCl:U are highly sensitive to water. In fact, the viscosity and the conductivity were 13 times lower and 10 times higher in the hydrated deep eutectic solvent, respectively, at only 6 wt% water content (Du et al. 2016). Likewise, Shah et al. found that within 10 wt% of water, the viscosity of ChCl:U was reduced by more than 80% and the conductivity was 3 times higher compared to the dried deep eutectic solvent. In contrast, the presence of water only gradually decreased the melting point and the density of the mixture (Shah and Mjalli 2014). According to Silva et al., the addition of 3 to 9 wt% water to ChCl:sugar deep eutectic solvents decreased their melting point, making them liquid at room temperature. The presence of water also contributes to a slight density reduction of 0.82 to 2.22%, a substantial drop in viscosity values, and an increase in the liquid's polarizability (Silva et al. 2018). For ChCl:carboxylic acids, a 5% difference in density was detected between a dried and water-saturated deep eutectic solvent (the water content of the hydrated ones varied between 9.88 and 19.40 wt% depending on the deep eutectic solvent following their exposition to air) at the same temperature which proves once again that this property is not very sensitive to the amount of water present in the sample. On the other hand, viscosity was strongly decreased in the hydrated deep eutectic solvents with 10 to 200 times lower values than the dried ones (Florindo et al. 2014). The presence of water (up to 5 wt%) also significantly decreased the viscosity and increased the conductivity of natural deep eutectic solvents based on ChCl and betaine as hydrogen bond acceptors and sugars and carboxylic acids as hydrogen bond donors (Aroso et al. 2017). Increasing the water content from 2 to 22 wt% in ChCl:lactic acid decreased the viscosity by two orders of magnitude and increased the electrical conductivity in a similar trend (Alcalde et al. 2019). Another study noticed a threefold and a tenfold decrease in the viscosity of glucose:ChCl:water natural deep eutectic solvent after the addition of 5 and 10% water (v/v), respectively (Dai et al. 2013). A significant decrease in the viscosity and a linear decrease in the density of several natural deep eutectic solvents were also observed in the presence of water, while the conductivity of five selected ternary natural deep eutectic solvents made of choline chloride, organic acids, sugars, and water firstly increased with the increasing water content and then decreased after reaching a peak value of around 10–100 times higher than that of pure natural deep eutectic solvent at 60–80 wt% water (Dai et al. 2015). Similar water effects on the density and viscosity of some organic acid-based natural deep eutectic solvents were obtained by Mitar et al. However, though clearly higher conductivities were seen in ChCl:organic acid-based natural deep eutectic solvents compared to sugar:organic acid in the studies of both Dai et al. and Mitar et al., no rise and fall was obtained by Mitar's group when studying the natural deep eutectic solvent's conductivity as a function of water content. The ionic conductivities simply increased with the water content, and this difference can be attributed to the narrower water range that was considered by Mitar et al. (10–50 wt%) (Mitar et al. 2019). The viscosity and conductivity of ChCl-based deep eutectic solvents with different glycols as hydrogen bond donor were also massively affected by the water content. In fact, the viscosity value was halved after the addition of about 7–10 wt% water. On the other hand, the conductivity of deep eutectic solvents firstly increased with the increasing water content, reached a maximum at 60 wt% water, at which values were 6–15 times higher than that of pure deep eutectic solvents, and then decreased. The ionic dissociation of deep eutectic solvents' components caused the initial increase in the conductivity which was later decreased owing to the dilution of the electrolytes at higher water content. The polarity linearly increased with the increasing water content for all three solvents (Gabriele et al. 2019). A similar non-monotonic behavior of the ionic conductivity was observed with aqueous solutions of ChCl:U and ChCl:EG via molecular dynamics (MD) simulations. The values reached the maximum also at 60 wt% water (Celebi et al. 2019). When varying the water content from 16 to 30 wt% in citric acid:sucrose deep eutectic solvent, a linear decrease of the density was observed, while a major impact was seen on the viscosity which decreased by up to 99.73% at the maximum studied water content (30 wt%) (Savi et al. 2019b). Likewise, the addition of 10 wt% water resulted in an 85.9% decrease in the viscosity of lactic acid:glucose natural deep eutectic solvent (Savi et al. 2019a).

Rublova et al. conducted an in-depth study about the effect of water on the surface tension of binary mixtures of "ChCl + water," "ChCl + ethylene glycol," and "ethylene glycol + water" and a ternary mixture of "ChCl + ethylene glycol + water." The interpretation of the variation of the surface tension and the thermodynamic characteristics of adsorption at the interface "solution/air" led to some interesting findings. When comparing the aqueous solution of ChCl to that of EG, stronger adsorption of choline cation was obtained owing to the choline cation-water hydrophobic interactions. The ternary mixture results revealed interactions between deep eutectic solvents' constituents in an adsorbed surface layer formed at the interface air/diluted solution of ChCl:EG which explains the way higher values of equilibrium adsorption constants for the ternary mixture compared to those related to "ChCl + water" and "ethylene glycol + water" mixtures (Rublova et al. 2020). When studying the effect of water on the surface tension of DL-menthol:octanoic acid deep eutectic solvent, Nunes et al. detected two consecutive behaviors: a decrease of the surface tension while increasing the water content reaching a minimum value at around 4000 ppm of water, followed by an increase of the surface tension (Nunes et al. 2019). The same behavior was observed by Sanchez-Fernandez et al. when studying the surface tension of ChCl:malonic acid as a function of water (Sanchez-Fernandez et al. 2017).

The papers dealing with the effect of water on the deep eutectic solvents' polarity are rather limited. The polarity of ChCl:U, ChCl:G, and ChCl:EG was investigated through the solvatochromic behavior of different absorbance and fluorescence probes in deep eutectic solvent-water mixtures. This approach would inform us about the interactions dominating these mixtures. The water addition happens to increase the dipolarity/polarizability and decrease the hydrogen bond basicity of all three solvents. The behavior of the fluorescent probes further revealed more relevant hydrogen bond interactions between added water and deep eutectic solvents' constituents for ChCl:G and ChCl:EG when compared to ChCl:U. The structural differences between the adopted hydrogen bond donors as well as the interstitial accommodation of water molecules within ChCl:U would explain the greater influence of water addition on ChCl:G and ChCl:EG rather than ChCl:U (Pandey and Pandey 2014).

1.5.2 Effect on Deep Eutectic Solvents' Network

As discussed above, water clearly impacts the physicochemical properties of the deep eutectic solvents, whether present in low or large amounts. Furthermore, investigating deep eutectic solvent-water interactions is crucial especially that binary mixtures of deep eutectic solvents and water have been commonly adopted in many applications already. In fact, the presence of water allows the circumvention of some of the shortcomings of deep eutectic solvents like their relative high viscosity while maintaining their unique and appealing properties, which explains the rising interest in deep eutectic solvent-water mixtures over the past few years. However, the high polarity of water and its propensity to interact with the hygroscopic components of the eutectic system makes it of paramount importance to check out if and how water affects the intra- and intermolecular bonds lying behind the supramolecular network of deep eutectic solvents. Despite their relevance, the investigations of the effect of water on deep eutectic solvent's system are rather restricted and mostly cover ChCl-based eutectics. Table 1.2 provides an overview of the reported studies dealing with the effect of water on various deep eutectic solvents' systems. This effect was examined via multiple techniques mainly NMR, Brillouin spectroscopies, and neutron total scattering, not to mention the MD simulations. Some studies proposed a passage from deep eutectic solvent to an aqueous solution of its individual components while adding water, and others suggested that a transition from "water-in-DES" to a "DES-in-water" system occurs at a certain hydration level. In the former system, water is seen as another hydrogen bond donor (Hammond et al. 2017b; López-Salas et al. 2019; Zhekenov et al. 2017), thus integrating into the deep eutectic solvent's network and subsequently strengthening the hydrogen bonds taking place between the hydrogen bond acceptor and the hydrogen bond donor at a low water content (Hammond et al. 2017a; Weng and Toner 2018). However, further dilution results in the weakening of the interactions that usually dominate in a deep eutectic solvent supramolecular structure owing to the tendency of water to interact with the deep eutectic solvent's forming compounds. The preferential hydration of chloride anions was reported in numerous papers dealing with different ChCl-based deep eutectic solvents like ChCl:U, ChCl:G, ChCl:EG, and ChCl:LA (Alcalde et al. 2019; Fetisov et al. 2018; Kaur et al. 2020; Kumari et al. 2018; Weng and Toner 2018). Yet, when it comes to the hydration level at which the transition happens, the values are not always consistent for the same deep eutectic solvent. For instance, the transition point varied between 15 and 51 wt% for ChCl:U (Hammond et al. 2017a; Kumari et al. 2018; Posada et al. 2017; Shah and Mjalli 2014). There are not enough studies to compare between the transition points of other deep eutectic solvents. Few studies also proved that temperature does not affect the structure of deep eutectic solvent-water mixtures (Celebi et al. 2019; Weng and Toner 2018). Further studies must be conducted on other deep eutectic solvents because although this transition is likely to occur in all the aqueous mixtures of deep eutectic solvents, the changeover water content obviously depends on the hydrogen bond acceptor and hydrogen bond donor types as well as their molar

Table 1.2 Investigations of the effect of water on deep eutectic solvents' systems (in chronological order)

Studied system(s)	Water content	Technique(s)	Comments	References
ChCl:U ChCl:thiourea 1:2	14–57–90 and 95 wt%	NMR	The characteristic chloride-HBD complex was maintained at 14 wt% water. Yet, a disruption was noticed at or above 57 wt% water content.	Gutiérrez et al. (2009)
ChCl:U 1:2	0–90 wt%	MD simulations	The water addition caused a decrease in the number of urea-urea and urea-Cl hydrogen bonds owing to the preferential binding of chloride anions to water and an increase in the ions' diffusion. The effect of water became notable beyond 25–30 wt% water.	Shah and Mjalli (2014)
ChCl:U ChCl:EG ChCl:G 1:2	2.5–20 wt%	Pulsed field gradient NMR	Unlike the case of ChCl:G, a difference between the diffusion coefficients of the aliphatic and hydroxyl groups of choline or the HBD was observed when adding water to both ChCl:U and ChCl:EG, suggesting an interaction with water.	D'Agostino et al. (2015)
1,2-Propanediol:ChCl:water 1:1:1	0–100% (v/v)	NMR	At a water content below 50% (v/v), the NADES structure is maintained. However, further dilution leads to an aqueous solution of the free components of the NADES.	Dai et al. (2015))
ChCl:G 1:3	25 and 50 wt%	NMR	The interactions between DES components are weakened up to 25 wt% water and new interactions with water are observed. However, in the presence of 50 wt% water, the interactions between ChCl and glycerol are completely disrupted.	Hadj-Kali et al. (2015)

ChCl:U 1:2	0.25 wt% (initial water content)	Measurement of water absorption as a function of the exposure time to the open air by Karl Fischer titration method at a temperature of 298 ± 2 K and a humidity of 60%	The water uptake by the DES is very high during the first hours with a rate of around 6.8 wt% per day. The rate decreased to 3.4 wt% after 10 hours and then was zeroed after 65 days' exposure to air with 40 wt% water content. On the other hand, water content can be easily tailored through heating. It went down from 40 wt% to 1.5 wt% after heating for 6 to 18 h.	Du et al. (2016)
ChCl:U ChCl:EG ChCl:G 1:2	0–1 mole fraction	MD simulations	Water molecules are integrated within the DES network below 30% mole fraction of water while retaining the intermolecular interactions occurring in the three DES. However, important changes were observed in fluidity of the different species and the number of hydrogen bonds above 50% mole fraction.	Zhekenov et al. (2017)
ChCl:U 1:2	0–100 wt%	Brillouin and NMR spectroscopies	Both techniques perfectly agreed on the dilution range at which structural rearrangements occur in DES-water mixtures. Two distinct regions intersecting at 85 wt% DES were observed for sound propagation via Brillouin spectroscopy, while hydroxyl protons of choline were completely exchanged with water at the same hydration level, as proved by NMR.	Posada et al. (2017)

(continued)

Table 1.2 (continued)

Studied system(s)	Water content	Technique(s)	Comments	References
ChCl:U 1:2	ChCl:U:water mixtures at molar ratios 1:2:1, 1:2:2, 1:2:5, 1:2:10, 1:2:15, and 1:2:20 (corresponding to 6.48–67.53 wt% water)	Neutron total scattering and empirical potential structure refinement	The hydrogen bonds between choline and urea are slightly reinforced at low water levels (≤1 mole fraction). The DES then resisted hydration between 2 and 10 mole fractions of water. At 15 mole fraction of water (83 mole % or 51 wt%), a transition from "water in DES" to "DES in water" is witnessed since the DES structural motifs disappeared and the system is rather considered as an aqueous solution of the DES components.	Hammond et al. (2017a)
ChCl:malic acid 1:1	2 mole equivalents of water (11.62 wt%)	Neutron total scattering and quasi-elastic neutron scattering	Water contributes to the hydrogen bonding structure of the DES by acting as a second smaller HBD at this hydration level. Water is sequestered into the DES by mostly interacting with the ammonium group of choline while preserving the interactions between the DES components.	Hammond et al. (2017b)
ChCl:U 1:2 ChCl:U:water 1:2:3	17 wt% for the hydrated DES	MD simulations	The anhydrous DES mainly consists of hydrogen bonds between urea and the chloride anions. These hydrogen bonds were disrupted in the hydrated DES since water tends to interact with chloride ions. In addition, some of urea-chloride hydrogen bonds are replaced by urea-water hydrogen bonds.	Fetisov et al. (2018)

Binary systems of 1:2 ChCl:G DES/water or glycerol/water	0–100% mole fraction (in 5% intervals) or 0.9 mole fraction for MD simulations	FTIR and Raman spectroscopies and MD simulations	Results analysis by a multivariate method detected five species in the DES-water mixture: pure water, DES, DES-water association, glycerol, and glycerol-water association. The DES-water interactions dominate over DES-DES interactions at 0.9 mole fraction of water. Yet, DES was identified as a highly stable distinct solvent from its constituents. MD simulations pictured the mixture of DES-water at 0.9 mole fraction of water as three different regions: a similar structure to that of pure DES, ChCl solvated by water, and self-assembled glycerol molecules (Fig. 1.8).	Ahmadi et al. (2018)

(continued)

Table 1.2 (continued)

Studied system(s)	Water content	Technique(s)	Comments	References
ChCl:U 1:2	0–100% mole fraction	MD simulations	The ion pairing and ionic hydration within the (DES + water) mixtures were evaluated. Results showed that both choline and chloride ions were not completely hydrated even at 0.9 mole fraction of water. However, the interactions between chloride anion and the surrounding water molecules are stronger than those between choline and water. Besides, the ion pairing between choline and chloride and the ionic hydration coexist in the studied mixtures. While the ion pairing decreases with the increasing water content, the ionic hydration increases and the curves meet at 0.8 mole fraction of water.	Gao et al. (2018)

ChCl:G 1:2	0–73.6 wt%	MD simulations	Water played both destructuring and structuring roles in ChCl:G DES. The water decreases the major type of interactions in the DES network which is the ChCl:G complex, with chloride anion serving as a bridge between the cation and the HBD. This effect was observed owing to the tendency of water to solvate chloride anions. On the other hand, the direct choline-glycerol hydrogen bonds were enhanced at low water content (10.1 wt%) compared to the anhydrous DES. Water molecules were also able to replace Cl^- by linking choline and glycerol. These types of interactions (choline-water-glycerol) reached a maximum at 35.8 wt% water. Overall, the chloride-bridge interactions between choline and glycerol were predominant below 21.8 wt% water.	Weng and Toner (2018)
ChCl:U 1:2	3.4–58.1 wt%	MD simulations	The DES structure is qualitatively preserved below 41 wt% of water. Above this water content, water tends to solvate chloride anions as well as choline cations, thus leading to a transition from DES to an aqueous solution of ChCl and urea.	Kumari et al. (2018)

(continued)

Table 1.2 (continued)

Studied system(s)	Water content	Technique(s)	Comments	References
ChCl:U ChCl:methylurea ChCl:EG ChCl:G 1:2 ChCl:glucose ChCl:xylitol 2:1 ChCl:oxalic acid ChCl:glutaric acid 1:1	Measured water absorption from air as a function of time (up to 4–8 h) at a temperature of 22 ± 2 °C and a humidity of 32%	Attenuated total reflectance infrared spectroscopy (ATR-IR) and perturbation-correlation moving-window two-dimensional correlation IR spectra techniques (PCMW2D-COS)	All the studied DES are hygroscopic, but their order for the absorption capacity is not in line with that of the absorption rate. Thus, a high absorption rate does not necessarily mean a high capacity. ChCl:glucose and ChCl:xylitol were considered for the investigation of the water absorption process. Results have shown that water molecules are attracted into the bulk of DES first, followed by their surface absorption. The early water absorption increases with the number of hydrophilic groups. Further water absorption depends on the strength of the hydrogen bonds between DES components. This explains why ChCl:glucose, with a higher number of hydrophilic groups, presents a higher rate but a lower absorption capacity when compared to ChCl:xylitol.	Chen et al. (2019b)
Resorcinol:hexylresorcinol:tetraethyl ammonium bromide 1.25:1:1	0–100 wt%	Brillouin and NMR spectroscopies	Both techniques led to the observation of two distinct behaviors, below and above 70 wt% DES, thus reflecting the presence of two hydrogen bond networks: a DES network at low water content and another water network at high water content.	Posada et al. (2019)

ChCl:malic acid 1:1	4–60 wt%	Brillouin and NMR spectroscopies	Two different behaviors were observed when varying the ratio of ChCl:malic acid to water. These two regimes are explained by the transition from "water-in-DES" to "DES-in-water" system. This transition point occurred at 60–70 wt% DES, as supported by Brillouin spectroscopy results, chemical shifts, self-diffusion coefficients, and the relaxation times obtained via NMR.	Roldán-Ruiz et al. (2019)
ChCl:U:water 1:1.95:1.67 (equivalent to 10.5 wt% water) ChCl:EG:water 1:1.89:1.33 (equivalent to 8.5 wt% water) ChCl:G:water 1:1.96:0.97 (equivalent to 5.3 wt% water)	Further addition of 10 wt% or 90 wt% of water to the ternary DES	NMR	The most marked hydrogen bonds in urea-based DES are the ones between hydroxyl protons of choline and water as well as between methyl groups of choline and urea. No correlation has been detected between urea and water. On the other hand, interactions between all species were observed in ChCl:G:water, but glycerol-water interactions were the weakest. In ChCl:EG:water, and unlike the other two DES, strong ethylene glycol-water interactions were detected. After 10 wt% dilution, the DES structures were preserved. However, after 90 wt% dilution, the mixtures are best described as aqueous solutions of individual compounds.	Delso et al. (2019)

(continued)

Table 1.2 (continued)

Studied system(s)	Water content	Technique(s)	Comments	References
ChCl:LA 1:1; 1:1.5; 1:2; 1:2.5	1 ChCl: 1 lactic acid: *j* $H_2 0$ (*j* = 1-10) for DFT and up to 22 wt% for MD simulations	DFT calculations and MD simulations	The DES consists of HBA:HBD clusters which are diluted upon water absorption. Water tends to interact with Cl and to a lesser extent with choline and lactic acid. Yet, the presence of water in the studied composition range does not disrupt the HBA:HBD hydrogen bonding of the DES.	Alcalde et al. (2019)
ChCl:DEG (n = 2) ChCl:TEG (n = 3) ChCl:PEG 200 (n ~ 4) 1:3 (with n the number of ethylene glycol units)	0–75 wt% $D_2 0$	FTIR and NMR	Both NMR and FTIR results showed that the interactions between DES components are gradually weaken by the addition of up to 50 wt% water but the supramolecular structures were somehow maintained. Further dilution led to the complete dissociation and hydration of the DES at 75 wt%.	Gabriele et al. (2019)
Resorcinol:U:ChCl 3:2:1	0–75 wt%	Brillouin spectroscopy, NMR, neutron scattering, MD simulations, and DSC	Water acts as another HBD or HBA at 75 wt% DES by strongly interacting with the compounds forming the ternary DES resorcinol:ChCl:U without highly affecting the hydrogen bonds between them. At this point, the system is known as "water in DES" with the lowest DES content or highest dilution point. Further, this mixture results in a new deeper quaternary eutectic, presenting a lower melting point than the corresponding ternary DES and water.	López-Salas et al. (2019)

ChCl:U ChCl:EG 1:2	0–100 wt%	MD simulations	Slight changes in hydrogen bond numbers and radial distribution functions were observed at a water content lower than 5 wt%, while the microscopic structure of both DES is massively affected between 5 and 40 wt% of water. Above 40 wt%, both DES lose their intermolecular structure due to dissolution in water and thus the studied parameters barely change. Besides, the structure of DES-water is not affected by the temperature.	Celebi et al. (2019)
RAMEB:Lev 1:27	2.5–60 wt%	NMR	The diffusion coefficients of the mixture's components revealed two well-differentiated behaviors, at a cross-over composition of 72 wt% DES, as the water content varied. This was explained by the predominance of the "water-in-DES" system above 72 wt% DES and a prevalence of the "DES-in-water" system below that same composition.	El Achkar et al. (2020a)

(continued)

Table 1.2 (continued)

Studied system(s)	Water content	Technique(s)	Comments	References
ChCl:EG 1:2	3.7–60.3 wt%	MD simulations	The nanostructure of ChCl:EG DES is maintained up to 40 mol% water (13.2 wt%). Between 27.6 and 43.2 wt%, a transition from DES-like structure to water-like structure is noticed. This transition is accompanied by the preferred hydration of chloride, the tendency of water to interact with choline's ammonium group, and the increased hydrogen bonding between ethylene glycol molecules.	Kaur et al. (2020)
ChCl:EG 1:2 and 1:3	0–87 wt%	DSC	When studying the phase behavior of ChCl:EG as a function of water content, two distinct thermal phase behaviors were detected at a cross-over composition of 30 wt% water content. When the latter is below 30 wt%, no water crystallization is observed and the glassy phase behavior of the pure DES is preserved. However, when the water content is above 30 wt%, there is a coexistence of ice which is reflected by the crystallization and a residual liquid phase, made of water and rich in DES components, called maximally freeze-concentrated solution. In this latter case, water is thought to be trapped via strong interactions between water and DES components.	Jani et al. (2020)

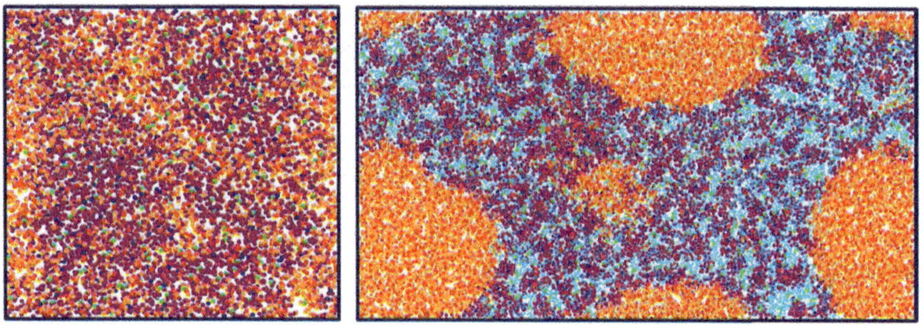

Fig. 1.8 Snapshots from molecular dynamics simulations of choline chloride:glycerol (left) and choline chloride:glycerol-water system (right) at 0.9 mole fraction of water. Purple, dark blue and green points represent the carbon, nitrogen and chlorine atoms of choline chloride, respectively. Orange points represent the carbon atoms of glycerol. Red and light blue points represent oxygen atoms and water molecules, respectively. (Reproduced from (Ahmadi et al. 2018) with permission from the PCCP Owner Societies)

ratio (Gabriele et al. 2019). Besides, a deep understanding of the impact of water on the deep eutectic solvent's structural arrangement will surely expand the possibility of tuning the deep eutectic solvent-water mixtures according to the desired applications (Fig. 1.8).

1.6 Conclusion

The quest for green solvents is a rising topic contributing to the goals of green chemistry. Deep eutectic solvents constitute the most considered and investigated solvents nowadays. These solvents possess quite interesting properties, thus increasing their possibility to replace other conventional solvents in numerous academic and industrial sectors. This chapter provides a general and comprehensive overview on the basics and properties of deep eutectic solvents. Their updated definition, classification, and preparation methods were elucidated. At a second stage, their main physicochemical properties particularly their phase behavior, density, viscosity, ionic conductivity, surface tension, and polarity are summarized. This segment also highlights the great tunability of these systems which properties can be modeled by several parameters. From the choice of the hydrogen bond donor, hydrogen bond acceptor, and their molar composition to the temperature and water content, these solvents can be designed to meet the requirements of the targeted applications. Finally, given the ubiquity of water and the wide consideration of binary mixtures of deep eutectic solvents and water in plenty of applications, the impact of water on both the physicochemical properties and the structure of the deep eutectic systems was emphasized. The presence of water significantly affects the properties of deep eutectic solvents. Even though it can present some beneficial outcomes, the presence of water may disturb the supramolecular network of the eutectic mixtures.

Several studies engaged in understanding the behavior of aqueous mixtures of deep eutectic solvents have recently emerged. A transition from one system to another upon the addition of water is proposed by several studies. A deeper understanding of deep eutectic solvents still requires further studies especially that there are a countless number of combinations that can lead to their formation.

References

Abbott AP, Capper G, Davies DL, Munro HL, Rasheed RK, Tambyrajah V (2001) Preparation of novel, moisture-stable, Lewis-acidic ionic liquids containing quaternary ammonium salts with functional side chains. Chem Commun (19):2010–2011. https://doi.org/10.1039/B106357J

Abbott AP, Capper G, Davies DL, Rasheed RK, Tambyrajah V (2003) Novel solvent properties of choline chloride/urea mixtures. Chem Commun 1:70–71. https://doi.org/10.1039/B210714G

Abbott AP, Boothby D, Capper G, Davies DL, Rasheed RK (2004a) Deep eutectic solvents formed between choline chloride and carboxylic acids: versatile alternatives to ionic liquids. J Am Chem Soc 126(29):9142–9147. https://doi.org/10.1021/ja048266j

Abbott AP, Capper G, Davies DL, Rasheed RK (2004b) Ionic liquid analogues formed from hydrated metal salts. Chem Eur J 10(15):3769–3774. https://doi.org/10.1002/chem.200400127

Abbott AP, Capper G, Gray S (2006) Design of improved deep eutectic solvents using hole theory. ChemPhysChem 7(4):803–806. https://doi.org/10.1002/cphc.200500489

Abbott AP, Barron JC, Ryder KS, Wilson D (2007a) Eutectic-based ionic liquids with metal-containing anions and cations. Chem Eur J 13(22):6495–6501. https://doi.org/10.1002/chem.200601738

Abbott AP, Harris RC, Ryder KS (2007b) Application of hole theory to define ionic liquids by their transport properties †. J Phys Chem B 111(18):4910–4913. https://doi.org/10.1021/jp0671998

Abbott AP, Harris RC, Ryder KS, D'Agostino C, Gladden LF, Mantle MD (2011) Glycerol eutectics as sustainable solvent systems. Green Chem 13(1):82–90. https://doi.org/10.1039/C0GC00395F

Abranches DO, Martins MAR, Silva LP, Schaeffer N, Pinho SP, Coutinho JAP (2019) Phenolic hydrogen bond donors in the formation of non-ionic deep eutectic solvents: the quest for type V DES. Chem Commun 55(69):10253–10256. https://doi.org/10.1039/C9CC04846D

Agieienko V, Buchner R (2019) Densities, viscosities, and electrical conductivities of pure anhydrous reline and its mixtures with water in the temperature range (293.15 to 338.15) K. J Chem Eng Data 64(11):4763–4774. https://doi.org/10.1021/acs.jced.9b00145

Ahmadi R, Hemmateenejad B, Safavi A, Shojaeifard Z, Shahsavar A, Mohajeri A, Dokoohaki MH, Zolghadr AR (2018) Deep eutectic–water binary solvent associations investigated by vibrational spectroscopy and chemometrics. Phys Chem Chem Phys 20(27):18463–18473. https://doi.org/10.1039/C8CP00409A

Alcalde R, Gutiérrez A, Atilhan M, Aparicio S (2019) An experimental and theoretical investigation of the physicochemical properties on choline chloride – lactic acid based natural deep eutectic solvent (NADES). J Mol Liq 290:110916. https://doi.org/10.1016/j.molliq.2019.110916

Aroso IM, Paiva A, Reis RL, Duarte ARC (2017) Natural deep eutectic solvents from choline chloride and betaine – physicochemical properties. J Mol Liq 241:654–661. https://doi.org/10.1016/j.molliq.2017.06.051

Celebi AT, Vlugt TJH, Moultos OA (2019) Structural, thermodynamic, and transport properties of aqueous reline and ethaline solutions from molecular dynamics simulations. J Phys Chem B 123(51):11014–11025. https://doi.org/10.1021/acs.jpcb.9b09729

Chen Z, Ludwig M, Warr GG, Atkin R (2017) Effect of cation alkyl chain length on surface forces and physical properties in deep eutectic solvents. J Colloid Interface Sci 494:373–379. https://doi.org/10.1016/j.jcis.2017.01.109

Chen W, Bai X, Xue Z, Mou H, Chen J, Liu Z, Mu T (2019a) The formation and physicochemical properties of PEGylated deep eutectic solvents. New J Chem 43(22):8804–8810. https://doi.org/10.1039/C9NJ02196E

Chen Y, Yu D, Chen W, Fu L, Mu T (2019b) Water absorption by deep eutectic solvents. Phys Chem Chem Phys 21(5):2601–2610. https://doi.org/10.1039/C8CP07383J

Choi YH, van Spronsen J, Dai Y, Verberne M, Hollmann F, Arends IWCE, Witkamp G-J, Verpoorte R (2011) Are natural deep eutectic solvents the missing link in understanding cellular metabolism and physiology? Plant Physiol 156(4):1701–1705. https://doi.org/10.1104/pp.111.178426

Coutinho JAP, Pinho SP (2017) Special issue on deep eutectic solvents: a foreword. Fluid Phase Equilib 448:1. https://doi.org/10.1016/j.fluid.2017.06.011

Cui Y, Li C, Yin J, Li S, Jia Y, Bao M (2017) Design, synthesis and properties of acidic deep eutectic solvents based on choline chloride. J Mol Liq 236:338–343. https://doi.org/10.1016/j.molliq.2017.04.052

D'Agostino C, Harris RC, Abbott AP, Gladden LF, Mantle MD (2011) Molecular motion and ion diffusion in choline chloride based deep eutectic solvents studied by 1H pulsed field gradient NMR spectroscopy. Phys Chem Chem Phys 13(48):21383. https://doi.org/10.1039/c1cp22554e

D'Agostino C, Gladden LF, Mantle MD, Abbott AP, Ahmed EI, Al-Murshedi AYM, Harris RC (2015) Molecular and ionic diffusion in aqueous – deep eutectic solvent mixtures: Probing inter-molecular interactions using PFG NMR. *Phys Chem Chem Phys* 17(23):15297–15304. https://doi.org/10.1039/C5CP01493J

Dai Y, van Spronsen J, Witkamp G-J, Verpoorte R, Choi YH (2013) Natural deep eutectic solvents as new potential media for green technology. Anal Chim Acta 766:61–68. https://doi.org/10.1016/j.aca.2012.12.019

Dai Y, Witkamp G-J, Verpoorte R, Choi YH (2015) Tailoring properties of natural deep eutectic solvents with water to facilitate their applications. Food Chem 187:14–19. https://doi.org/10.1016/j.foodchem.2015.03.123

Delso I, Lafuente C, Muñoz-Embid J, Artal M (2019) NMR study of choline chloride-based deep eutectic solvents. J Mol Liq 290:111236. https://doi.org/10.1016/j.molliq.2019.111236

Du C, Zhao B, Chen X-B, Birbilis N, Yang H (2016) Effect of water presence on choline chloride-2urea ionic liquid and coating platings from the hydrated ionic liquid. Sci Rep 6(1). https://doi.org/10.1038/srep29225

Duarte ARC, Ferreira ASD, Barreiros S, Cabrita E, Reis RL, Paiva A (2017) A comparison between pure active pharmaceutical ingredients and therapeutic deep eutectic solvents: solubility and permeability studies. Eur J Pharm Biopharm 114:296–304. https://doi.org/10.1016/j.ejpb.2017.02.003

Dwamena AK, Raynie DE (2020) Solvatochromic parameters of deep eutectic solvents: effect of different carboxylic acids as hydrogen bond donor. J Chem Eng Data 65(2):640–646. https://doi.org/10.1021/acs.jced.9b00872

El Achkar T, Fourmentin S, Greige-Gerges H (2019) Deep eutectic solvents: an overview on their interactions with water and biochemical compounds. J Mol Liq 288:111028. https://doi.org/10.1016/j.molliq.2019.111028

El Achkar T, Moufawad T, Ruellan S, Landy D, Greige-Gerges H, Fourmentin S (2020a) Cyclodextrins: from solute to solvent. Chem Commun 56:3385–3388. https://doi.org/10.1039/D0CC00460J

El Achkar T, Moura L, Moufawad T, Ruellan S, Panda S, Longuemart S, Legrand F-X, Costa Gomes M, Landy D, Greige-Gerges H, Fourmentin S (2020b) New generation of supramolecular mixtures: characterization and solubilization studies. Int J Pharm:119443. https://doi.org/10.1016/j.ijpharm.2020.119443

Fahri F, Bacha K, Chiki FF, Mbakidi J-P, Panda S, Bouquillon S, Fourmentin S (2020) Air pollution: new bio-based ionic liquids absorb both hydrophobic and hydrophilic volatile organic compounds with high efficiency. Environ Chem Lett 18:1403–1411. https://doi.org/10.1007/s10311-020-01007-8

Fetisov EO, Harwood DB, Kuo I-FW, Warrag SEE, Kroon MC, Peters CJ, Siepmann JI (2018) First-principles molecular dynamics study of a deep eutectic solvent: choline chloride/urea and its mixture with water. J Phys Chem B 122(3):1245–1254. https://doi.org/10.1021/acs.jpcb.7b10422

Florindo C, Oliveira FS, Rebelo LPN, Fernandes AM, Marrucho IM (2014) Insights into the synthesis and properties of deep eutectic solvents based on Cholinium chloride and carboxylic acids. ACS Sustain Chem Eng 2(10):2416–2425. https://doi.org/10.1021/sc500439w

Florindo C, McIntosh AJS, Welton T, Branco LC, Marrucho MI (2018) A closer look into deep eutectic solvents: Exploring intermolecular interactions using solvatochromic probes. Phys Chem Chem Phys 20(1):206–213. https://doi.org/10.1039/C7CP06471C

Florindo C, Branco LC, Marrucho IM (2019) Quest for green-solvent design: from hydrophilic to hydrophobic (deep) eutectic solvents. ChemSusChem 12(8):1549–1559. https://doi.org/10.1002/cssc.201900147

Francisco M, van den Bruinhorst A, Kroon MC (2012) New natural and renewable low transition temperature mixtures (LTTMs): screening as solvents for lignocellulosic biomass processing. Green Chem 14(8):2153–2157. https://doi.org/10.1039/C2GC35660K

Gabriele F, Chiarini M, Germani R, Tiecco M, Spreti N (2019) Effect of water addition on choline chloride/glycol deep eutectic solvents: characterization of their structural and physicochemical properties. J Mol Liq 291:111301. https://doi.org/10.1016/j.molliq.2019.111301

Gao Q, Zhu Y, Ji X, Zhu W, Lu L, Lu X (2018) Effect of water concentration on the microstructures of choline chloride/urea (1:2) /water mixture. Fluid Phase Equilib 470:134–139. https://doi.org/10.1016/j.fluid.2018.01.031

García G, Aparicio S, Ullah R, Atilhan M (2015) Deep eutectic solvents: physicochemical properties and gas separation applications. Energy Fuel 29(4):2616–2644. https://doi.org/10.1021/ef5028873

Germani R, Orlandini M, Tiecco M, Del Giacco T (2017) Novel low viscous, green and amphiphilic N-oxides/phenylacetic acid based deep eutectic solvents. J Mol Liq 240:233–239. https://doi.org/10.1016/j.molliq.2017.05.084

Gomez FJV, Espino M, Fernández MA, Silva MF (2018) A greener approach to prepare natural deep eutectic solvents. Chem Select 3(22):6122–6125. https://doi.org/10.1002/slct.201800713

González CG, Mustafa NR, Wilson EG, Verpoorte R, Choi YH (2018) Application of natural deep eutectic solvents for the "green"extraction of vanillin from vanilla pods. Flavour Fragr J 33(1):91–96. https://doi.org/10.1002/ffj.3425

Gutiérrez MC, Ferrer ML, Mateo CR, del Monte F (2009) Freeze-drying of aqueous solutions of deep eutectic solvents: a suitable approach to deep eutectic suspensions of self-assembled structures. Langmuir 25(10):5509–5515. https://doi.org/10.1021/la900552b

Hadj-Kali MK, Al-khidir KE, Wazeer I, El-blidi L, Mulyono S, AlNashef IM (2015) Application of deep eutectic solvents and their individual constituents as surfactants for enhanced oil recovery. Colloids Surf A Physicochem Eng Asp 487:221–231. https://doi.org/10.1016/j.colsurfa.2015.10.005

Häkkinen R, Willberg-Keyriläinen P, Ropponen J, Virtanen T (2019) Effect of composition and water content on physicochemical properties of choline chloride-boric acid low-melting mixtures. J Mol Liq 280:104–110. https://doi.org/10.1016/j.molliq.2019.02.011

Hammond OS, Bowron DT, Edler KJ (2017a) The effect of water upon deep eutectic solvent nanostructure: an unusual transition from ionic mixture to aqueous solution. Angew Chem Int Ed 56(33):9782–9785. https://doi.org/10.1002/anie.201702486

Hammond OS, Bowron DT, Jackson AJ, Arnold T, Sanchez-Fernandez A, Tsapatsaris N, Garcia Sakai V, Edler KJ (2017b) Resilience of malic acid natural deep eutectic solvent nanostructure to solidification and hydration. J Phys Chem B 121(31):7473–7483. https://doi.org/10.1021/acs.jpcb.7b05454

Haraźna K, Walas K, Urbańska P, Witko T, Snoch W, Siemek A, Jachimska B, Krzan M, Napruszewska BD, Witko M, Bednarz S, Guzik M (2019) Polyhydroxyalkanoate-derived hydrogen-bond donors for the synthesis of new deep eutectic solvents. Green Chem 21(11):3116–3126. https://doi.org/10.1039/C9GC00387H

Hayyan A, Mjalli FS, AlNashef IM, Al-Wahaibi T, Al-Wahaibi YM, Hashim MA (2012) Fruit sugar-based deep eutectic solvents and their physical properties. Thermochim Acta 541:70–75. https://doi.org/10.1016/j.tca.2012.04.030

Hayyan A, Mjalli FS, AlNashef IM, Al-Wahaibi YM, Al-Wahaibi T, Hashim MA (2013) Glucose-based deep eutectic solvents: physical properties. J Mol Liq 178:137–141. https://doi.org/10.1016/j.molliq.2012.11.025

Ibrahim RK, Hayyan M, AlSaadi MA, Ibrahim S, Hayyan A, Hashim MA (2019) Physical properties of ethylene glycol-based deep eutectic solvents. J Mol Liq 276:794–800. https://doi.org/10.1016/j.molliq.2018.12.032

Jani A, Sohier T, Morineau D (2020) Phase behavior of aqueous solutions of ethaline deep eutectic solvent. J Mol Liq 304:112701. https://doi.org/10.1016/j.molliq.2020.112701

Kamlet MJ, Taft RW (1976) The solvatochromic comparison method. I. the .Beta.-scale of solvent hydrogen-bond acceptor (HBA) basicities. J Am Chem Soc 98(2):377–383. https://doi.org/10.1021/ja00418a009

Kamlet MJ, Abboud JL, Taft RW (1977) The solvatochromic comparison method. 6. The .Pi.* scale of solvent polarities. J Am Chem Soc 99(18):6027–6038. https://doi.org/10.1021/ja00460a031

Kareem MA, Mjalli FS, Hashim MA, AlNashef IM (2010) Phosphonium-based ionic liquids analogues and their physical properties. J Chem Eng Data 55(11):4632–4637. https://doi.org/10.1021/je100104v

Kaur S, Gupta A, Kashyap HK (2020) How hydration affects the microscopic structural morphology in a deep eutectic solvent. J Phys Chem B 124(11):2230–2237. https://doi.org/10.1021/acs.jpcb.9b11753

Kim SH, Park S, Yu H, Kim JH, Kim HJ, Yang Y-H, Kim YH, Kim KJ, Kan E, Lee SH (2016) Effect of deep eutectic solvent mixtures on lipase activity and stability. J Mol Catal B Enzym 128:65–72. https://doi.org/10.1016/j.molcatb.2016.03.012

Kumari P, Shobhna, Kaur S, Kashyap HK (2018) Influence of hydration on the structure of reline deep eutectic solvent: a molecular dynamics study. *ACS Omega* 3(11):15246–15255. https://doi.org/10.1021/acsomega.8b02447

Lapeña D, Lomba L, Artal M, Lafuente C, Giner B (2019) The NADES glyceline as a potential green solvent: a comprehensive study of its thermophysical properties and effect of water inclusion. J Chem Thermodyn 128:164–172. https://doi.org/10.1016/j.jct.2018.07.031

López-Salas N, Vicent-Luna JM, Imberti S, Posada E, Roldán MJ, Anta JA, Balestra SRG, Madero Castro RM, Calero S, Jiménez-Riobóo RJ, Gutiérrez MC, Ferrer ML, del Monte F (2019) Looking at the "water-in-deep-eutectic-solvent" system: a dilution range for high performance eutectics. ACS Sustain Chem Eng 7(21):17565–17573. https://doi.org/10.1021/acssuschemeng.9b05096

Martins MAR, Crespo EA, Pontes PVA, Silva LP, Bülow M, Maximo GJ, Batista EAC, Held C, Pinho SP, Coutinho JAP (2018) Tunable hydrophobic eutectic solvents based on terpenes and monocarboxylic acids. ACS Sustain Chem Eng 6(7):8836–8846. https://doi.org/10.1021/acssuschemeng.8b01203

Martins MAR, Pinho SP, Coutinho JAP (2019) Insights into the nature of eutectic and deep eutectic mixtures. J Solut Chem 48(7):962–982. https://doi.org/10.1007/s10953-018-0793-1

Meng X, Ballerat-Busserolles K, Husson P, Andanson J-M (2016) Impact of water on the melting temperature of urea + choline chloride deep eutectic solvent. New J Chem 40(5):4492–4499. https://doi.org/10.1039/C5NJ02677F

Mitar A, Panić M, Prlić Kardum J, Halambek J, Sander A, Zagajski Kučan K, Radojčić Redovniković I, Radošević K (2019) Physicochemical properties, cytotoxicity, and antioxidative activity of natural deep eutectic solvents containing organic acid. Chem Biochem Eng Q 33(1):1–18. https://doi.org/10.15255/CABEQ.2018.1454

Mjalli FS (2016) Mass connectivity index-based density prediction of deep eutectic solvents. Fluid Phase Equilib 409:312–317. https://doi.org/10.1016/j.fluid.2015.09.053

Mjalli FS, Shahbaz K, AlNashef IM (2015) Modified Rackett equation for modelling the molar volume of deep eutectic solvents. Thermochim Acta 614:185–190. https://doi.org/10.1016/j.tca.2015.06.026

Mulia K, Fauzia F, Krisanti EA (2019) Polyalcohols as hydrogen-bonding donors in choline chloride-based deep eutectic solvents for extraction of Xanthones from the pericarp of Garcinia mangostana L. Molecules 24(3):636. https://doi.org/10.3390/molecules24030636

Nunes RJ, Saramago B, Marrucho IM (2019) Surface tension of dl-menthol:octanoic acid eutectic mixtures. J Chem Eng Data 64(11):4915–4923. https://doi.org/10.1021/acs.jced.9b00424

Paiva A, Craveiro R, Aroso I, Martins M, Reis RL, Duarte ARC (2014) Natural deep eutectic solvents – solvents for the 21st century. ACS Sustain Chem Eng 2(5):1063–1071. https://doi.org/10.1021/sc500096j

Pandey A, Pandey S (2014) Solvatochromic probe behavior within choline chloride-based deep eutectic solvents: effect of temperature and water. J Phys Chem B 118(50):14652–14661. https://doi.org/10.1021/jp510420h

Pandey A, Rai R, Pal M, Pandey S (2013) How polar are choline chloride-based deep eutectic solvents? Phys Chem Chem Phys 16(4):1559–1568. https://doi.org/10.1039/C3CP53456A

Posada E, López-Salas N, Riobóo RJJ, Ferrer ML, Gutiérrez MC, Monte F d (2017) Reline aqueous solutions behaving as liquid mixtures of H-bonded co-solvents: microphase segregation and formation of co-continuous structures as indicated by Brillouin and 1H NMR spectroscopies. Phys Chem Chem Phys 19(26):17103–17110. https://doi.org/10.1039/C7CP02180A

Posada E, Roldán-Ruiz MJ, Jiménez Riobóo RJ, Gutiérrez MC, Ferrer ML, del Monte F (2019) Nanophase separation in aqueous dilutions of a ternary DES as revealed by Brillouin and NMR spectroscopy. J Mol Liq 276:196–203. https://doi.org/10.1016/j.molliq.2018.11.139

Reichardt C (1994) Solvatochromic dyes as solvent polarity indicators. Chem Rev 94(8):2319–2358. https://doi.org/10.1021/cr00032a005

Ribeiro BD, Florindo C, Iff LC, Coelho MAZ, Marrucho IM (2015) Menthol-based eutectic mixtures: hydrophobic low viscosity solvents. ACS Sustain Chem Eng 3(10):2469–2477. https://doi.org/10.1021/acssuschemeng.5b00532

Rodriguez Rodriguez N, van den Bruinhorst A, Kollau LJBM, Kroon MC, Binnemans K (2019) Degradation of deep-eutectic solvents based on choline chloride and carboxylic acids. ACS Sustain Chem Eng 7(13):11521–11528. https://doi.org/10.1021/acssuschemeng.9b01378

Roldán-Ruiz MJ, Jiménez-Riobóo RJ, Gutiérrez MC, Ferrer ML, del Monte F (2019) Brillouin and NMR spectroscopic studies of aqueous dilutions of malicine: determining the dilution range for transition from a "water-in-DES" system to a "DES-in-water" one. J Mol Liq 284:175–181. https://doi.org/10.1016/j.molliq.2019.03.133

Rublova Y, Kityk A, Danilov F, Protsenko V (2020) Mechanistic study on surface tension of binary and ternary mixtures containing choline chloride, ethylene glycol and water (components of aqueous solutions of a deep eutectic solvent, ethaline). Z Phys Chem 234(3):399–413. https://doi.org/10.1515/zpch-2019-1492

Sanchez-Fernandez A, Hammond OS, Jackson AJ, Arnold T, Doutch J, Edler KJ (2017) Surfactant–solvent interaction effects on the micellization of cationic surfactants in a carboxylic acid-based deep eutectic solvent. Langmuir 33(50):14304–14314. https://doi.org/10.1021/acs.langmuir.7b03254

Santana APR, Mora-Vargas JA, Guimarães TGS, Amaral CDB, Oliveira A, Gonzalez MH (2019) Sustainable synthesis of natural deep eutectic solvents (NADES) by different methods. J Mol Liq 293:111452. https://doi.org/10.1016/j.molliq.2019.111452

Savi LK, Carpiné D, Waszczynskyj N, Ribani RH, Haminiuk CWI (2019a) Influence of temperature, water content and type of organic acid on the formation, stability and properties of functional natural deep eutectic solvents. Fluid Phase Equilib 488:40–47. https://doi.org/10.1016/j.fluid.2019.01.025

Savi LK, Dias MCGC, Carpine D, Waszczynskyj N, Ribani RH, Haminiuk CWI (2019b) Natural deep eutectic solvents (NADES) based on citric acid and sucrose as a potential green technology: a comprehensive study of water inclusion and its effect on thermal, physical and rheological properties. Int J Food Sci Technol 54(3):898–907. https://doi.org/10.1111/ijfs.14013

Scientific opinion on safety and efficacy of choline chloride as a feed additive for all animal species (2011) EFSA J 9(9):2353. https://doi.org/10.2903/j.efsa.2011.2353

Shah D, Mjalli FS (2014) Effect of water on the thermo-physical properties of reline: an experimental and molecular simulation based approach. Phys Chem Chem Phys 16(43):23900–23907. https://doi.org/10.1039/C4CP02600D

Shahbaz K, Mjalli FS, Hashim MA, AlNashef IM (2011a) Prediction of deep eutectic solvents densities at different temperatures. Thermochim Acta 515(1):67–72. https://doi.org/10.1016/j.tca.2010.12.022

Shahbaz K, Mjalli FS, Hashim MA, AlNashef IM (2011b) Using deep eutectic solvents based on methyl triphenyl phosphunium bromide for the removal of glycerol from palm-oil-based biodiesel. Energy Fuel 25(6):2671–2678. https://doi.org/10.1021/ef2004943

Shahbaz K, Baroutian S, Mjalli FS, Hashim MA, AlNashef IM (2012a) Densities of ammonium and phosphonium based deep eutectic solvents: prediction using artificial intelligence and group contribution techniques. Thermochim Acta 527:59–66. https://doi.org/10.1016/j.tca.2011.10.010

Shahbaz K, Mjalli FS, Hashim MA, AlNashef IM (2012b) Prediction of the surface tension of deep eutectic solvents. Fluid Phase Equilib 319:48–54. https://doi.org/10.1016/j.fluid.2012.01.025

Shahbaz K, Bagh FSG, Mjalli FS, AlNashef IM, Hashim MA (2013) Prediction of refractive index and density of deep eutectic solvents using atomic contributions. Fluid Phase Equilib 354:304–311. https://doi.org/10.1016/j.fluid.2013.06.050

Silva LP, Fernandez L, Conceição JHF, Martins MAR, Sosa A, Ortega J, Pinho SP, Coutinho JAP (2018) Design and characterization of sugar-based deep eutectic solvents using conductor-like screening model for real solvents. ACS Sustain Chem Eng. https://doi.org/10.1021/acssuschemeng.8b02042

Smith EL, Abbott AP, Ryder KS (2014) Deep eutectic solvents (DESs) and their applications. Chem Rev 114(21):11060–11082. https://doi.org/10.1021/cr300162p

Smith PJ, Arroyo CB, Hernandez FL, Goeltz JC (2019) Ternary deep eutectic solvent behavior of water and urea–choline chloride mixtures. J Phys Chem B. https://doi.org/10.1021/acs.jpcb.8b12322

Tang B, Row KH (2013) Recent developments in deep eutectic solvents in chemical sciences. Monatshefte Für Chem – Chem Month 144(10):1427–1454. https://doi.org/10.1007/s00706-013-1050-3

Teles ARR, Capela EV, Carmo RS, Coutinho JAP, Silvestre AJD, Freire MG (2017) Solvatochromic parameters of deep eutectic solvents formed by ammonium-based salts and carboxylic acids. Fluid Phase Equilib 448:15–21. https://doi.org/10.1016/j.fluid.2017.04.020

Valvi A, Dutta J, Tiwari S (2017) Temperature-dependent empirical parameters for polarity in choline chloride based deep eutectic solvents. J Phys Chem B 121(50):11356–11366. https://doi.org/10.1021/acs.jpcb.7b07754

van Osch DJGP, Zubeir LF, van den Bruinhorst A, Rocha MAA, Kroon MC (2015) Hydrophobic deep eutectic solvents as water-immiscible extractants. Green Chem 17(9):4518–4521. https://doi.org/10.1039/C5GC01451D

Weng L, Toner M (2018) Janus-faced role of water in defining nanostructure of choline chloride/glycerol deep eutectic solvent. Phys Chem Chem Phys 20(35):22455–22462. https://doi.org/10.1039/C8CP03882A

Yusof R, Abdulmalek E, Sirat K, Rahman MBA (2014) Tetrabutylammonium bromide (TBABr)-based deep eutectic solvents (DESs) and their physical properties. Molecules 19(6):8011–8026. https://doi.org/10.3390/molecules19068011

Zhang Q, De Oliveira Vigier K, Royer S, Jérôme F (2012) Deep eutectic solvents: syntheses, properties and applications. Chem Soc Rev 41(21):7108–7146. https://doi.org/10.1039/C2CS35178A

Zhekenov T, Toksanbayev N, Kazakbayeva Z, Shah D, Mjalli FS (2017) Formation of type III deep eutectic solvents and effect of water on their intermolecular interactions. Fluid Phase Equilib 441:43–48. https://doi.org/10.1016/j.fluid.2017.01.022

Chapter 2
Deep Eutectic Solvents for Innovative Pharmaceutical Formulations

Canh-Hung Nguyen, Luc Augis, Sophie Fourmentin, Gillian Barratt, and François-Xavier Legrand

Contents

C.-H. Nguyen
Université Paris-Saclay, CNRS UMR 8612, Institut Galien Paris-Saclay, Châtenay-Malabry, France

Department of Pharmaceutics, Hanoi University of Pharmacy, Hanoi, Vietnam

L. Augis · G. Barratt (✉) · F.-X. Legrand (✉)
Université Paris-Saclay, CNRS UMR 8612, Institut Galien Paris-Saclay, Châtenay-Malabry, France
e-mail: gillian.barratt@universite-paris-saclay.fr; francois-xavier.legrand@universite-paris-saclay.fr

S. Fourmentin
Unité de Chimie Environnementale et Interactions sur le Vivant (UCEIV, UR 4492), SFR Condorcet FR CNRS 3417, Université du Littoral Côte d'Opale, Dunkerque, France

S. Fourmentin et al. (eds.), *Deep Eutectic Solvents for Medicine, Gas Solubilization and Extraction of Natural Substances*, Environmental Chemistry for a Sustainable World 56, https://doi.org/10.1007/978-3-030-53069-3_2

Abstract Finding alternative solvents for industrial processes, such as chemical synthesis or extraction of biologically active molecules that are less toxic and more environmentally friendly than the organic solvents used up to now, is a major societal issue. These alternative solvents are often described as "green, biodegradable" solvents. Deep eutectic solvents, which are mixtures of simple and often naturally occurring compounds, have been extensively studied in this regard. Among their possible applications, there has been increasing interest in their use for the preparation of pharmaceutical formulations. Indeed, by changing the nature and ratio of their components, deep eutectic solvents can be adapted to a wide range of active molecules, from poorly soluble small molecules to labile macromolecules. The use of deep eutectic solvents to solubilize active molecules with low aqueous solubility and/or low permeability could be an alternative approach to increase their dissolution and in vivo absorption. This could result in significant increases of bioavailability or enhanced therapeutic efficacy of currently marketed drugs. Moreover, deep eutectic solvents can be used to limit phenomena like polymorphism or degradation which present a challenge to drug formulation. However, despite being generally described as biodegradable and nontoxic due to the nature of their constituents, the safety of deep eutectic solvents, which possess both novel physicochemical and biological properties, cannot be taken for granted and must therefore be carefully studied during development stages. Therefore, this chapter presents not only recent progress in the application of deep eutectic solvents in the development of formulations for improving therapeutic efficacy by different routes of administration but also studies that have been undertaken to investigate the toxicity of deep eutectic solvents to both living organisms and the environment.

Keywords Solubility · Macromolecules · Antimicrobial · Topical route · Oral route · Nasal route

Abbreviations

A	Acetamide
AA	Acrylic acid
ACA	Aconitic acid
Arg	Arginine
β-ala	β-alanine
B	Betaine
C	Choline
CA	Citric acid
CC	Choline chloride
CH	Cholinium hydroxide
DEEAC	*N,N*-diethylethanolammonium chloride
EG	Ethylene glycol

F	Fructose
G	Glucose
GA	Glycolic acid
GE	Geranate
GL	Glycerol
GLA	Glutamic acid
GLN	Glutamine
HBA	Hydrogen bond acceptor
HBD	Hydrogen bond donor
LA	Lactic acid
LAU	Lauric acid
MA	Malic acid
MEA	Maleic acid
MOA	Malonic acid
OA	Oxalic acid
P	Proline
PD	1,2-Propanediol (propylene glycol)
PHE	Phenylalanine
S	Sucrose
TA	Tartaric acid
TEAC	Tetraethylammonium chloride
TEG	Triethylene glycol
TBAB	Tetrabutylammonium bromide
TPAB	Tetrapropylammonium bromide
U	Urea
W	Water
X	Xylitol
ZC	Zinc chloride

2.1 Introduction

Recent progress in synthetic organic chemistry, molecular modelling, and high-throughput screening has led to the production of a large number of new chemical entities with specific pharmacological activity, as well as discovering new indications for existing molecules. However, for a promising molecule to become an effective medicine, it is necessary to select an adequate solvent for its formulation that will both preserve its integrity and be compatible with the requirements for administration to human patients. Many molecules that are revealed to have high specificity for pharmacological targets have low water solubility that limits both their formulation and their bioavailability; however, traditional organic solvents are not acceptable for pharmaceutical applications and also present considerable constraints for manufacturing. At the same time, alternative routes of administration avoiding intravenous injection, such as oral and topical routes, are attracting more and more interest because of their ease of use and better acceptability. This brings new challenges to formulation and a need for suitable additives.

Ionic liquids (IL) and deep eutectic solvents (DES) have been studied for over a century, but interest in their pharmaceutical applications is more recent (Agatemor et al. 2018; Adawiyah et al. 2016). Both solvents are mixtures of low-molecular-weight organic salts, containing a hydrogen bond donor (HBD) and a hydrogen bond acceptor (HBA), that are liquid below 100 °C. In 2011, Choi et al. put forward the concept of natural deep eutectic solvents (NADES) composed of amino acids, sugars, sugar alcohols, and polyalcohols that are found in living cells (Choi et al. 2011). They postulated that these compounds could represent a "missing link" between aqueous and lipophilic media and provide an environment that would allow cells to survive in extreme conditions such as dehydration. These compositions can be considered as "green solvents" that do not carry the same risks of toxicity and handling as traditional organic solvents (Durand et al. 2016; Vanda et al. 2018). As such, they have numerous applications in pharmaceutical formulations, which will be detailed in this chapter.

2.1.1 Definitions and Components

The terms "ionic liquid" and "deep eutectic solvent" are often used interchangeably, but there are differences between the two systems (see correspondence arising from Banerjee et al. 2018a; from Rogers and Gurau 2018; and from Banerjee et al. 2018b). Although both can be considered as low transition temperature mixtures, there are differences in the way in which they are obtained (Durand et al. 2016). Deep eutectic solvents are prepared by mixing two molecules, and the driving force for their formation is hydrogen bonding. On the other hand, ionic liquids are organic or inorganic salts whose behavior is mainly defined by ionic interactions. Within the category of deep eutectic solvents, some authors have proposed a subclass called natural deep eutectic solvents (NADES) composed of molecules that are found within cells, as mentioned above (Choi et al. 2011; Vanda et al. 2018). In this chapter, we will concentrate on deep eutectic solvents, although ionic liquids will be included when they have been compared with deep eutectic solvents in the same experimental system.

As can be seen in Tables 1, 2, 3, 4, 5 and 6, by far, the most frequently used hydrogen bond acceptor is choline or choline chloride, while the hydrogen bond donors are often organic acids or sugars, or urea.

2.1.2 Physicochemical Properties: Characterization of Deep Eutectic Solvents

The physicochemical characterization of deep eutectic solvents systems is based around thermal and rheological properties. The low transition temperature can be checked by simple measurement of melting point, but differential scanning calorimetry gives a more accurate picture of the thermal behavior.

As far as rheology is concerned, viscosity can be an issue for deep eutectic solvents (Moura et al. 2017). On one hand, for topical applications, a more viscous preparation would be an advantage as long as it can be easily spread on the skin. On the other hand, for other routes, excessive viscosity would hamper administration by other routes, such as oral, and can be a cause of toxicity.

2.2 Toxicity of Deep Eutectic Solvents

For any new pharmaceutical ingredient, toxicity profiling is extremely important. Many of the studies described above included an assessment of the effect of the deep eutectic solvents on mammalian tissues. A few systematic studies of toxicity have been made. Since deep eutectic solvents are proposed as "green solvents," emphasis has often been placed on their toxicity to the environment, and in particular, aquatic organisms have often been used as indicators. The studies that have been published have usually focused on bacterial cells, human cell lines in culture, and in vivo studies in rodents, although other organisms such as fungal cells, plants, invertebrates, and fish have also figured in the literature.

2.2.1 Toxicity Toward Microorganisms

In 2013, Hayyan et al. first studied the toxicity of deep eutectic solvents toward bacterial and eukaryotic cells (Hayyan et al. 2013). They chose two Gram-negative bacteria species (*Escherichia coli*, *Pseudomonas aeruginosa*) and two Gram-positive ones (*Bacillus subtilis*, *Staphylococcus aureus*) as well as brine shrimp (*Artemia saliva*) larvae as an indicator of ecotoxicity. They used deep eutectic solvents prepared from choline chloride combined with urea, glycerol, ethylene glycol, and triethylene glycol. In this study, they observed no inhibition of bacterial growth. When Mao et al. investigated the toxicity of deep eutectic solvents composed of choline chloride and urea, ethylene glycol, and glycerol toward *Arthrobacter simplex*, using growth, metabolic activity, and membrane permeability as the criteria of toxicity, they observed that the toxicity of the deep eutectic solvents was lower than that of the individual components (Mao et al. 2018). The ability of the bacteria to perform biotransformations of steroids was also improved in the deep eutectic solvents.

Wen et al. attempted to assess deep eutectic solvent toxicity and biodegradability toward several organisms including a bacterium, *Escherichia coli* (Wen et al. 2015). The deep eutectic solvents were composed of choline chloride or choline acetate with urea, glycerol, acetamide, or ethylene glycol at different molar ratios. Biodegradability was determined in closed bottles with microorganisms from a water treatment plant. In their experiments, studied solvents were toxic to bacteria at concentrations above 75 μM, and they inhibited bacterial growth much more than

Table 1 Studies on deep eutectic solvent toxicity

Type of deep eutectic solvent studied	Active agent, if any	Toxicity criteria	Conclusions	References
CC with U, GL, EG, and triethylene glycol		Growth of bacteria and brine shrimp larvae	Some toxicity toward shrimp larvae	Hayyan et al. (2013)
CC with U, GL, and EG		Growth of *Arthrobacter simplex*	Lower toxicity of deep eutectic solvents compared to the individual components	Mao et al. (2018)
CC and choline acetate with U, acetamide, G, and EG		Growth of a bacterium (*Escherichia coli*), garlic, (*Allium sativum*) and hydra (*Hydra sinensis*)	Deep eutectic solvents inhibit bacterial growth more than the sum of their component, but toxicity to higher organisms is less than that of individual components.	Wen et al. (2015)
Various		Growth of bacteria and yeast. Human cells lines (HeLa, MCF-7 and HEK293T) – mitochondrial activity	Toxicity of deep eutectic solvents containing organic acids	Radošević et al. (2018)
Various		Ecotoxicity, marine bacterium *Aliivibrio fischeri*, and inhibition of bioluminescence	Tetramethylammonium chloride less toxic than tetrapropylammonium chloride and CC safe	Macário et al. (2018a, b)
CC with alcohols, sugars, organic acids, U, and ZC		Ecotoxicity, fungal cells, and carp	Only deep eutectic solvents with ZC, MOA, or *para*-toluene sulfonic acid are toxic.	Juneidi et al. (2016)
Acetylcholine chloride (AcChCl): acetamide (1:2)		Growth of a bacterium (*Escherichia coli*)	Toxicity above 300 nM caused by acidification of the culture medium	Torregrosa-Crespo et al. (2020)

(continued)

Table 1 (continued)

Type of deep eutectic solvent studied	Active agent, if any	Toxicity criteria	Conclusions	References
Various ammonium-based		Human cell lines – mitochondrial activity, microscopic examination, apoptosis, lactate dehydrogenase release, reactive oxygen species production, and radical scavenging. Acute toxicity in mice	CC:U 1:3 lethal to mice	Hayyan et al. (2015)
CC with G, F, S, GL, MOA, and W		Human cell lines – mitochondrial activity, modelling of interactions with phospholipids	Organic acids main source of toxicity	Hayyan et al. (2016)
Various ammonium-based		Human cell lines – mitochondrial activity, membrane permeability, reactive oxygen species production, and radical scavenging. Acute toxicity in mice	CC-based less toxic than DEEAC-based	Mbous et al. (2017)
CC with G, F, X, GL, and MA	Grape skin extract	Human cell lines – mitochondrial activity		Radošević et al. (2016)
CC with G, GL, and OA		Fish (CCO) and human (MCF-7) cell lines – mitochondrial activity and wheat seed germination	Toxicity of OA-containing deep eutectic solvent	Radošević et al. (2015)
Deep eutectic solvents containing organic acids		Human cells lines (HeLa, MCF-7 and HEK293T) – mitochondrial activity	Deep eutectic solvents are safe and show antioxidative activity.	Mitar et al. (2019)
CA with L-arginine or ethambutol	Therapeutic deep eutectic solvents	Toxicity to human intestinal Caco2 cells – mitochondrial activity	Deep eutectic solvent toxicity intermediate between components – CA is toxic due to pH lowering.	Santos et al. (2019)

(continued)

Table 1 (continued)

Type of deep eutectic solvent studied	Active agent, if any	Toxicity criteria	Conclusions	References
Choline- and imidazole-based ionic liquids	Caffeine and salicylic acid	Human keratinocyte lines – mitochondrial activity	Choline-based less toxic than imidazole-based	Santos de Almeida et al. (2017)
Deep eutectic solvents and ionic liquid, including C:GE		Normal human bronchial epithelial cells – lactic acid release and interleukin-α1 secretion. Fourier transform infrared spectroscopy in porcine skin	CAGE formulation has very low toxicity.	Zakrewsky et al. (2014)
C:GE	Insulin	Fourier transform infrared spectroscopy in porcine skin	Lipid loss from skin	Banerjee et al. (2017) and Tanner et al. (2018)
Various		Human HaCaT and MNT-1 cells lines – mitochondrial activity	CC and tetramethylammonium chloride-based deep eutectic solvents less toxic than tetrabutylammonium chloride-based ones	Macário et al. (2019)
CC with EG, GL, or U		Human A549 lung cells – mitochondrial activity	Toxicity in low millimolar range	Shekaari et al. (2019)
CC with sugars or alcohols		HEK-293 human embryonic kidney cells – mitochondrial activity and quantitative structure-activity relationship analysis	Deep eutectic solvents slightly more toxic than their components and less toxic than imidazolium ionic liquids	Ahmadi et al. (2018)
Bacteria-derived hydroxylic acids and various hydrogen bond acceptor, including CC		Mouse embryonic fibroblast (MEF) 3T3 cell line, live/dead by fluorescence	CC-based deep eutectic solvents had low toxicity.	Haraźna et al. (2019)
B:GL 1:2	Green coffee bean extract	Repeated oral administration to rats	Enlarged stomach and might be consequence of viscosity	Benlebna et al. (2018)

(continued)

Table 1 (continued)

Type of deep eutectic solvent studied	Active agent, if any	Toxicity criteria	Conclusions	References
CC:GL 1:2	Salvianolic acid B	Acute oral toxicity in mice	Some mice died immediately.	Chen et al. (2017)
C:GE	Insulin	Oral bioavailability study	No reported toxicity	Banerjee et al. (2018a)
C:GE 1:2		Repeated oral administration to rats on high-fat diet	No observed toxicity	Nurunnabi et al. (2019)
P:GLA 2:1	Rutin	Oral bioavailability study	No reported toxicity	Faggian et al. (2016)
P:MA:LA:W 1:0.2:0.3:0.5	Berberine	Oral bioavailability study	No reported toxicity	Sut et al. (2017)
Various, from the literature		Multitasking-Quantitative Structure Toxicity Relationship analysis	Importance of hydrogen bond donor for toxicity sugar alcohols and straight-chain alcohols > sugars and amides > organic and inorganic acids (most toxic)	Halder et al. (2019)

Abbreviations: *AA* acrylic acid, *ACA* aconitic acid, *Arg* arginine, *β-ala* β-alanine, *B* betaine, *C* choline, *CA* citric acid, *CC* choline chloride, *CH* cholinium hydroxide, *DEEAC* *N,N*-diethylethanolammonium chloride, *EG* ethylene glycol, F fructose, G glucose, *GA* glycolic acid, *GE* geranate, *GL* glycerol, *GLA* glutamic acid, *GLN* glutamine, *HBA* hydrogen bond acceptor, *HBD* hydrogen bond donor, *LA* lactic acid, *LAU* lauric acid, *MA* malic acid, *MEA* maleic acid, *MOA* malonic acid, *OA* oxalic acid, P proline, *PD* 1,2-propanediol (propylene glycol), *PHE* phenylalanine, S sucrose, *TA* tartaric acid, *TEAC* tetraethylammonium chloride, *TBAB* tetrabutylammonium bromide, *TPAB* tetrapropylammonium bromide, U urea, W water, X xylitol, *ZC* zinc chloride

the individual components. As far as biodegradability was concerned, only the choline chloride:urea and choline chloride:acetamide deep eutectic solvents could be considered as fully biodegradable in their system (Wen et al. 2015).

In a study carried out by Radošević et al. (Radošević et al. 2018), the toxicity of natural deep eutectic solvents toward several strains of bacteria (*Escherichia coli*, *Proteus mirabilis*, *Salmonella typhimurium*, *Pseudomonas aeruginosa*, *Staphylococcus aureus*) as well as the fungal species *Candida albicans* was tested. They observed toxicity for compositions containing organic acids but that the solvents possessed antioxidative activity.

More recently, Macário et al. approached the question of the ecotoxicology of deep eutectic solvents using mixture theory in an attempt to predict their toxicity as a function of their components (Macário et al. 2018a & b). Toxicity was assessed using the inhibition of bioluminescence produced by the marine bacterium *Aliivibrio fischeri*. In one study, tetramethylammonium chloride, tetraethylammonium chloride, and tetrapropylammonium chloride where used as hydrogen bond acceptors with ethylene glycol and 1-propanol as hydrogen bond donors. Mixture toxicity theory was used to analyze the results obtained with the deep eutectic solvents and with their components separately. Neither the individual components nor the solvents showed excessive toxicity. A model in which the hydrogen bond donor and hydrogen bond acceptor acted on different sites was found to give reasonable agreement with the data and allow the toxicity of deep eutectic solvents to be predicted. Among the hydrogen bond acceptor, toxicity increased with the length of the carbon chain: tetramethylammonium chloride being the least toxic and tetrapropylammonium chloride the most toxic (Macário et al. 2018a). In another study, the same approach was applied to deep eutectic solvents based on choline chloride as hydrogen bond donor with a number of hydrogen bond acceptors including ethylene glycol and urea (Macário et al. 2018b). In this case, they observed an antagonistic effect between the components of the deep eutectic solvents so that the mixture was less toxic than the sum of the parts. In particular, choline chloride with urea or 1-propanol gave products with very low toxicity, making them suitable for use as alternative "green" solvents.

Juneidi et al. used four fungal species (*Phanerochaete chrysosporium*, *Aspergillus niger*, *Lentinus tigrinus,* and *Candida cylindracea*) as their reporter organisms (Juneidi et al. 2016). The deep eutectic solvents tested were based on choline chloride with alcohols, sugars, organic acids, urea, and zinc chloride as the hydrogen bond donor component. Only zinc chloride and organic acids showed growth-inhibiting effects on fungi, which were reduced slightly when these components were in the form of a deep eutectic solvent rather than presented as individual compounds.

A recent review suggested that antimicrobial testing of deep eutectic solvents was better performed in liquid suspension culture than by antibiograms so that toxicity could be monitored as a function of time (Torregrosa-Crespo et al. 2020). Using a novel deep eutectic solvent composition, acetylcholine chloride:acetamide 1:2, they observed toxicity to *Escherichia coli* at concentrations above 300 nM and that this was partially due to acidification of the medium by degradation products of deep eutectic solvent components. This highlights the necessity of studying the stability of deep eutectic solvents over time as well as working with freshly prepared material.

2.2.2 Toxicity Toward Human Cells Lines and Ex Vivo Studies on Skin

The team of Hayyan has studied the toxicity of deep eutectic solvents based on choline chloride with four different hydrogen bond donors toward a panel of human cell lines using a variety of criteria to assess toxicity (Hayyan et al. 2015). Again, the deep eutectic solvents studied were choline chloride with glycerol, ethylene glycol, triethylene glycol, and urea, always in a 1:3 ratio. The cells lines were PC3

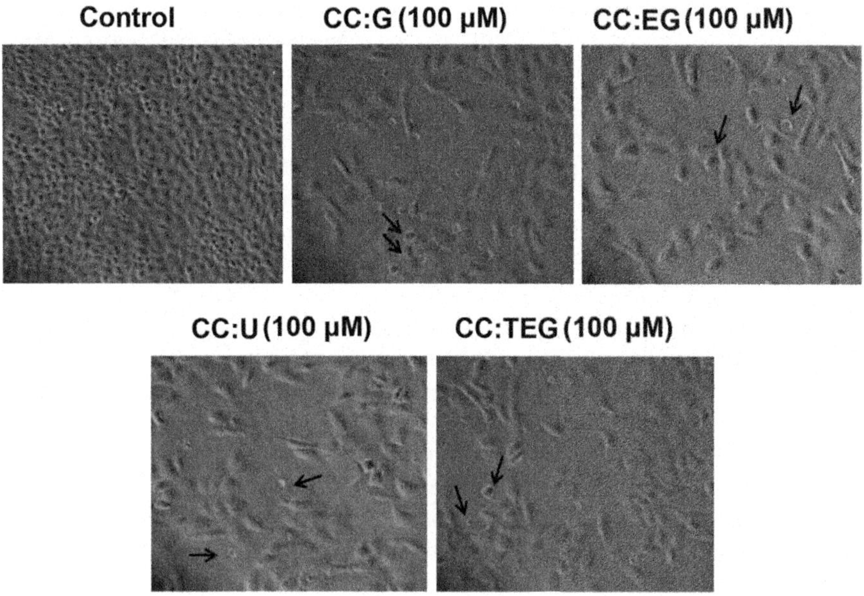

Fig. 2.1 Morphological alterations of MCF-7 cells treated with deep eutectic solvents. Cells were treated with the half maximal inhibitory concentration of each solvent for 24 h, and their morphology was analyzed using light microscopy. Arrow shows shrunken or apoptotic cells. (Hayyan et al. 2015)

(prostate cancer), A375 (malignant melanoma), HepG2 (hepatocellular cancer), MCF-7 (breast cancer), H413 (oral carcinoma), and OKF6 (normal oral keratinocytes). The deep eutectic solvents, at different dilutions in culture medium, were compared with their individual components using the 3-(4,5-dimethylthiazol-2-yl)-2,5-diphenyltetrazolium bromide (MTT) reduction assay. The results showed appreciable toxicity of the solvents, which was generally intermediate between that of the components, except for those based on urea, which were less toxic than the constituent chemicals. When the toxicity toward the cancer cell lines was compared with that toward the normal keratinocytes, only a small degree of selectivity toward malignant cells was observed.

Microscopic examination of MCF-7 cells after treatment with deep eutectic solvents at 100 µM, approximately equal to the half maximal inhibitory concentration (IC_{50}), showed reduced confluence compared with control cells and some rounded cells characteristic of apoptosis (Fig. 2.1). Therefore, flow cytometry was used to determine the proportion of annexin-positive cells, which was slightly increased with the deep eutectic solvents. On the other hand, no propidium iodide-positive (necrotic) cells were observed. Furthermore, no deoxyribonucleic acid (DNA) fragmentation was observed (Hayyan et al. 2015).

Some evidence of membrane damage by deep eutectic solvents was provided by measuring the release of a cytoplasmic enzyme, lactic dehydrogenase, and the penetration of a fluorescent marker into the MCF-7 cells. The deep eutectic solvents containing ethylene glycol and triethylene glycol had more effect that the other two. The redox properties of the deep eutectic solvents were also studied. Firstly, the generation of reactive oxygen species (ROS) by MCF-7 cells was measured by a fluorogenic technique. The deep eutectic solvents caused a dose-dependent increase in reactive oxygen species production, which was particularly marked with choline chloride:triethylene glycol. Somewhat paradoxically, the authors also measured the antioxidant activity of the deep eutectic solvents, using quercetin as a positive control. Their radical scavenging capacity was found to be less than 1% of that of quercetin (Hayyan et al. 2015).

In a follow-up study, Hayyan et al. investigated the effect of introducing water into the deep eutectic solvents on the toxicity toward human cancer cell lines as described above (Hayyan et al. 2016). Deep eutectic solvents that contained choline chloride with fructose, glucose, sucrose, or glycerol with water as a third component showed half maximal inhibitory concentration values above 100 mM, while a deep eutectic solvent composed of choline chloride and malonic acid alone, without water, was much more toxic. Since the viscosity of this solvent was not higher than some of the water-containing ones, and taking into account results from the literature, it was concluded that the presence of organic acids is a main source of deep eutectic solvent toxicity. They also used molecular modelling to investigate the possible interactions between deep eutectic solvents and membrane phospholipids. This predicted that deep eutectic solvents could interact with the cell surface.

Somewhat different in vitro results were obtained by Radošević et al. who examined the cytotoxicity of deep eutectic solvents composed of choline chloride and five different hydrogen bond donors (glucose, fructose, xylose, glycerol, and malic acid) toward two human cell lines (HeLa and MCF-7) using the 2-(4-iodophenyl)-3-(4-nitrophenyl)-5-(2,4-disulfophenyl)-2H-tetrazolium (WST-1) reduction assay (Radošević et al. 2016). They found that the concentration that reduced 2-(4-iodophenyl)-3-(4-nitrophenyl)-5-(2,4-disulfophenyl)-2H-tetrazolium (WST-1) metabolism by 50% was more than 2 mg/mL for all compositions. However, a deep eutectic solvent containing oxalic acid, although efficient for the application under consideration (extract of antioxidants from grape skin), was discarded because of slight toxicity (Radošević et al. 2015). In this study, in contrast to the results of Wen et al. (Wen et al. 2015), a high level of biodegradability was found for the tested solvents, with the lowest level being recorded for the solvent with oxalic acid. In a further study, a third human cell line, HEK293T, was added to the panel tested (Radošević et al. 2018). Only deep eutectic solvents containing organic acids (for example, citric acid) showed appreciable toxicity. These results were confirmed by Mitar et al. (Mitar et al. 2019), who also demonstrated the antioxidative properties of the natural deep eutectic solvents. In this work, a further criterion was added – corrosiveness toward metal (steel). This was found to be extremely low for all the deep eutectic solvents tested, meaning that they could be used in pipelines and reactors without any problem.

With a view to the oral administration of deep eutectic solvent-based formulations, Santos et al. tested the cytotoxicity of deep eutectic solvents based on citric acid against the intestinal cell line Caco-2 (Santos et al. 2019). Citric acid itself is quite toxic due to the acidic pH that it imparts to the culture medium, while the two active molecules included (ethambutol and L-arginine) were not toxic, with the deep eutectic solvents showing intermediate values.

Given the interest of deep eutectic solvents for topical applications, it is also important to determine whether irritation or any other toxicity is observed in the skin. An investigation was carried out in vitro by Santos de Almeida et al. on a human keratinocyte cell line, HaCat, using the 3-(4,5-dimethylthiazol-2-yl)-2,5-diphenyltetrazolium bromide (MTT) reduction assay (Santos de Almeida et al. 2017). They compared deep eutectic solvents based on choline combined with glutamine or phenylalanine with ionic liquids based on imidazoles. The choline-based solvents were found to be less cytotoxic than the imidazole-based ones. The results for the two choline-based solvents were very similar, with half maximal inhibitory concentrations for a 24-h exposure of about 0.4% v/v in the culture medium.

Zakrewsky et al. chose a primary cell line, normal human bronchial epithelial cells to test the in vitro toxicity of ionic liquids destined for transdermal delivery, and combined this with the antimicrobial activity of the formulations to select the most promising for in vivo applications (Zakrewsky et al. 2014). Of the 12 formulations tested, the "CAGE" formulation of choline and geranic acid and a choline-hexanoate formulation combined low cytotoxicity with high activity against biofilms. The potential for skin irritation was assessed by measuring interleukin-α1 secretion from a multilayer human skin model and by Fourier transform infrared (FTIR) spectroscopy to look at changes in bands characteristic of skin lipids in porcine skin. Although its individual components caused significant interleukin-α1 release 4 h after application, the level observed with the choline:geranic acid solvent gave a result similar to the negative control. No changes in the band of the Fourier transform infrared spectrum between 1650 and 1660 cm^{-1} were observed for the choline:geranic acid solvent and two other choline-based deep eutectic solvents, while the individual components caused significant changes. However, in their study of protein penetration through skin, they observed stretching of the peaks between 2850 and 2920 cm^{-1} that are characteristic of lipid extraction (Banerjee et al. 2017). In a later study, the influence of the ratio of choline to geranic acid was investigated (Tanner et al. 2018). The decrease in the peak, indicating lipid loss, was proportional to the geranic acid content, that is, the hydrophobicity of the mixture. However, pure geranic acid did not promote insulin delivery through the skin, probably because it was too hydrophobic to dissolve the protein.

Macário et al. examined the potential skin toxicity of a number of deep eutectic solvents using two cells lines, (HaCaT, a human keratinocyte line, and MNT-1, a human melanoma line) with 3-(4,5-dimethylthiazol-2-yl)-2,5-diphenyltetrazolium bromide (MTT) reduction as the indicator of cell viability (Macário et al. 2019). In general, they observed that solvents containing choline chloride or tetramethylammonium chloride as hydrogen bond donor were not toxic to these cell lines, while those containing tetrabutylammonium chloride were toxic. As far as the hydrogen bond acceptors tested were concerned (butanoic acid, hexanoic acid, 1-propanol, ethylene glycol, and urea), only butanoic acid was toxic as an individual agent, while the deep eutectic solvents formed between this and tetramethylammonium chloride were only slightly toxic and choline chloride-butanoic acid was nontoxic.

Shekaari et al. measured the cytotoxicity of three deep eutectic solvents (choline chloride with ethylene glycol, glycerol, and urea) against A549 lung cells using 3-(4,5-dimethylthiazol-2-yl)-2,5-diphenyltetrazolium bromide (MTT) reduction as the criterion of toxicity and obtained half maximal inhibitory concentration values in the low millimolar range (Shekaari et al. 2019). Choline chloride-based deep eutectic solvents were evaluated for toxicity against HEK-293 human embryonic kidney cells by Ahmadi et al. (Ahmadi et al. 2018). The hydrogen bond donors were different sugars or alcohols. They observed that the solvents were more toxic than their starting components but that this toxicity was moderate, with half maximal inhibitory concentration values derived from 3-(4,5-dimethylthiazol-2-yl)-2,5-diphenyltetrazolium bromide (MTT) reduction between 4 and 75 millimolar. The toxicity observed was lower than that of a classical imidazolium-based ionic liquid used in comparison. Of course, to obtain the half maximal inhibitory concentration values, the formulations must have been diluted to different extents in cell culture medium, so the interactions between the components of the deep eutectic solvents will have been altered.

Haraźna et al. recently proposed a new type of hydrogen bond donor: a mixture of (*R*)-3-hydroxyheptanoic and (*R*)-3-hydroxynonanoic acids derived from a bacterial cell wall polymer (Haraźna et al. 2019). These were used to form ionic liquids or deep eutectic solvents with a number of hydrogen bond acceptor, including choline chloride. Toxicity was assessed against the mouse embryonic fibroblast (MEF) 3 T3 cell line using a fluorescence microscopy technique to distinguish between live and dead cells. No reduction in cell viability was observed for concentrations up to 500 μg/mL of the choline chloride-based deep eutectic solvents. This deep eutectic solvent was also found to be completely biodegradable in 5 days in a standard test with activated sludge.

2.2.3 Toxicity Toward Plants, Invertebrates, and Fish

The study of Wen et al. included a plant (garlic, *Allium sativum*) and an invertebrate (hydra, *Hydra sinensis*) (Wen et al. 2015). Deep eutectic solvents containing choline chloride or choline acetate with urea, glycerol, acetamide, or ethylene glycol in various proportions were tested. Some deep eutectic solvents or their components reduced root growth in garlic bulbs, in particular choline chloride and glycerol both had strong effects, but lower toxicity was observed when they were combined as a deep eutectic solvent. Similarly, the choline salts, particularly the chloride, were found to be detrimental to hydra growth, but the deep eutectic solvents were less toxic than the sum of their parts (Fig. 2.2).

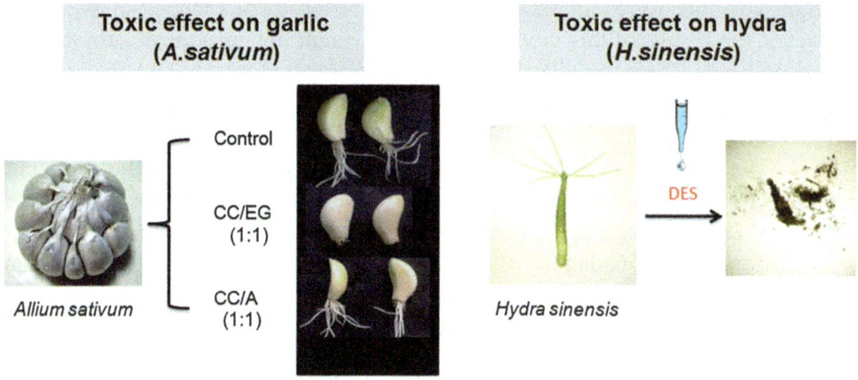

Fig. 2.2 Effect of DES on garlic bulbs and on hydra. (Adapted from Wen et al. 2015 with permission of Elsevier)

Brine shrimp (*Artemia saliva*) larvae were chosen as an indicator of ecotoxicity by Hayyan et al. (Hayyan et al. 2013). Deep eutectic solvents prepared from choline chloride combined with urea, glycerol, ethylene glycol, and triethylene glycol exerted some toxicity toward the shrimp larvae. This toxicity was more than the additive effect of the individual components, and the viscosity of the deep eutectic solvents was suggested as one possible explanation.

A more complex marine organism, the common carp (*Cyprinus carpio*) was used by Juneidi et al. to evaluate deep eutectic solvents based on choline chloride combined with alcohols, sugars, organic acids, urea, and zinc chloride (Juneidi et al. 2016). Acute toxicity toward the fish was performed by adding the solvents to the water in the aquarium and adjusting the pH if necessary. Mortality was measured after 16 days. In agreement with their data obtained with fungi, only containing zinc chloride, malonic acid, or *para*-toluenesulfonic acid showed median lethal dose values below 100 mg/L.

2.2.4 *In Vivo Toxicity in Rodents*

Hayyan et al. determined the acute toxicity toward mice of a range of choline chloride-based deep eutectic solvents. The median lethal dose (LD_{50}) of the deep eutectic solvents was determined in ICR mice after a single oral administration. Values of 5 to 6 g/kg were obtained for the solvents with glycerol, ethylene glycol, and triethylene glycol, while the choline chloride:urea 1:3 caused immediate death of the animals. However, a 1:2 composition had similar toxicity to the other solvents. Blood biochemistry showed some elevation of transaminases indicative of liver damage. The conclusion of this study was that choline-based deep eutectic solvents are not completely devoid of toxicity and that the compositions need to be carefully considered (Hayyan et al. 2015). In a later study, some natural deep eutectic solvents were assessed using the same protocol (Mbous et al. 2017). However, the results of this in vivo study are difficult to interpret because they are quoted in g/mL. The natural deep eutectic solvents were found to be less toxic than other deep eutectic solvents, despite the high viscosity of some natural deep eutectic solvents.

The safety of deep eutectic solvents for use in the extraction of phenolic compounds was considered by Benlebna et al. focusing on their possible effects after ingestion. They studied the toxicity of a green coffee bean extract in deep eutectic solvents composed of betaine:glycerol (1:2 molar ratio) after repeated oral administration to rats (Benlebna et al. 2018). They encountered difficulties in administration due to the viscosity of the solvents and the deaths of two rats on days 5 and 10 after the first gavage. These rats were found to have a greatly enlarged stomach (Fig. 2.3), which might have been related to the increased water consumption observed in the treated group. Increases in plasma and liver lipids were also recorded. Unfortunately, a control group with deep eutectic solvents without extract was not tested in this study. The results reinforce the conclusions of Hayyan et al. that deep eutectic solvents are not necessarily innocuous by the oral route (Hayyan et al. 2015).

On the other hand, Chen et al. tested the acute oral toxicity of a choline chloride:glycerol (1:2) solvent in rats and obtained a median lethal dose of 7733 mg/kg (Chen et al. 2017). The mice that died did so within 4 hours of receiving the dose and were first excited and then showed reduced activity, breathlessness, convulsions, and tremor. In the studies of the oral bioavailability of insulin in

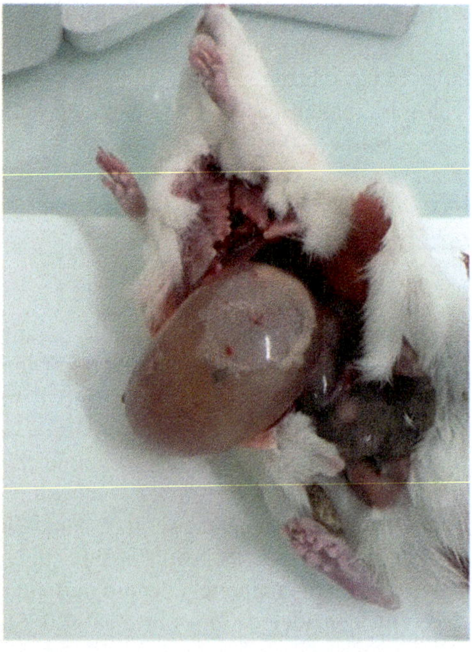

Fig. 2.3 Photograph of the stomach of a rat that died 5 days after gavage with a glycerol:betaine NADES in a molar ratio 2:1 containing 10% v/v of water. (Benlebna et al. 2018, reprinted with permission of American Chemical Society)

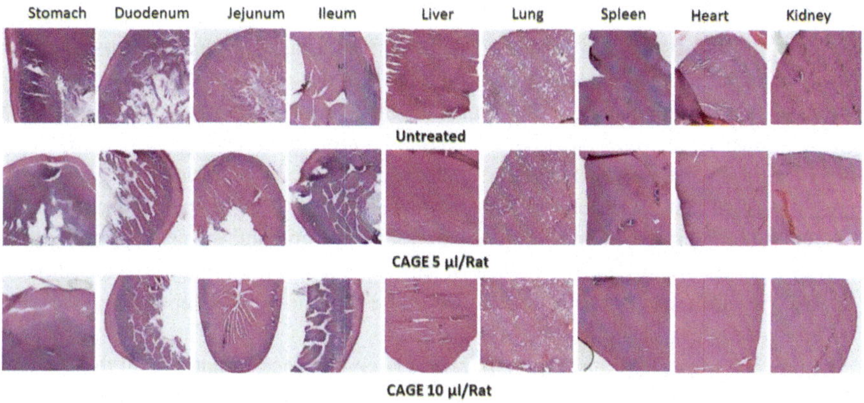

Fig. 2.4 Photomicrographs of hematoxylin and eosin-stained harvested tissue sections after 30 days oral administration of CAGE. (Nurunnabi et al. 2019; reprinted with permission of Proceedings of the National Academy of Sciences)

the "CAGE" formulation of choline and geranic acid, no toxic effects and no histological changes in the intestines of the treated rats were observed (Banerjee et al. 2018a). With the ability of the "CAGE" formulation to reduce obesity, daily oral administration to rats for 30 days was well tolerated (Nurunnabi et al. 2019). Blood parameters and biomarkers of toxicity were unchanged compared with controls, and no abnormalities could be detected in the internal organs of the treated animals (Fig. 2.4).

Faggian et al. did not mention any toxicity observed when they determined the oral pharmacokinetics of rutin administered to rats in a proline:glutamic acid 2:1 deep eutectic solvent (Faggian et al. 2016). Similarly, Sut et al. did not record any adverse events when measuring the oral pharmacokinetics of berberine in three different natural deep eutectic solvents (Sut et al. 2017).

2.2.5 *Attempts to Predict Toxicity of Deep Eutectic Solvents*

Table 1 summarizes the information above concerning the toxicity of deep eutectic solvents. It is clear that a wide range of different models and criteria of toxicity have been used and that there is a need for more systematic investigations. Recently, a few investigators have tried to analyze the results to determine the parameters underlying toxicity and thereby predict it. Thus, Macário et al. applied mixture theory to their results of toxicity toward a marine organism (Macário et al. 2018a & b). Ahmadi et al. performed a Quantitative Structure Activity Relationships (QSAR) analysis on their data concerning the toxicity of choline chloride-based systems toward HEK-293 cells (Ahmadi et al. 2018). They were able to identify important structural parameters: the number of rotatable bonds, the mean atomic van de Waals volume, and the C2 ratio of the hydrogen bond acceptor and hydrogen bond donor.

A recent article by Halder et al. has brought together data from a number of the studies reported above and subjected them to a multitasking-Quantitative Structure Toxicity Relationships (mtk-QSTR) analysis (Halder et al. 2019). This revealed a number of parameters that were important in determining toxicity: polarizability, electronegativity, nature of the hydrogen bond donor, and topological features. They were able to rank the main classes of hydrogen bond donor in terms of deep eutectic solvent toxicity: sugar alcohols and straight-chain alcohols giving low toxicity, sugars and amides giving intermediate toxicity, and organic and inorganic acids giving a high level of toxicity. More studies of this type, and more standardized protocols for toxicity studies, should in the future guide the choice of deep eutectic solvent compositions for pharmaceutical applications.

Table 1 summarizes the results that have been obtained to date about the toxicity of deep eutectic solvents. It is clear that a wide range of different models and criteria of toxicity have been used and that there is a need for more systematic investigations.

2.3 Pharmaceutical Applications of Deep Eutectic Solvents

It will be evident from the above that the main application of deep eutectic solvents in pharmaceutical science is as solvents for hydrophobic drugs. As well as a large number of small molecules, deep eutectic solvents have also been found to have advantages for solubilizing larger macromolecules such as proteins, nucleic acids, and polysaccharides. Not only do they increase solubility per se compared with aqueous media but may also increase the stability of the compound and favor particular conformations. Deep eutectic solvents have also been discovered to have intrinsic properties such as antimicrobial activity and as promoters of absorption.

2.3.1 Solubilization of Small Molecules in Deep Eutectic Solvents

Early Work and Model Drugs

Some early work on deep eutectic solvents as vehicles for drug solubilization was performed by Morrison et al. (Morrison et al. 2009). They studied the thermal properties of mixtures of choline chloride with urea or malonic acid using, among other techniques, hot stage microscopy to allow them to define eutectic compositions of urea:choline chloride 2:1 and malonic acid:choline chloride 1:1. The solubility of five poorly water-soluble compounds (benzoic acid, danazol, griseofulvin, AMG517, itraconazole) in these mixtures was tested, yielding improvements in all cases, notably an increase in the solubility of the antifungal drug itraconazole by 22,000 times compared with water.

Dai et al. studied the potential of a range of natural deep eutectic solvent compositions as "green" solvents for a number of natural products with pharmaceutical activity: rutin, quercetin, cinnamic acid, carthamin, 1,8-dihydroxyl anthraquinone, taxol, and ginkgolide B (Dai et al. 2013). A large array of combinations of compounds were screened using nuclear magnetic resonance (NMR) spectroscopy, thermogravimetric analysis, differential scanning calorimetry, and viscosity measurements in order to characterize the mixtures. Four compositions were chosen for solubility studies: 1,2-propanediol:choline chloride:water 1:1:1, glucose:choline chloride:water 2:5:5, lactic acid:glucose:water 5:1:3, and xylitol:choline chloride:water 1:2:3. For all the active compounds cited, at least one of the natural deep eutectic solvents allowed a considerable increase in solubility compared with pure water, the most spectacular being that of quercetin multiplied over 400,000 times in xylitol:choline chloride:water. They also observed that temperature had a strong effect on solubility in natural deep eutectic solvents, with an increase of 1.6 to 2.3 times at 50 °C compared with 40 °C. In a follow-up study, in 2015, they investigated the effect of adding water to the natural deep eutectic solvents containing quercetin and carthamin in an attempt to reduce the viscosity (Dai et al. 2015). Although the presence of water reduced hydrogen bonding in the deep eutectic solvents, a small proportion was able to reduce the viscosity significantly without much impact on the solubility of the active molecule.

Choi et al. reported that the solubility of the flavonoid rutin was 50–100 fold higher in natural deep eutectic solvents than in water, the highest solubility being attained in an aconitic acid:choline chloride mixture (Choi et al. 2011). They also reported good solubility of paclitaxel and ginkgolide B in a deep eutectic solvent composed of glucose and choline chloride. A naturally occurring alkaloid with antiparasitic activity, berberine, was formulated in deep eutectic solvents passed on proline and organic acids with the aim of improving its oral bioavailability (Sut et al. 2017).

In a study published in 2016, Li and Lee coined the term "deep eutectic solvent derivatives" for mixtures of choline chloride and carboxylic acids that, although not precisely eutectics, were liquid at room temperature (Li and Lee 2016). They determined the solubility of four model drugs (itraconazole, piroxicam, lidocaine, and posaconazole) in a choline chloride:glycolic acid 1:2 mixture. Again, the solubility of itraconazole and that of another antifungal from the same class, posaconazole, was greatly increased in these solvents, with smaller effects seen on piroxicam and lidocaine. The addition of a third component, oxalic acid, further increased the solubility of the antifungals but also increased the viscosity of the formulations.

Lidocaine (a local anesthetic) was also used as a model compound for investigating the potential of deep eutectic solvents by Gutiérrez et al. (2018). In a first, theoretical, study, they examined the molecular interactions that could occur between lidocaine and the selected deep eutectic solvent components: choline chloride, β-alanine (hydrogen bond acceptors), and lactic acid (hydrogen bond donor) (Gutiérrez et al. 2018). In the second study, they studied deep eutectic solvents composed of arginine combined in an equimolar ratio with tartaric acid, oxalic acid, and glutamic acid, both theoretically and experimentally, using density functional theory and molecular dynamics (Gutiérrez et al. 2019).

Thioflavin T is a fluorescent molecule that undergoes a change when bound to amyloid fibrils and can therefore be used as a sensor for neurogenerative diseases. Gautam et al. studied the fluorescence behavior of this molecule in two deep eutectic solvents: choline chloride:urea 1:2 and *N,N*-diethylethanolammonium chloride:urea 1:2 (Gautam et al. 2018). High quantum yield and fluorescence lifetime were obtained in both solvents.

Palmelund et al. (2019). performed a similar study comparing experimentally determined solubility with theoretical predictions from conductor-like screening model for real solvents for a range of eleven active molecules in six deep eutectic solvents compositions and three conventional solvents (water, ethanol, and polyethylene glycol 300). A good correlation was observed between the results, suggesting that predictive techniques could be used to reduce the number of experimental determinations. For most of the molecules tested, conventional solvents gave the best solubility, but often the best solvent was ethanol, which could bring toxicity and safety problems. However, for paracetamol and especially for celecoxib, their highest solubility was obtained in a deep eutectic solvent.

Anti-inflammatory and Analgesic Drugs

Nonsteroidal anti-inflammatory drugs suffer from low water solubility. Lu et al. determined the solubility of five such drugs (aspirin, acetaminophen, ketoprofen, naproxen, and ibuprofen) in a panel of deep eutectic solvents based on choline chloride, ethylammonium chloride, tetrapropylammonium bromide, betaine, or choline bitartrate as the hydrogen bond acceptor and a variety of sugars, alcohols, organic acids, and urea as the hydrogen bond donor (Lu et al. 2016). Good solubility was obtained in 17 selected deep eutectic solvents, but the authors were not able to predict solubility by looking at various physical properties of the deep eutectic solvents or at the hydrogen bond acceptor to hydrogen bond donor ratio. They also monitored the stability of aspirin in a choline chloride:1,2-propanediol (propylene glycol) 1:2 solvents compared with pure water and a 1:1 solvent:water mixture and observed a reduced rate of cleavage to salicylic acid at 80 °C in the deep eutectic solvent.

A later study by Mokhtarpour et al. focused on naproxen solubility in three deep eutectic solvent compositions: choline chloride:ethylene glycol 1:2, choline chloride:urea 1:2, and choline chloride:malonic acid 1:1 (Mokhtarpour et al. 2019a). All compositions increased the drug solubility more than 3300-fold with malonic acid as the hydrogen bond donor. They used several modelling procedures to attempt to predict solubility and performed density measurements to probe the interactions between the solvent and the solute. They also report a similar study with indomethacin in deep eutectic solvents composed of tetrabutylammonium bromide with ethylene glycol or glycerol (Mokhtarpour et al. 2019b). An increase of solubility of over

Table 2 Solubilization of small molecules in deep eutectic solvents

Compound	Pharmaceutical activity	Optimal deep eutectic solvent composition(s)	Therapeutic deep eutectic solvents	References
1,8-dihydroxyl anthraquinone	Natural product	LA:G:W 5:1:3		Dai et al. (2013)
Acetaminophen	Analgesic and antipyretic	TPAB:PD 1:2		Lu et al. (2016)
Acetaminophen	Analgesic and antipyretic	CC:U 1:2; CC:MOA 1:1; CC:OA 1:1,; CC:GL 1:2; CC:EG1:2		Shekaari et al. (2017, 2018a)
Adiphenine	Local anesthetic	Drug with GL, U, or aspirin	✓	Abbott et al. (2017)
AMG517	Transient receptor potential vanilloid 1 antagonist	MOA:CC 1:1		Morrison et al. (2009)
Antofloxacin	Antibiotic	CC:*p*-toluenesulfonic acid 1:2		Zhang et al. (2019)
Aspirin (acetylsalicylic acid)	Anti-inflammatory	CC:drug and menthol:drug	✓	Aroso et al. (2016)
Aspirin (acetylsalicylic acid)	Anti-inflammatory	TPAB:PD 1:2		Lu et al. (2016)
Aspirin (acetylsalicylic acid)	Anti-inflammatory	CC:drug	✓	Abbott et al. (2017)
Aspirin (acetylsalicylic acid)	Anti-inflammatory	CC:drug	✓	Abranches et al. (2019)
Benzalkonium chloride	Antiseptic	Drug:AA	✓	Wang et al. (2017)
Benzoic acid	Model drug	CC:U 1:2		Morrison et al. (2009)
Benzoic acid	Model drug	CC:drug and menthol:drug	✓	Aroso et al. (2016) and Duarte et al. (2017)
Berberine	Various (natural alkaloid)	P:MA:LA:W 1:0.2:0.3:0.5		Sut et al. (2017)
bis(2-ethylhexyl) 2-(4-hydroxy-3,5-dimethoxybenzyl) malonate (Bis-EHBm)	Antioxidant	PD:CC:W1:1:1		Durand et al. (2017)

(continued)

Table 2 (continued)

Compound	Pharmaceutical activity	Optimal deep eutectic solvent composition(s)	Therapeutic deep eutectic solvents	References
Caffeine	Model drug	CH:GLN 1:1; CH:PHE 1:1		Santos de Almeida et al. (2017)
Carthamin	Antioxidant	X:CC:W 1:2:3		Dai et al. (2013, 2015)
Cefadroxil	Antibiotic	C:GE 1:2		Zakrewsky et al. (2015, 2016b)
Ceftazidime	Antibiotic	C:GE 1:2		Zakrewsky et al. (2015, 2016b)
Cinnamic acid	Antioxidant and antimicrobial	PD:CC:W 1:1:1		Dai et al. (2013)
Clavulanic acid	Antibiotic	B:U 1:1.5		Olivares et al. (2018)
Chlorambucil	Anticancer	Conjugate with dimethylaminopropanol: 1,4, butanediol	✓	Pradeepkumar et al. (2018)
Coenzyme Q	Antioxidant	Coenzyme Q:LAU 2:1	✓	Tarate and Bansal (2015)
3,4-dihydro-6-hydroxy-7-methoxy-2,2-dimethyl-1(2H)-benzopyran (CR-6)	Antioxidant	PD:CC:W 1:1:1		Durand et al. (2017)
3,4-dihydro-6-hydroxy-7-methoxy-2,2-dimethyl-1(2H)-benzopyran palmitate (CR-6 palmitate)	Antioxidant	PD:CC:W 1:1:1		Durand et al. (2017)
Curcumin	Antimicrobial and photosensitizer	G:S 1:1; MEA:CC 1:3		Wikene et al. (2015a)
Danazol	Gonadotrophin inhibitor	MOA:CC 1:1		Morrison et al. (2009)
Decyl rosmarinate	Antioxidant	PD:CC:W 1:1:1		Durand et al. (2017)
Erythritol	Antibiotic	ZC:drug 1:3; B:drug 2:1	✓	Lim et al. (2018)
Ethambutol	Antibiotic	CA:ethambutol:W 2:1:10	✓	Santos et al. (2019)

(continued)

Table 2 (continued)

Compound	Pharmaceutical activity	Optimal deep eutectic solvent composition(s)	Therapeutic deep eutectic solvents	References
Ginkgolide B	Platelet-activating factor antagonist	PD:CC:W 1:1:1		Dai et al. (2013)
Ginkgolide B	Platelet-activating factor antagonist	CC:G		Choi et al. (2011)
Griseofulvin	Antifungal	CC:MOA 1:1		Morrison et al. (2009)
Ibuprofen	Anti-inflammatory	TPAB:PD 1:2		Lu et al. (2016)
Ibuprofen	Anti-inflammatory	Drug:menthol	✓	Aroso et al. (2015) and Duarte et al. (2017)
Ibuprofen	Anti-inflammatory	Drug:various HBA	✓	Wolbert et al. (2019)
Ibuprofen	Anti-inflammatory	CC:drug	✓	Abranches et al. (2019)
Imipenem	Antibiotic	B:U 1:1.5		Olivares et al. (2018)
Indomethacin	Anti-inflammatory	TBAB:EG; TBAB:GL		Mokhtarpour et al. (2019b)
Itraconazole	Anti-inflammatory	CC:GA 1:2; CC:GA:OA 1:1.7:0.3		Li and Lee (2016)
Itraconazole	Antifungal	CC:MOA 1:1		Morrison et al. (2009)
Ketoprofen	Anti-inflammatory	TPAB:PD 1:2		Lu et al. (2016)
Ketoprofen	Anti-inflammatory	CC:drug	✓	Abranches et al. (2019)
Lamotrigine	Antiepileptic	CC:EG		Shekaari et al. (2019)
L-arginine	Amino acid	CA:L-Arg:W various ratios	✓	Santos et al. (2019)
Lidocaine	Local anesthetic	CC:GA 1:2; CC:GA:OA 1:1.7:0.3		Li and Lee, (2016)
Lidocaine	Local anesthetic	CC:LA; β-ala:LA; Arg:GLA; Arg:OA; Arg:TA		Gutiérrez et al. (2018, 2019)
Lidocaine	Local anesthetic	Various HBA, including thymol	✓	Wolbert et al. (2019)
Lidocaine	Local anesthetic	Prilocaine	✓	Wojnarowska et al. (2018)

(continued)

Table 2 (continued)

Compound	Pharmaceutical activity	Optimal deep eutectic solvent composition(s)	Therapeutic deep eutectic solvents	References
Naproxen	Anti-inflammatory	CC:U 1:2; CC:GL 1:2		Lu et al. (2016)
Naproxen	Anti-inflammatory	CC:EG 1:2; CC:U 1:2; CC:MOA 1:1		Mokhtarpour et al. (2019a)
Naringin dihydrochalcone	Antioxidant	CA:G:W 1:1:12; CC:CA:W 1:1:11; CC:G:W 1:1:11		Tang et al. (2016)
Nobiletin	Glucose-lowering agent	C:GE 1:2		Hattori et al. (2019)
Paclitaxel	Anticancer	CC:G		Choi et al. (2011)
Paracetamol	Antalgic	CC:drug	✓	Abbott et al. (2017)
Paracetamol	Antalgic	Drug:phenol	✓	Potticary et al. (2019)
Paracetamol	Antalgic	CC:drug	✓	Abranches et al. (2019)
Phenylacetic acid	Model drug	CC:drug and menthol:drug	✓	Aroso et al. (2016) and Duarte et al. (2017)
Phenylacetic acid	Model drug	Various HBA	✓	Wolbert et al. (2019)
Phloretin	Antioxidant	CC:S:W 1:1:10; CC:CA:W 1:1:11		Tang et al. (2016)
Phlorizin	Antioxidant	CA:G:W 1:1:12; CC:CA:W 1:1:11; CC:G:W 1:1:11		Tang et al. (2016)
Piroxicam	Anti-inflammatory	CC:GA 1:2; CC:GA:OA 1:1.7:0.3		Li and Lee (2016)
Posaconazole	Antifungal	CC:GA 1:2; CC:GA:OA 1:1.7:0.3		Li and Lee, (2016)
Prilocaine	Local anesthetic	Lidocaine	✓	Wojnarowska et al. (2018)
Quercetin	Various (natural flavonoid)	X:CC:W 1:2:3		Dai et al. (2013, 2015)
Ranitidine	Antihistaminic	Drug with GL, U, or aspirin	✓	Abbott et al. (2017)
Resveratrol	Antioxidant	PD:CC:W 1:1:1		Shamseddin et al. (2017)
Rutin	Various (natural flavonoid)	G:CC:W 2:5:5		Dai et al. (2013)

(continued)

Table 2 (continued)

Compound	Pharmaceutical activity	Optimal deep eutectic solvent composition(s)	Therapeutic deep eutectic solvents	References
Rutin	Various (natural flavonoid)	P:GLA 2:1		Faggian et al. (2016)
Rutin	Various (natural flavonoid)	CC:ACA		Choi et al. (2011)
Salicylic acid	Model drug	CH:GLN 1:1; CH:PHE 1:1		Santos de Almeida et al. (2017)
Salicylic acid	Model drug	CC:drug	✓	Abbott et al. (2017)
Salsalate	Brown adipose tissue inducer	PD:CC:W 1:1:1		Rozema et al. (2015)
Salvianolic acid B	Antioxidant	CC:GL 1:2		Chen et al. (2016)
Sinapine	Antioxidant	PD:CC:W 1:1:1		Durand et al. (2017)
Sulfanilamide	Antibiotic	CC:EG 1:1		Jeliński et al. (2019a)
Sulfacetamide	Antibiotic	CC:EG 1:1		Jeliński et al. (2019a)
Taxol	Anticancer	LA:G:W 5:1:3		Dai et al. (2013)
Meso-tetra-(4-carboxyphenyl)-porphine (TCPP)	Photosensitizer	CA:S 1:1; MA:F:G 1:1:1; CC:X 5:2		Wikene et al. (2016)
Thioflavin T	Biological sensor	CC:U 1:2; DEAC:U 1:2		Gautam et al. (2018)
Meso-tetra(4-hydroxyphenyl) porphine (THPP)	Photosensitizer	CA:S 1:1; G: MA 1:1; G:F:MA 1:1:1		Wikene et al. (2015b, 2017) and Tønnesen and Wikene (2016)

For abbreviations, see Table 1

17,000-fold was obtained in the tetrabutylammonium bromide:ethylene glycol (TBAB:EG) solvent compared with water.

Salicylic acid itself was one of two model compounds chosen by Santos de Almeida to investigate the feasibility of deep eutectic solvents for cutaneous applications, the other being caffeine (Santos de Almeida et al. 2017). They found that the best solubility was obtained with choline-based solvents rather than those based on halogenated imidazole. As observed by Dai et al. for other drugs, solubility increased with increased temperature (Dai et al. 2013). Another salicylic acid derivative, salsalate, was formulated in a natural deep eutectic solvent composed of equimolar proportions of 1,2-propanediol, choline chloride, and water after testing a number of compositions. This provided a nontoxic alternative to dimethyl sulfoxide (DMSO) for studies of its physiological effects in brown adipose tissue cells (Rozema et al. 2015).

The group of Shekaari have characterized the solubility of acetaminophen in choline-based deep eutectic solvents in close detail. In a first work, they compared urea, oxalic acid, and malonic acid as hydrogen bond donors (Shekaari et al. 2017). The last named had the greatest impact on solubility. As might be expected, increasing concentrations of deep eutectic solvents in water and increased temperature increased the solubility. They measured the volumetric and compressibility properties in the drug in the different solvents to gain an insight into the interactions between the drug and the solvent, which were in fact found to be strong. In a follow-up study, they added some thermodynamic measurements and determined that the main contribution to the energy of dissolution was enthalpic (Shekaari et al. 2018a). In another work, they performed similar experiments using deep eutectic solvents with the compositions choline chloride:glycerol 1:2 and choline chloride:ethylene glycol 1:2. Choline chloride:ethylene glycol was the better solvent (Shekaari et al. 2018b). They also extended their observations to another active molecule, the anti-epileptic drug lamotrigine, again finding that choline chloride:ethylene glycol was the best solvent of three tested (Shekaari et al. 2019).

Phenols and Flavonoids

Salvianolic acid B is a natural product with applications for cardiovascular diseases. Although it is water-soluble, it is easily hydrolyzed with subsequent loss of activity. Chen et al. investigated the possible improvement in stability brought about by the use of deep eutectic solvents, composed of choline chloride with ethylene glycol, glycerol, 1,2-propanediol, and 1,4-butanediol (Chen et al. 2016). The degradation of salvianolic acid B was reduced in all tested solvents compared with water, with maximum stability observed in choline chloride:glycerol with an optimal molar ratio of 1:2. Infrared spectroscopy showed interactions between the deep eutectic solvents and the carbonyl group of this phenolic acid. The authors then determined the pharmacokinetics and acute toxicity of salvianolic acid B after oral administration within this optimized deep eutectic solvent (see below).

As well as their use as solvents for phenolic compounds, deep eutectic solvents have also been employed for the extraction of this type of molecule from plants. Thus, Benlebna et al. used a natural deep eutectic solvent composed of betaine and glycerol (1:2) to extract polyphenols from green coffee beans (Benlebna et al. 2018). However, some toxicity was observed when this extract was administered orally to rats (see section on toxicity above). Radošević et al. also used natural deep eutectic solvents to extract phenols and anthocyanins from grape skin (Radošević et al. 2016). Formulations combining choline chloride with glucose, fructose, xylose, glycerol, and malic acid, in mixtures containing 30% of water, were prepared. The best extraction efficiency was achieved with the deep eutectic solvent containing malic acid, judged not only by assay of the different molecules extracted but also by the antioxidant capacity. The toxicity of the deep eutectic solvents determined in vitro toward two human tumor cell lines was low, while the antiproliferative activity of the choline chloride:malic acid extract was much higher than that of a traditional methanolic extract.

In a work designed to evaluate the antioxidant capacity of a number of phenolic compounds, Durand et al. used a natural deep eutectic solvent consisting of 1,2-propanediol (propylene glycol), choline chloride, and water in equimolar proportions as a vehicle to add the active molecules to cultured fibroblasts (Durand et al. 2017). For compounds with low water solubility, this provided a better method for obtaining dose-response curves than ethanol or dimethyl sulfoxide. The same group used this natural deep eutectic solvent to formulate resveratrol and test its ability to inhibit the enzyme matrix metalloprotease-9 in cultures of activated macrophages, thereby avoiding the toxicity associated with dissolving the molecule in dimethyl sulfoxide (Shamseddin et al. 2017).

Flavonoids are another class of antioxidant products for which deep eutectic solvents can bring considerable advantages in extraction and solubilization. To this end, Tang et al. made a detailed study of the solubility of three plant-derived flavonoids: phloretin, phlorizin, and naringin dihydrochalcone (naringin DC) in natural deep eutectic solvents that combined citric acid, choline chloride, glucose, and sucrose in various combinations with water (Tang et al. 2016). The density and viscosity of the solvents were measured, and the solubility parameters were analyzed. As observed in many studies, solubility in the natural deep eutectic solvents increased with increasing temperature. Taking all the factors into consideration, the best solvents for the extraction and solubilization of flavonoids were reported to be citric acid:glucose:water 1:1:12 and choline chloride:citric acid:water 1:1:11.

The solubility of nobiletin, a glucose-lowering flavonoid, was enhanced 450 times compared with water in the so-called "CAGE" solvent consisting of geranic acid and choline in a 2:1 molar ratio. Nucleic magnetic resonance studies showed overlapping spectra, suggesting that the flavonoid might be able to act as a hydrogen bond acceptor within the solvent (Hattori et al. 2019).

Photosensitizers

Photosensitizing compounds have great potential for the treatment of both infections and cancer but are generally hydrophobic and thus difficult to administer. Photodynamic therapy (PDT) consists of first giving the photosensitizer which accumulates, with variable degrees of specificity, in the target (bacteria or tumor cell). On illumination, the photosensitizer produces reactive oxygen species that destroy the target. It is clear that the success of this therapy depends on fast and accurate accumulation of the photosensitizer in the target. A group at the University of Oslo has been studying the use of natural deep eutectic solvents as solvents for agent used in antimicrobial photodynamic therapy. They first evaluated a panel of solvents for a natural compound, curcumin (Wikene et al. 2015a). Of 17 solvents tested, 2 gave acceptable solubility: D-(+)-glucose:sucrose 1:1 and maleic acid:choline chloride 1:3. Hydrolysis of curcumin was also reduced in these solvents. The phototoxicity of these formulations toward bacteria is discussed in the section on the antimicrobial activity of deep eutectic solvents. They then turned their attention to an anionic porphyrin, *meso*-tetra-(4-carboxyphenyl)-porphine (TCPP) (Wikene et al. 2016). From absorption and fluorescence spectroscopy, *meso*-tetra-(4-carboxyphenyl)-porphine was found to be soluble in a dicationic form in acidic deep eutectic solvents formed from citric acid:sucrose 1:1, DL-malic acid:D-fructose:D-glucose 1:1:1, and choline chloride:xylitol 5:2. Phototoxicity was observed toward several bacterial species (see below).

The same group also investigated a neutral porphyrin, *meso*-tetra(4-hydroxyphenyl)porphine (THPP) (Wikene et al. 2015b). Two natural deep eutectic solvents were selected for their ability to increase the solubility and stability: citric acid:sucrose 1:1 and D-glucose:DL-malic acid 1:1. They were able to show increased phototoxicity toward *Enterococcus faecalis* and *Escherichia coli* compared with phosphate-buffered saline. A follow-up study (Wikene et al. 2017) extended these observations to four bacterial and one fungal species. In 2016, a patent was filed entitled "Eutectic solvents and uses thereof," describing the use of deep eutectic solvents and natural deep eutectic solvents for bacterial killing by photodynamic therapy (Tønnesen and Wikene 2016).

Antimicrobial Drugs

Other patent applications, filed by Zakrewsky et al. claim that a wide range of drugs can be incorporated into deep eutectic solvents to improve their transdermal penetration, with treatment of skin infections as a particular example (see below) (Zakrewsky et al. 2015; Zakrewsky et al. 2016b).

Several more specific reports have described the solubilization of antimicrobial molecules in deep eutectic solvents. Olivares et al. have formulated two β-lactam antibiotics, clavulanic acid and imipenem, in the composition of betaine and urea in a molar ratio of 1:1.5 (Olivares et al. 2018). They used spectroscopic techniques (infrared (IR) and nuclear magnetic resonance) to characterize the deep eutectic solvents, with the results highlighting the importance of hydrogen bonds in their microstructure and the disruption of this structure by addition of water. Both the chemical stability and the antimicrobial activity of the two antibiotics were increased severalfolds by incorporation into the solvents compared with an aqueous solution. The solubility of antofloxacin hydrochloride, an antibiotic in the fluoroquinolone family, was found to be increased in a deep eutectic solvent composed of choline chloride:*para*-toluenesulfonic acid 1:2 compared with ethanol:water and ethanol:acetonitrile mixtures (Zhang et al. 2019).

Two other antibacterial molecules – sulfanilamide and sulfacetamide – were solubilized in natural deep eutectic solvents based on choline chloride paired with glycerol, sugars, or sugar alcohols (Jeliński et al. 2019a). They reported both experimental results and theoretical determinations of thermodynamic parameters using the conductor-like screening model for real solvents (COSMO-RS) protocol. The solvent composed of choline chloride and glycerol in a 1:1 molar ratio gave the highest solubility, and in general, equimolar proportions were the best solvents. In the theoretical model, these mixtures were found to have the lowest enthalpy.

Therapeutic Deep Eutectic Solvents

It is possible that an active molecule can itself act as a component of a deep eutectic solvent. As early as 1998, Stott et al. combined the nonsteroidal anti-inflammatory drug ibuprofen with a number of terpenes, including menthol and thymol, to form eutectic mixtures in an attempt to improve the transdermal penetration of the drug (Stott et al. 1998). In 2015, Su and Klibanov investigated the possibility of forming deep eutectic solvents with aspirin as the hydrogen bond donor (Su and Klibanov 2015). They observed that in a mixture with choline chloride (choline chloride:aspirin 2:1), the solubility of aspirin was greatly increased, and it was also more resistant to degradation in the deep eutectic solvent. Tarate and Bansal described a deep eutectic solvent formed between coenzyme Q and lauric acid in a 2:1 molar ratio (Tarate and Bansal 2015). Abbott et al. also described the formation of deep eutectic solvents between choline chloride and active pharmaceutical ingredients that could act as hydrogen bond donors, such as salicylic acid, paracetamol, and aspirin, or between hydrogen bond acceptors including adiphenine and ranitidine with glycerol, urea, or aspirin (Abbott et al. 2017).

A similar approach was taken with erythritol, a sugar alcohol that is able to inhibit the growth of biofilms and dental plaque. Lim et al. mixed this with a zwitterion and betaine and formed a complex that was stable in water and capable of dispersing biofilms (Lim et al. 2017). Although this may not be strictly considered as a deep eutectic solvent, they obtained similar results with a mixture of erythritol and zinc chloride (Lim et al. 2018). The cationic complex formed was found to disrupt the interactions within bacterial exopolysaccharides. In a similar vein, Wang et al. created an antibacterial deep eutectic solvent using the cationic antiseptic benzalkonium chloride with acrylic acid (Wang et al. 2017). The resulting mixture could be incorporated into a dental composite that showed good mechanical properties and biocompatibility.

In a study undertaken by Santos et al. in 2019, two molecules that could have activity in the treatment of tuberculosis, the antibiotic ethambutol and L-arginine, which can relieve the symptoms of tuberculosis by stimulating the immune system, were combined with citric (Santos et al. 2019). The systems were characterized by polarized optical microscopy, differential scanning calorimetry (DSC), and nuclear magnetic resonance. The solubility of ethambutol in the deep eutectic solvent was considerably higher than that in water. Some toxicity was observed toward Caco-2 cells, which seemed to be linked to the acid pH of the solvents (Santos et al. 2019).

Aroso et al. were the first to refer to this sort of system as therapeutic deep eutectic solvents (THEDES). They performed a study in which they associated three drugs that can act as hydrogen bond donors (acetylsalicylic acid, benzoic acid, and phenylacetic acid) with choline chloride or menthol (Aroso et al. 2016). The antibacterial activity of benzoic and phenylacetic acids was maintained in the solvents. They also showed increased permeability through an artificial membrane when the drugs were formulated in this way (Duarte et al. 2017).

Wolbert et al. attempted to use thermodynamic modelling to predict the formation of therapeutic deep eutectic solvents (Wolbert et al. 2019). Lidocaine, ibuprofen, and phenylacetic acid were used as examples of drugs that could act as hydrogen bond donors, while the potential hydrogen bond acceptors were thymol, vanillin, lauric acid, *para*-toluic acid, benzoic acid, and cinnamic acid. Predictions were made based on melting temperature and melting enthalpy, and differential scanning calorimetry was used to confirm the formation of a deep eutectic solvent. In most cases, the melting point of the mixture could be predicted to +/− 3 °C. They were also able to model the behavior of a ternary system of lidocaine, thymol, and water. This revealed that a small amount of water did not necessarily perturb the formation of a therapeutic deep eutectic solvent. Other solvents formed from lidocaine and prilocaine, and their hydrochloride salts were studied by Wojnarowska et al. both experimentally by differential scanning calorimetry and crystallography as well as in silico methods. Both the ionic (hydrochloride) and the nonionic combinations were able to form eutectic mixtures. The importance of hydrogen bonding in the nonionic deep eutectic solvents was highlighted (Wojnarowska et al. 2018).

Pereira et al. investigated the formation of therapeutic deep eutectic solvents between a monoterpene with anticancer activity, limonene, and capric acid, malic acid, ibuprofen, and menthol. In particular, the combination with ibuprofen increased the solubility of the anti-inflammatory drug and reinforced its activity against reactive oxygen species while also possessing antiproliferative activity against HT29 cells (Pereira et al. 2019).

After using their own observations and data from the literature on choline chloride-based deep eutectic solvents, Abranches et al. used the conductor-like screening model for real solvents method to generate a model and attempt to predict the formation of therapeutic deep eutectic solvents from acetylsalicylic acid, ibuprofen, paracetamol, and ketoprofen (Abranches et al. 2019). In general, they found good agreement between the eutectic temperature calculated from the model and that observed experimentally. The results of all these studies using in silico techniques to predict eutectic formation should allow a more rational approach to formulation compared with the empirical one that has been used up to now.

All these observations concerning the solubility of pharmaceutically active low-molecular-weight compounds in deep eutectic solvents are summarized in Table 2.

2.3.2 Solubilization and Stabilization of High-Molecular-Weight Molecules

As well as their applications for formulation of small molecules, deep eutectic solvents have also been found to be good solvents for biological macromolecules. As early as 2011, Choi et al. observed good solubility of salmon deoxyribonucleic acid (DNA), albumin, amylase, and starch in some natural deep eutectic solvents (Choi et al. 2011).

In their 2013 study, Dai et al. included three high-molecular-weight natural products: gluten (a plant protein), deoxyribonucleic acid, and starch (Dai et al. 2013). For each of these compounds, a deep eutectic solvent composition was found that could increase their solubility compared with water. For starch, which is not soluble in pure water even at 100 °C, 7.55 g/mole solvent were detected in glucose:choline chloride:water 2:5:5 at this temperature. The solubilities of both gluten and deoxyribonucleic acid were both increased in a solvent consisting of lactic acid, glucose, and water in a 5:1:3 molar ratio by factors of 88 and 34, respectively. Recently, Haraźna et al. showed a good solubility of lignin, a biopolymer from wood, in a solvent formed from choline chloride and a mixture of bacterially derived organic acids (Haraźna et al. 2019).

Table 3 gives a summary of the applications of deep eutectic solvents to biomacromolecules. Some more detailed results for several classes of macromolecule are given below.

Cage Molecules

Cyclodextrins and curcurbit[*n*]urils are smaller macromolecules that have the particularity of being cyclic cage molecules that can complex a wide range of guest molecules. As such, they have a number of applications in the pharmaceutical and food industries, but their water solubility can be limiting, especially for curcurbit[*n*]urils and β- and some substituted cyclodextrins. In a patent application filed by Fourmentin et al. in 2016 and published in 2018 (Fourmentin et al. 2018), it was reported that cyclodextrins could be dissolved in deep eutectic solvents composed of choline chloride and urea. In follow-up studies (Moufawad et al. 2019), the dissolution process of cyclodextrins in this deep eutectic solvent was described in detail, and it was shown that the ability of the cyclodextrin to form inclusion complexes with methyl orange was retained. Furthermore, the effect of adding water to the system was studied, and a complex formed between β-cyclodextrin and piroxicam in the deep eutectic solvent was characterized by nuclear magnetic resonance (NMR) (Colombo Dugoni et al. 2019). More recently, El Achkar et al. have demonstrated the possibility of preparing a low melting mixture composed of levunilic acid and a cyclodextrin: the randomly methylated β-cyclodextrin that is able to include trans-anethole (Achkar et al. 2020).

Table 3 Effect of deep eutectic solvents on macromolecules

Compound	Class	Optimal deep eutectic solvent composition(s)	Main observations	References
Curcurbit[*n*]urils	Cage molecules	CC:U 1:2	Solubility increased 3.9 to 10 times	Scherman and McCune (2018) and McCune et al. (2017)
Cyclodextrins	Cage molecules	CC:U 1:2	Solubility increased 3.5 to 55 times (best result for β-cyclodextrin)	Scherman and McCune (2018) and McCune et al. (2017)
Cyclodextrins	Cage molecules	CC:U 1:2	Solubility; ability to include guest molecules retained	Fourmentin et al. (2018), Moufawad et al. (2019) and Colombo Dugoni et al. (2019)
Salmon deoxyribonucleic acid	Nucleic acid	MA:P 1:1	Solubility higher than in water	Choi et al. (2011)
Deoxyribonucleic acid	Nucleic acid	LA:G:W5:1:3	Solubility increased by a factor of 34	Dai et al. (2013)
Nucleic acids	Nucleic acid	CC:U and ionic liquids	Stabilization of triplexes and quadruplexes	Tateishi-Karimata and Sugimoto (2014)
Deoxyribonucleic acid	Nucleic acid	CC:GL 1:4	Self-assembly into complex structures	Gállego et al. (2015)
Starch	Polysaccharide	CC:G 1:1	Solubility 17 g/L (not soluble in water)	Choi et al. (2011)
Starch	Polysaccharide	G:CC:W 2:5:5	Solubility 7.55 g/mole solvent (not soluble in water)	Dai et al. (2013)
Lignin	Polyphenol from wood	CC:organic acids	Good solubility	Haraźna et al. (2019)
Gluten	Protein (storage)	LA:G:W 5:1:3	Solubility increased by a factor of 88	Dai et al. (2013)
Zein	Plant protein	CC:GL 1:2	Plasticizing properties	Qu et al. (2019)
Soy protein	Plant protein	CC:GL 1:2	Plasticizing properties	Qu et al. (2019)
Gelatin	Structural protein	CC:GL 1:2 Imipramine:GL	Plasticizing properties; high ductile strength	Qu et al. (2019)

(continued)

Table 3 (continued)

Compound	Class	Optimal deep eutectic solvent composition(s)	Main observations	References
Bovine serum albumin	Protein (blood)	C:GE	Transport through porcine skin	Banerjee et al. (2017)
Bovine serum albumin	Protein (blood)	CC:GL; CC:U	Preservation of folding as measured by circular dichroism	Sanchez-Fernandez et al. (2017)
Bovine serum albumin	Protein (blood)	B:LA	Good solubility	Su and Klibanov, (2015)
Albumin	Protein (blood)	F:G:S 1:1:1	Solubility 30 g/L	Choi et al. (2011)
β-lactoglobulin	Protein (milk)	Molecularly imprinted polymer	Extraction from complex mixture	Fu et al. (2019)
Interferon α2	Protein (cytokine)	CC:F	Better stability at room temperature compared with phosphate-buffered saline	Lee et al. (2018)
Lysozyme	Protein (enzyme)	CC:GL; CC:U	Structural changes compared with buffer; reduced antibacterial activity	Esquembre et al. (2013)
Lysozyme	Protein (enzyme)	CC:U	Prevention of aggregation and better refolding after denaturation	Niknaddaf et al. (2018)
Lysozyme	Protein (enzyme)	Ionic liquids; CC:carboxylic acid	Rapid formation of fibrils and nanofibers	Silva et al. (2018a, b)
Lysozyme	Protein (enzyme)	CC:GL; CC:U	Preservation of folding as measured by circular dichroism	Sanchez-Fernandez et al. (2017)
Lysozyme	Protein (enzyme)	TEAC:EG; CC:LA; CC:GL	Good solubility, enzyme activity retained, and thermal stability	Su and Klibanov, (2015)
α-chymotrypsinogen	Protein (enzyme)	ZC:PD	Good solubility	Su and Klibanov (2015)
Subtilisin	Protein (enzyme)	CC:GL	Good solubility	Su and Klibanov, (2015)

(continued)

Table 3 (continued)

Compound	Class	Optimal deep eutectic solvent composition(s)	Main observations	References
Versatile peroxidase	Protein (oxidizer)	CC:GL	Increased stability and activity compared with aqueous buffer	Mamashli et al. (2018)
Catalase	Protein (enzyme)	CC:GL 1:2; CC:U 1:2	Enzyme activity preserved	Harifi-Mood et al. (2017)
Laccase	Protein (enzyme)	B:GL 1:2	Stability and enzyme activity preserved	Khodaverdian et al. (2018)
Lipase	Protein (bacterial enzyme)	Cholinium acetate	Stability and enzyme activity preserved	Nascimento et al. (2019)
Insulin	Protein (hormone)	C:GE	Transport through porcine skin; hypoglycemic effect in normal rats	Banerjee et al. (2017)
Insulin	Protein (hormone)	C:GE	Transport through porcine skin; changes in skin structure detected by Fourier transform infrared spectroscopy	Tanner et al. (2018)
Insulin	Protein (hormone)	C:GE 1:2	Lowering of blood glucose in normal rats after oral administration	Banerjee et al. (2018a)
Insulin	Protein (hormone)	CC:MA 2:1	Increased passage, lowering of blood glucose in normal rats after nasal administration, and no visible toxicity	Li et al. (2019)

For abbreviations, see Table 1

Another patent (Scherman and McCune 2018) concerns the use of deep eutectic solvents for the solubilization of cage molecules including cyclodextrins and curcurbit[*n*]urils. These authors used a deep eutectic solvent of choline chloride and urea in a 1:2 molar ratio to dissolve these compounds and found an increase in solubility compared with water for all classes that was as high as 55-fold for β-cyclodextrin (McCune et al. 2017). With the aid of a colored reporter molecule, methylviologen, they were able to show that dissolution in the deep eutectic solvent did not upset the cage-guest equilibrium. These results with cage molecules open up many opportunities for pharmaceutical formulation.

Nucleic Acids

As a result of their specific base-pairing properties, nucleic acids have many applications in biomedical science, both in therapeutic and diagnostic scenarios. However, nucleic acid structures are not extremely stable in aqueous solution. Tateishi-Karimata and Sugimoto have reviewed work in which nucleic acids have been dissolved in a hydrated ionic liquid composed of choline dihydrogen phosphate and a deep eutectic solvent made of choline chloride and urea (Tateishi-Karimata and Sugimoto 2014). They reported differences in base-pairing compared with sodium chloride solution and a stabilization of triplexes in ionic liquid and quadruplexes in deep eutectic solvent. Since these structures are used in a number of devices in nanomedicine, this is an important result. Gállego et al. studied the formation of deoxyribonucleic acid nanostructures in deep eutectic solvents using atomic force microscopy, mobility on agarose gels, and circular dichroism (Gállego et al. 2015). They observed that a deep eutectic solvent composed of choline chloride and glycerol in a 1:4 ratio allowed folding into complex structures that could be transferred into aqueous solution.

Proteins

Ionic liquids and deep eutectic solvents are also attracting much interest as alternative solvents for proteins, for both fundamental studies and applications in medicine and biotechnology. The three-dimensional structure of protein molecules is essential for their function but is only marginally stable, being held together by a large number of weak interactions: hydrogen bonds, electrostatic interactions, van de Waals forces, and hydrophobic interactions. It follows that the solvent has a profound influence on protein conformation and activity. Given that, as discussed by Choi et al. many components of deep eutectic solvents are found in living cells and may be involved in survival of stress conditions, they could be considered as a potential alternative to aqueous solvents for proteins (Choi et al. 2011). This hypothesis was investigated by Sanchez-Fernandez et al. with respect to two model proteins: lysozyme and bovine serum albumin (BSA) (Sanchez-Fernandez et al. 2017). Two main physiochemical techniques were used to probe the protein structure: circular dichroism and small-angle neutron scattering (SANS). In particular, circular dichroism in the far-ultraviolet (far-UV) region gives an indication of the secondary structure of the protein and allows the proportion of α-helix, β-sheet, and random coil to be estimated. On the other hand, the near-UV region gives information on the tertiary structure.

The solubility of bovine serum albumin, bovine pancreatic α-chymotrypsinogen, bacteria subtilisin, and hen egg white lysozyme in a panel of low transition temperature mixtures was tested by Su and Klibanov (Su and Klibanov 2015). The solubilities depended on the DES composition and the structures of the proteins themselves and in some cases the solubility approached that observed in water. The enzymatic activity of lysozyme was tested in a number of mixtures and found to be almost completely conserved. The thermal stability of the enzyme was also conserved or even increased.

As indicated by circular dichroism, the secondary structure of bovine serum albumin was not affected by the use of a deep eutectic solvent composed of choline chloride and glycerol either pure or mixed with water compared with phosphate-buffered saline. On the other hand, changes in the near-ultraviolet region suggested a modification of tertiary structure in the pure deep eutectic solvent (Sanchez-Fernandez et al. 2017). Bovine serum albumin in pure solvents was irreversibly denatured by heating to 80 °C, but the results in water are not reported. As far as lysozyme was concerned, its secondary structure of a mixture of α-helix and β-sheet was not modified in the chosen deep eutectic solvents: choline chloride with glycerol or urea. Small-angle neutron scattering was used to get a better impression of the protein conformation in solution. This showed that the globular shape of bovine serum albumin became less symmetrical in pure deep eutectic solvent, indicating that the solvation of the protein was modified in this solvent. On the other hand, when a hydrated deep eutectic solvent was used, the results were not different from pure water. This was interpreted as water forming a shell around the protein, preventing direct interaction with the deep eutectic solvent. In the case of lysozyme, the protein was found to be only partially folded in the deep eutectic solvent, irrespective of whether it contained glycerol or urea. This might not have been expected because high concentrations of urea denature proteins.

Another study of lysozyme structure in deep eutectic solvents was carried out by Esquembre et al. (Esquembre et al. 2013). They used circular dichroism and also followed the intrinsic fluorescence of tryptophan residues that gives information about folding and unfolding of the protein. In this study, the deep eutectic solvents were composed of choline chloride with either glycerol or urea in a 2:1 molar ratio. Tryptophan fluorescence of lysozyme was slightly blue shifted in solvents compared with buffer solution, whereas denatured lysozyme shows a red shift. Heating of lysozyme caused unfolding in all solvents, but monitoring of the mean fluorescence energy indicated different processes in buffer and in deep eutectic solvent solution: in the latter, the protein appeared to pass through a number of intermediate states. These states were examined by means of circular dichroism spectra. Differences in the near-ultraviolet region were recorded as a function of temperature and revealed that modifications in the tertiary structure began to occur at a lower temperature for the deep eutectic solvent containing urea than for the deep eutectic solvent with glycerol or with buffer. While the thermally induced unfolding of lysozyme was completely reversible in buffer, it was only partially so in the deep eutectic solvent with glycerol and irreversible in the urea-containing deep eutectic solvent. The biological activity of lysozyme was also tested in media containing various proportions of the choline chloride:glycerol solvent by its capacity to lyse a Gram-positive bacterium. The speed of bacterial degradation decreased as the proportion of the solvent increased. The authors concluded that a detailed characterization of the system is necessary before deep eutectic solvent can be used as a solvent for biocatalysis (Esquembre et al. 2013).

Harifi-Mood et al. studied the structure and activity of the enzyme catalase from bovine liver in two deep eutectic solvents: choline chloride:glycerol 1:2 and choline chloride:urea 1:2 (Harifi-Mood et al. 2017). Fluorescence spectroscopy and circular dichroism indicated that the secondary structure of the protein was modified in the deep eutectic solvents, more in the glycerol-containing one than in the urea-based one. The enzymatic activity was slightly reduced in the deep eutectic solvents compared with phosphate buffer, but the binding affinity of the substrate was increased. The use of deep eutectic solvents as solvents for the bacterial enzyme laccase was investigated by Khodaverdian et al. (Khodaverdian et al. 2018). They used natural deep eutectic solvents based on choline or betaine, the latter leading to better preservation of enzyme activity. In particular, a deep eutectic solvent composed of sorbitol, betaine, and water in equimolar proportions improved the stability of the enzyme at high temperature compared with aqueous buffer. Once again, the intrinsic fluorescence of tryptophan residues was used to monitor structural changes in the protein.

The influence of choline-based ionic liquids on the catalytic activity of lipase from *Aspergillus niger* was investigated by Nascimento et al. (Nascimento et al. 2019). At low concentrations, all the ionic liquids tested (choline, choline acetate, choline propanoate, choline butanoate, choline pentanoate, and choline hexanoate) allowed the activity to be maintained or even enhanced. However, higher concentrations of the anions with longer alkyl chains (pentanoate and hexanoate) inhibited enzyme activity, suggesting that these formulations could be useful in modulating lipase activity in vivo.

In living cells, protein folding and refolding is facilitated by other proteins known as chaperones that play an important role in the response of the cell to stress. The fact that many of the components of deep eutectic solvents are found in cells and are related to survival under extreme condition led Niknaddaf et al. to put forward the hypothesis that these solvents could act as chemical chaperones (Niknaddaf et al. 2018). They conducted experiments to test this using a model of lysozyme aggregation by the reducing agent dithiothreitol. They observed that 20% of choline chloride:urea solvent in water completely inhibited this aggregation, which is brought about by the reduction of disulfide bonds in the protein. This might appear to be contradictory with the results of Esquembre et al. above (Esquembre et al. 2013); however, the mechanism of denaturation was different. They also looked at refolding by dilution after denaturation with guanidinium chloride. The presence of the choline chloride:urea solvent in the dilution buffer reduced aggregation compared with simple buffer. Monitoring of the intrinsic fluorescence of tryptophan confirmed that the solvents promoted correct refolding. The biological activity of the enzyme, again measured by the ability to lyse bacteria, was also recovered better when the solvent was present in the renaturation solution (Niknaddaf et al. 2018).

In two articles published in 2018 by Silva et al. another aspect of lysozyme behavior was considered. In this work, they were attempting to produce nanofibers from the protein, that is, promoting its denaturation and aggregation. Protein nanofibers have a number of applications in biotechnology, for example, in tissue engineering and biosensors, particularly when combined with polymers. In one study, they observed that hen egg white lysozyme fibrillation was considerably faster in ionic liquids than in water or pH 2 buffer. Various spectroscopic and microscopic techniques revealed nanofibers of up to 1 μm in length, with a diameter depending on the ionic liquid used (Silva et al. 2018a). The presence of the acetate anion was found to favor the formation of β-sheet structures which are more inclined to aggregate and produce fibrils. In another work, acetate and other carboxylic acids were used in combination with choline chloride to form deep eutectic solvents that were then used as a medium for the formation of lysozyme nanofibers in acidic solution (Silva et al. 2018b). The dimensions of the fibers depended on the nature of the carboxylic acid, the longest fibers being obtained with lactic acid. Again, the formation was rapid, complete within a few hours. The nanofibers so generated were incorporated into carbohydrate-based films where they increased the tensile strength.

Qu et al. have investigated the possibility of creating materials for drug delivery from proteins dissolved in deep eutectic solvents (Qu et al. 2019). They tested the plasticizing properties of deep eutectic solvents, in particular "Glyceline," choline chloride:glycerol 1:2, with two plant-derived globular proteins, zein from maize and soy protein. Although the solvents modified the properties of the proteins, the resulting materials were not as strong and ductile as those using PEG or fatty acid modifiers. On the other hand, when gelatin, a fibrous protein, was incorporated into Glyceline mixtures, a material with high ductile strength was obtained. The choline chloride could be replaced by imipramine hydrochloride, forming a therapeutic deep eutectic solvent, and patches could be prepared that increased the penetration of the drug into bovine skin.

Another enzyme whose activity has been improved by incorporation into a deep eutectic solvent is versatile peroxidase, a heme peroxidase of fungal origin. It has potential industrial applications for which an alternative to organic solvents would be desirable. Thus, a deep eutectic solvent composed of choline chloride and glycerol (1:2) was tested (Mamashli et al. 2018). The activity of the enzyme at neutral pH was higher in the solvent-buffer mixtures than in buffer alone. A number of techniques, including circular dichroism and fluorescence spectroscopy, showed changes in protein structure in the deep eutectic solvent, with exposure of hydrophobic regions and rearrangement of the heme group. Furthermore, the thermal stability of the enzyme was improved in deep eutectic solvent.

Interferon-α2 (IFN-α2) is a cytokine with several therapeutic applications in the treatment of cancer and liver disease. However, therapeutic proteins are notoriously difficult to formulate, store, and handle. Lee et al. investigated the formulation of this protein in several natural deep eutectic solvents with particular focus on its thermal stability (Lee et al. 2018). An equimolar mixture of choline chloride and fructose was found to preserve the biological activity of interferon-α2 during a 2-h incubation at temperatures of 37 °C, 50 °C, and even 70 °C. Circular dichroism and fluorescence spectroscopy experiments showed that the protein conformation was preserved when it was heated in the choline chloride:fructose deep eutectic solvents, in contrast to the modifications observed when it was formulated in phosphate-buffered saline (PBS). This opens up the way to formulations that do not need a strict cold chain to preserve their activity.

As discussed below, proteins, and in particular insulin, have been included in deep eutectic solvents in order to improve their cutaneous, oral, or nasal bioavailability (Banerjee et al. 2017; Tanner et al. 2018; Banerjee et al. 2018a; Li et al. 2019).

2.3.3 *Antimicrobial Applications*

As can be seen from the section on solubilization of small molecules, a number of antimicrobial agents have been incorporated into deep eutectic solvents. However, beyond this, it has been observed that some deep eutectic solvents formulations themselves have antiseptic or antimicrobial activity in their own right, especially by the topical route. The antimicrobial applications of deep eutectic solvents are summarized in Table 4.

Deep Eutectic Solvents with Intrinsic Antimicrobial Activity

Zakrewsky et al. provide some convincing evidence that a deep eutectic solvent composed of choline and geranic acid in a 1:2 molar ratio, named CAGE, could act as an efficient antiseptic agent on the skin (Zakrewsky et al. 2016a). They have shown antimicrobial activity against a wide range of bacteria, fungi, and viruses, combined with low skin toxicity. An in vivo study was carried out on a skin infection with *Propionibacterium acnes* injected intradermally in rat ears. The choline:geranic acid solvent was much more effective than clindamycin in saline in reducing ear thickness and reaching bacteria in deeper skin layers. This development of a cheap, effective, and nontoxic topical antimicrobial formulation has considerable importance given the rise of resistance to current agents and the withdrawal of some compounds because of safety concerns.

This research group has filed several patents with claims concerning the potential of ionic liquids for use on the skin: their lack of irritability and their ability to transport substances through the skin (Zakrewsky et al. 2015; Zakrewsky et al. 2016b; Kellar et al. 2018). In particular, the latest patent application (Kellar et al. 2018) refers to the role of ionic liquids in wound healing as a result of their ability to inhibit biofilm formation and to control pathogen growth. The "CAGE" formulation of choline and geranic acid figures prominently in the examples cited. A recent study has used some physicochemical techniques to probe the mechanism of action of the choline:geranic acid solvent (Ibsen et al. 2018). Molecular dynamic simulations were used to understand how the "CAGE" components might interact with the membrane of *Escherichia coli*. The choline:geranate pair seems to be able to penetrate through the lipopolysaccharide (LPS) layer and bind to the negatively charged membrane. Membrane disruption could be observed in "CAGE"-treated cells by scanning electron microscopy. Fourier transform infrared spectroscopy with

Table 4 Antimicrobial activity of deep eutectic solvent formulations

Active agent	Deep eutectic solvent composition	Results in vitro	Results in vivo	References
None	C:GE 1:2	Broad-spectrum antimicrobial activity	Treatment of *Cutibacterium acnes* infection in mice	Zakrewsky et al. (2016a)
None	CC:MOA 1:2	Inhibition of growth and β-lactamase production		Sadaf et al. (2018)
None	Various including C:GE 1:2	Inhibition of biofilms	Reduced bacterial growth in wound dressings	Kellar et al. (2018)
None	Various, containing organic acids	Activity against yeast and cancer cells; antioxidant properties		Radošević et al. (2018)
None	C:GE 1:2	Activity against *Escherichia coli*; studies of mechanism		Ibsen et al. (2018)
None	CC:AA polymerized with molecular imprinting	Inhibition of bacterial growth; extraction of β-lactoglobulin		Fu et al. (2019)
Various	Various including CC:carboxylic acid	Inhibition of biofilms	Skin permeation	Zakrewsky et al. (2015)
Benzalkonium chloride	BC:AA	Incorporation into dental composite		Wang et al. (2017)
Erythritol	B:E 2:1; ZC:E 1:3	Penetration into and detachment of biofilms		Lim et al. (2018)
β-lactams: clavulanic acid and imipenem	B:U 1:1.5	Increased stability and maintenance of activity		Olivares et al. (2018)
Photosensitizers	Various, including CA:S 1:1	Phototoxicity to bacteria	Antifungal effect	Tønnesen and Wikene (2016) and Wikene et al. (2015a, 2016, 2017)
L-arginine and ethambutol	Drug:CA	Toxicity to Caco2 cells		Santos et al. (2019)
Benzoic acid and phenylacetic acid	Drug:menthol	Activity against Gram - and Gram + bacteria		Aroso et al. (2016)

For abbreviations, see Table 1

attenuated total reflection demonstrated increased lipid content in treated cells. Importantly, *Escherichia coli* cells did not develop resistance after multiple or prolonged exposures to the choline:geranic acid solvent.

Another mechanism of activity of deep eutectic solvents against bacteria was investigated by Sadaf et al. (Sadaf et al. 2018). In in vitro studies, they looked at the ability of a formulation that they named "maline," composed of malonic acid and choline chloride in a 2:1 molar ratio to inhibit the activity of an extracellular enzyme, β-lactamase from *Bacillus cereus*. As in the studies on lysozyme reported above, circular dichroism and intrinsic fluorescence were used to monitor protein structure. This showed a transition of α-helix to random coil in the presence of maline, together with fluorescence quenching that could have been due to loss of the tertiary structure of the protein.

The antimicrobial activity of a number of natural deep eutectic solvents was tested toward several bacterial species, as well as a yeast strain and three human cell lines by Radošević et al. (Radošević et al. 2018). The highest activity toward bacteria was obtained with deep eutectic solvents containing organic acids – oxalic, malic, and citric. The yeast *Candida albicans* was quite resistant to these solvents. They also recorded cytotoxic effects toward human tumor cells lines HeLa and MCF-7 with a choline chloride:oxalic acid deep eutectic solvent while a normal cell line, HEK293T, was more resistant. The antioxidant capacity of the deep eutectic solvents was measured by their capacity to absorb oxygen radicals. Most of the solvents tested had oxygen scavenging properties. One deep eutectic solvent, composed of betaine, malic acid, and proline, stimulated mammalian cell growth while having strong antioxidant properties.

Recently, Silva et al. investigated the antimicrobial properties of deep eutectic solvents formed entirely from fatty acids, combining capric acid with other saturated fatty acids of longer chain length (Silva et al. 2019). They observed activity against Gram-positive bacteria and *Candida albicans*, particularly with a capric acid:lauric acid formulation. A significant result is that these deep eutectic solvents could promote the detachment of *Escherichia coli* biofilms (Fig. 2.5).

Therapeutic Deep Eutectic Solvents for Applications in the Oral Cavity

As mentioned above, Lim et al. formed a complex between erythritol and betaine in an attempt to treat biofilms in the oral cavity (Lim et al. 2017). The extracellular matrix secreted by the bacteria in these biofilms makes them extremely resistant to antibiotics. Detachment of *Streptococcus mutans* biofilms observed in an in vitro assay was observed after application of the betaine:erythritol (2:1) mixture but not with the components separately. Measurements by atomic force microscopy showed a reduction in the adhesive strength of the biofilm after contact with the betaine-erythritol mixture. It was observed that the water-insoluble extracellular polysaccharide could be dissolved in this mixture. These results are encouraging for the treatment of dental plaque because the constituent molecules are nontoxic. In a related study, the same group observed activity against biofilms of a complex

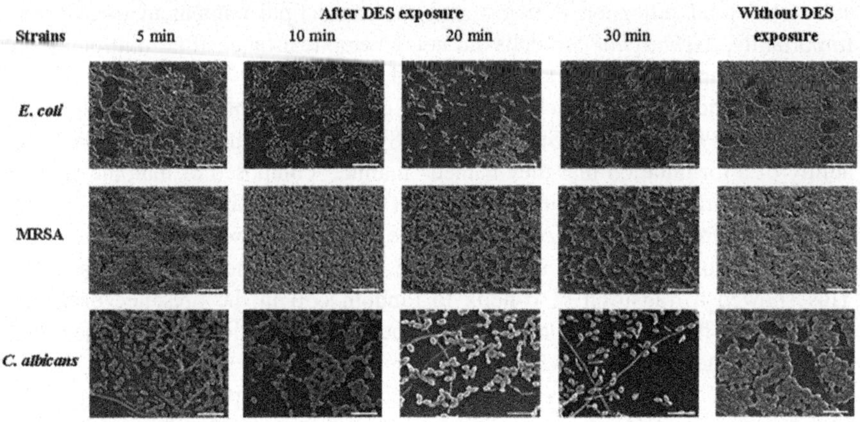

Fig. 2.5 SEM images of biofilms from three different microorganisms (*Escherichia coli*, methicillin-resistant *Staphylococcus aureus*, and *Candida albicans*) seeded onto coverslips after various times of exposure to a capric acid:lauric acid formulation. (Silva et al. 2019; reprinted with permission of Elsevier)

formed between erythritol and zinc chloride (Lim et al. 2018). The complex was able to interfere with intra- and intermolecular bonds in the exopolysaccharide.

Wang et al. also used a deep eutectic solvent-based approach to dental treatment (Wang et al. 2017). A mixture of benzalkonium chloride and acrylic acid could be incorporated into dental resins without altering their mechanical properties, while benzalkonium chloride alone compromised the strength of the resin during storage. The antibacterial activity of these materials was tested by placing discs in agar plates inoculated with *Streptococcus mutans* or *Staphylococcus aureus*. The deep eutectic solvent was as effective as benzalkonium chloride alone in controlling bacterial growth. The toxicity toward host cells was tested using the osteoblast line MC3T3-E1 and the 3-(4,5-dimethylthiazol-2-yl)-2,5-diphenyltetrazolium bromide (MTT) assay. Cells cultivated with the composite containing the solvent did not show any decrease in viability compared with control cells, while the composite containing benzalkonium chloride alone reduced their viability by about 50%.

Formulation of Conventional Antibiotics in Deep Eutectic Solvents

Santos et al. formulated two molecules that are potentially active in tuberculosis, L-arginine and ethambutol, as deep eutectic solvents with citric acid but did not study the antimicrobial activity (Santos et al. 2019). On the other hand, Aroso et al. showed that the activity of benzoic acid and phenylacetic acid against both Gram-negative and Gram-positive bacteria was comparable to that of the free drug when they were prepared as deep eutectic solvents with menthol (Aroso et al. 2016).

The stability of two β-lactam antibiotics, clavulanic acid and imipenem, was improved by dissolution in a deep eutectic solvent composed of betaine and urea

(Olivares et al. 2018). Their antimicrobial activity was determined in vitro by both a limiting dilution method to obtain the minimum inhibitory concentration (MIC) and a disk diffusion assay where the formulations were incorporated into cellulose disks and placed on agar plates containing the microorganisms. The test bacteria used were *Escherichia coli* and *Pseudomonas aeruginosa*. When freshly prepared solutions were used, the minimum inhibitory concentration of both antibiotics was identical in aqueous solution and in the solvents. However, when the formulations were kept at 25 °C for 7 days before adding them to the bacteria, the minimum inhibitory concentration of the water solution of imipenem increased at least 32-fold and that of aqueous clavulanic acid doubled, whereas no changes were observed for the deep eutectic solvents compared to a fresh preparation. These results are in agreement with stability studies. The disk assay was used because it allowed the deep eutectic solvent to be added to the bacterial cultures without dilution that could perturb the solvent structure. The size of the ring of growth inhibition of *Pseudomonas aeruginosa* was similar for antibiotics in aqueous solution and in the deep eutectic solvents, showing that they remained fully bioavailable in the latter formulation. This opens up possibilities for the use of these solvents in sustained release formulations without loss of antibiotic activity.

Formulation of Photosensitizers in Deep Eutectic Solvents for Antimicrobial Activity

As described above, deep eutectic solvents have been used to improve the solubility and stability of photosensitizing compounds, and the resulting formulations have been used for antimicrobial photodynamic therapy. This is reported in a series of articles and a patent application from the group of Tønnesen and Wikene (Tønnesen and Wikene 2016). The first compound to be used was the small, naturally occurring molecule curcumin (Wikene et al. 2015a). Curcumin in a deep eutectic solvent composed of maleic acid and choline chloride in 1:3 molar ratio was phototoxic toward *Escherichia coli* at a concentration of 1.25 μM, lower than any previously reported result. At this concentration, there was no contribution of the toxicity of the deep eutectic solvent itself.

Two porphyrin-based photosensitizers were formulated in a similar way. *Meso*-tetra-(4-carboxyphenyl)-porphine (TCPP) is an anionic porphyrin that could be solubilized in acidic deep eutectic solvents, citric acid:sucrose 1:1, DL-malic acid:D-fructose:D-glucose 1:1:1, and choline chloride:xylitol 5:2, with improved photostability compared with methanolic solution (Wikene et al. 2016). The phototoxicity of these preparations were tested against both Gram-positive (*Enterococcus faecalis*, *Staphylococcus aureus*) and Gram-negative (*Escherichia coli*) bacterial species. Before testing the phototoxicity of *meso*-tetra-(4-carboxyphenyl)-porphine, it was necessary to establish the photo- and dark toxicity of the deep eutectic solvent and define the dilution at which the effects of the medium were not significant. In these conditions, *meso*-tetra-(4-carboxyphenyl)-porphine in citric acid:sucrose or malic acid:fructose:glucose was much more phototoxic to all three bacterial species than a solution in ethanol. A neutral porphyrin, *meso*-tetra(4-hydroxyphenyl)porphine (THPP), was also formulated in this way (Wikene et al. 2015b). Two deep eutectic solvents, citric acid:sucrose 1:1 and D-glucose:DL-malic acid 1:1, were chosen for tests of phototoxicity against *Enterococcus faecalis* and *Staphylococcus aureus*. These solutions showed much higher antibacterial phototoxicity than a solution of *meso*-tetra(4-hydroxyphenyl) porphine in phosphate-buffered saline. In a subsequent study, Wikene et al. (2017) extended the observations to four bacterial species (*Escherichia coli, Staphylococcus epidermidis, Pseudomonas aeruginosa, Klebsiella pneumoniae*) as well as a fungal pathogen (*Candida albicans*). In these experiments, citric acid:sucrose 1:1 and DL-malic acid:D-fructose:D-glucose 1:1:1 were the chosen deep eutectic solvents. *Meso*-tetra(4-hydroxyphenyl)porphine in the deep eutectic solvents was much more phototoxic toward *Escherichia coli* than *meso*-tetra(4-hydroxyphenyl)porphine in phosphate-buffered saline. Similar results were obtained with *Pseudomonas aeruginosa* and *Staphylococcus epidermidis*. *Klebsiella pneumoniae* and *Candida albicans* were less sensitive.

Molecular Imprinting

A different approach was recently described by Fu et al. (Fu et al. 2019). They used a deep eutectic solvent composed of choline chloride and acrylic acid to produce a molecularly imprinted polymer designed to selectively extract β-lactoglobulin from complex mixtures. The polymer was formed on a base of molybdenum disulfide (MoS_2) containing some iron oxide nanoparticles (Fe_3O_4) to allow the polymer to be manipulated using an external magnetic field. The deep eutectic solvent was polymerized in the presence of the protein using ethylene glycol dimethacrylate to cross-link acrylic acid. The resulting poly(ChCl-AA deep eutectic solvent)@Fe_3O_4@ MoS_2 was able to selectively extract β-lactoglobulin from milk but also showed antibiotic properties. The growth of *Staphylococcus aureus*, *Escherichia coli,* and *Bacillus subtilis* was retarded in the presence of the polymeric material. However, an inverse effect was obtained with *Pseudomonas fluorescens*. The growth of bacterial on biomaterials always poses a problem for their use, so these results are encouraging.

2.3.4 Cutaneous Applications

Ionic liquids and deep eutectic solvents have great potential for topical application to the skin because of their low toxicity and their rheological properties. Furthermore, there is evidence that these formulations can act as promoters of absorption across the outer layers of the skin and help to deliver active substance into and across the skin. Early work by Stott et al. showed that forming eutectics between ibuprofen and various terpenes could enhance their penetration across human epidermis (Stott et al. 1998).

Transdermal penetration enhancement by deep eutectic solvents was claimed by Zakrewsky et al. in 2014 and in several patents (Zakrewsky et al. 2015; Zakrewsky et al. 2016b; Kellar et al. 2018), with particular reference to treating infections with microbial biofilms in the skin (Zakrewsky et al. 2014). Skin penetration was first determined using radiolabelled mannitol as a model drug and pig skin in Franz diffusion cells. One formulation in particular, a mixture of choline chloride and the geranate anion, was able to promote penetration into the epidermis and dermis. A similar result was obtained with the antibiotic cefadroxil incorporated into the deep eutectic solvent. The efficacy of this deep eutectic solvent, with or without antibiotic, was tested on a biofilm formed by *Pseudomonas aeruginosa* in simulated wounds in an in vitro human skin model. The choline chloride:geranate deep eutectic solvent alone caused a very significant reduction in bacterial viability that was further decreased by the inclusion of a related antibiotic (ceftazidime), whereas the ceftazidime in saline had only a small effect. Finally, it was important to assess the tolerance of the skin to deep eutectic solvent application. This was done using Fourier transform infrared spectroscopy to monitor changes in stratum corneum structure in full thickness pig skin. The use of the choline chloride:geranate deep eutectic solvent did not provoke any significant changes (Zakrewsky et al. 2014).

Santos de Almeida et al. also investigated the potential of deep eutectic solvents (ionic liquids as they call them) for cutaneous applications and in particular compared imidazole and choline as the hydrogen bond acceptor (Santos de Almeida et al. 2017). They chose two model drugs, caffeine and salicylic acid, and studied their permeation through full-thickness pig skin in Franz cells under occlusive conditions as well as the toxicity toward the human HaCaT keratinocyte cell line, using the 3-(4,5-dimethylthiazol-2-yl)-2,5-diphenyltetrazolium bromide (MTT) reduction assay as the endpoint. As reported above, an increase in caffeine solubility of two to three times was observed with choline-based formulations while the increases in salicylic acid solubility were more modest. On the other hand, imidazole-based formulations did not increase drug solubility. Permeation across pig skin from saturated solutions of the drug in 5:95 deep eutectic solvent:water was studied. The imidazole-based formulations increased passage across the skin, while choline-based ones had no promoting effect. This might have been due to a surfactant effect of the imidazole side chain. Reduced passage across the skin would be an advantage for an activity in the skin without systemic toxicity. The choline-based formulations were less toxic than the imidazole-based ones.

Choline:glutamine and choline:phenylalanine deep eutectic solvents loaded with caffeine were formulated as oil-in-water emulsions and gels suitable for skin applications. Emulsions containing the deep eutectic solvents had lower viscosity than a simple caffeine-containing emulsion and needed more surfactant for stability. However, the presence of the deep eutectic solvents prevented the crystallization of caffeine within the emulsion. A similar observation was made when gels containing caffeine were prepared from Carbopol® 940. Gel stability was not compromised by the presence of the deep eutectic solvent (Santos de Almeida et al. 2017).

As well as their possible applications for the cutaneous delivery of small molecules, deep eutectic solvents have potential for the delivery of proteins to the skin. Banerjee et al. observed that the choline:geranate (1:2) solvent, which they refer to as CAGE formulation, could promote the topical delivery of proteins as varied as bovine serum albumin, ovalbumin, and insulin (Banerjee et al. 2017). Labelled proteins were found to penetrate pig skin, reaching the epidermis and the dermis and passing into the acceptor compartment. Confocal microscopy allowed the penetration to be visualized. Fourier transform infrared spectroscopy revealed that the choline:geranic acid formulations modified some peaks in a way that could be interpreted as lipid extraction from the skin. Circular dichroism studies of the secondary structure of insulin showed that the α-helical content of the protein was not affected by incorporation into the deep eutectic solvents. Finally, the effect of insulin solubilized in choline:geranic acid solvent on the blood glucose levels of nondiabetic rats was tested. Application of insulin in choline:geranic acid formulation onto the skin resulted in a fall in blood glucose that was more sustained than that obtained after an injection of insulin (although the injected dose was 1 U/kg while the topical dose was 25 U/kg). Choline:geranic acid solvent alone or insulin in buffer applied to the skin had no effect.

In a later study, the solvent formulation was studied in more detail by varying the ratio of choline to geranic acid from 1:4 to 2:1 (Tanner et al. 2018). Nuclear magnetic resonance spectroscopy was used to map the interactions between the protons in the two components, and this was found to differ as the proportions changed. This would modify the ability of the solvent to accommodate solutes and allow the composition to be tailored for different types of molecules. Both the viscosity and conductivity of the formulation decreased as the proportion of geranate increased. The thermal stability of the formulations was quite similar over the different ratios, with temperatures of decomposition of 169 to 210 °C. As in previous articles, Fourier transform infrared spectroscopy was used to assess the impact of the deep eutectic solvent on stratum corneum structure, looking at peak height between 2800 and 3000 cm^{-1} that shows the vibrations in lipid bonds. A decrease in peak height suggests that deep eutectic solvent can extract lipids, and as the solvent becomes more hydrophobic, with a higher proportion of geranate, this effect is more pronounced. Finally, the different formulations were tested for their ability to promote the passage of fluorescently labelled insulin across pig skin in a diffusion cell. The formulations with a higher proportion of geranate (1:2 and 1:4) showed the presence of the protein in the stratum corneum, epidermis, and dermis, while for the 1:1 and 2:1 formulations, and for phosphate-buffered saline, it showed fluorescence restricted to the stratum corneum (Fig. 2.6).

The "CAGE" formulation was also employed for a low-molecular-weight glucose-lowering agent, the flavonoid nobiletin (Hattori et al. 2019). The transdermal passage of this molecule was significantly increased by the use of this solvent, and plasma concentrations were enhanced compared with the drug formulated in aqueous buffer or ethanol. Lowering of blood glucose concentrations was observed after transdermal administration to normal rats (Hattori et al. 2019).

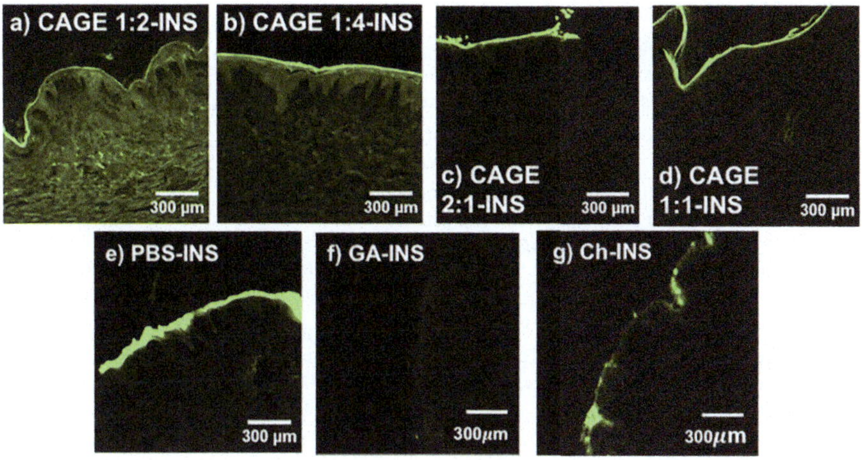

Fig. 2.6 Confocal microscopy of porcine skin after application of FITC insulin dissolved in (**a**) 1:2 CAGE, (**b**) 1:4 CAGE, (**c**) 2:1 CAGE, (**d**) 1:1 CAGE, (**e**) PBS, (**f**) geranic acid, and (**g**) choline bicarbonate. (Tanner et al. 2018; reprinted with permission of Elsevier)

A further work aimed at elucidating general principles for transdermal deep eutectic solvent formulation employed two model molecules, acarbose and ruxolitinib, as examples of hydrophilic and hydrophobic drugs, respectively (Tanner et al. 2019). Different choline-to-geranic acid proportions were used, as well as alternative carboxylic acids to replace geranic acid. The ability of the solvent systems to deliver the active molecules across skin in vitro was correlated with 2D nuclear magnetic resonance studies. It was observed that the transdermal transport capacity of the deep eutectic solvent was inversely related to the strength of the interionic interactions in the solvent. As a result, a deep eutectic solvent containing citronellic acid was found to be the optimal vehicle for ruxolitinib.

Based on the body of work accumulated with the "CAGE" solvent, Qi and Mitragotri have recently published a mechanistic study (Qi and Mitragotri 2019). They measured the skin penetration of fluorescent dextran of different molecular weights and concluded that for molecules up to 150 kilodaltons, transport was enhanced irrespective of molecular size. Combined with their observations on lipid extraction, they concluded that the "CAGE" formulation creates a more favorable environment for diffusion in the skin that does not rely on physiological transport pathways (Qi and Mitragotri 2019).

The use of ionic liquids and deep eutectic solvents for transdermal applications was recently comprehensively reviewed by Sidat et al. (Sidat et al. 2019). In particular, they list some studies in which ionic liquids have been used in synergy with conventional chemical penetration enhancers.

Table 5 provides a summary of work done on the transdermal applications of deep eutectic solvents.

Table 5 Cutaneous applications of deep eutectic solvent formulations

Active agent	Deep eutectic solvent composition	Results in vitro	Results in vivo	References
Ibuprofen	Drug:terpenes	Transport through human epidermis		Stott et al. (1998)
Cefadroxil and deep eutectic solvent alone	Deep eutectic solvent and ionic liquid, including C:GE	Antibiotic transport into skin; activity against biofilms		Zakrewsky et al. (2014) and Zakrewsky et al. (2015, 2016b)
Caffeine and salicylic acid	Choline and imidazole-based ionic liquids	Transport through porcine skin		Santos de Almeida et al. (2017)
Labelled bovine serum albumin and insulin	C:GE	Transport through porcine skin	Hypoglycemic effect in normal rats with insulin	Banerjee et al. (2017)
Insulin	C:GE, various ratios	Transport through porcine skin; changes in skin structure detected by Fourier transform infrared spectroscopy		Tanner et al. (2018)
Nobiletin	C:GE 1:2	Transport through porcine skin	Increased plasma concentrations and hypoglycemic effect in normal rats	Hattori et al. (2019)
Fluorescent dextrins	C:GE 1:2	Transport through porcine skin; changes in skin structure detected by Fourier transform infrared spectroscopy (mechanistic study)		Qi and Mitragotri (2019)
Acarbose, ruxolitinib	CAGE and similar	Transport through skin; correlation with bidimensional nuclear magnetic resonance spectroscopy		Tanner et al. (2019)

For abbreviations, see Table 1

2.3.5 *Oral Applications*

After the observations on the transcutaneous passage of insulin, the effect of formulation in deep eutectic solvents on its oral bioavailability was investigated (Banerjee et al. 2018a). Administration of insulin by the oral route has been a goal of pharmaceutical technology for decades, with all the advantages that this would bring in terms of patient compliance. However, the high molecular weight, hydrophilic nature, and susceptibility to digestive enzymes of proteins seriously limit its oral bioavailability. Any formulation must be able both to protect the protein in the gastrointestinal tract and promote its absorption across the intestinal epithelium.

The stability of insulin in the choline:geranic acid ("CAGE") deep eutectic solvent was assessed by examining its circular dichroism spectrum after storage at 4 °C and 25 °C. Insulin formulated in choline:geranic acid solvent was also less susceptible to degradation by trypsin. The protein was found to be stable for 4 months (Banerjee et al. 2018a). The formulation was administered orally to nondiabetic rats in gastro-resistant capsules and showed a dose-dependent glucose-lowering effect, which was not observed with insulin in saline. Histological studies of intestinal samples taken 5 hours after insulin-"CAGE" administration showed no structural damage (Fig. 2.7). This is in accordance with observations made on Caco-2 monolayers in vitro. The transport of fluorescently labelled insulin across these monolayers was also measured and found to be enhanced tenfold by the presence of choline:geranic acid solvent. Since this was accompanied by a reduction in the transepithelial electrical resistance, it could be concluded that the deep eutectic solvents could temporarily open tight junctions between the epithelial cells (supplementary material, Banerjee et al. 2018a). Another observation that could shed light on "CAGE" mechanism(s) of action is that it reduced the viscosity of the intestinal mucus, which would facilitate contact between the protein and the brush border of the enterocytes.

The "CAGE" formulation has also been found to exert a biological activity in its own right. Nurunnabi et al. have reported that the formulation can reduce the absorption of fat from the intestine and could therefore counteract obesity. A model fat molecule, docosahexaenoic acid, formed particle in the solvent that were too large to be adsorbed. Rats fed a high-fat diet in conjunction with the "CAGE" formulation gained less weight than those not given "CAGE" (Nurunnabi et al. 2019).

There are also examples of the use of deep eutectic solvents to improve the oral bioavailability of small molecules. Rutin, a disaccharide of quercetin, was formulated in a deep eutectic solvent composed of proline and glutamic acid 2:1. A pharmacokinetic study was performed in Balb/c mice, comparing the solvent with a suspension in water. The solvent allowed a twofold increase in the maximum blood concentration (C_{max}) and the area under the curve (AUC) while retarding the time corresponding to the maximum plasma concentration (T_{max}) from 15 minutes with the aqueous suspension to an hour with the deep eutectic solvent (Faggian et al. 2016).

The same team also formulated the naturally occurring alkaloid berberine in deep eutectic solvents for oral applications (Sut et al. 2017). After a solubility study, three deep eutectic solvents were chosen for a pharmacokinetic study: proline:malic acid 1:2, proline:urea 2:1, and proline:malic acid:lactic acid:water 1:0.2:0.3:0.5. All three increased the area under the curve compared with an aqueous suspension, the most effective being the proline:malic acid:lactic acid:water composition (Fig. 2.8).

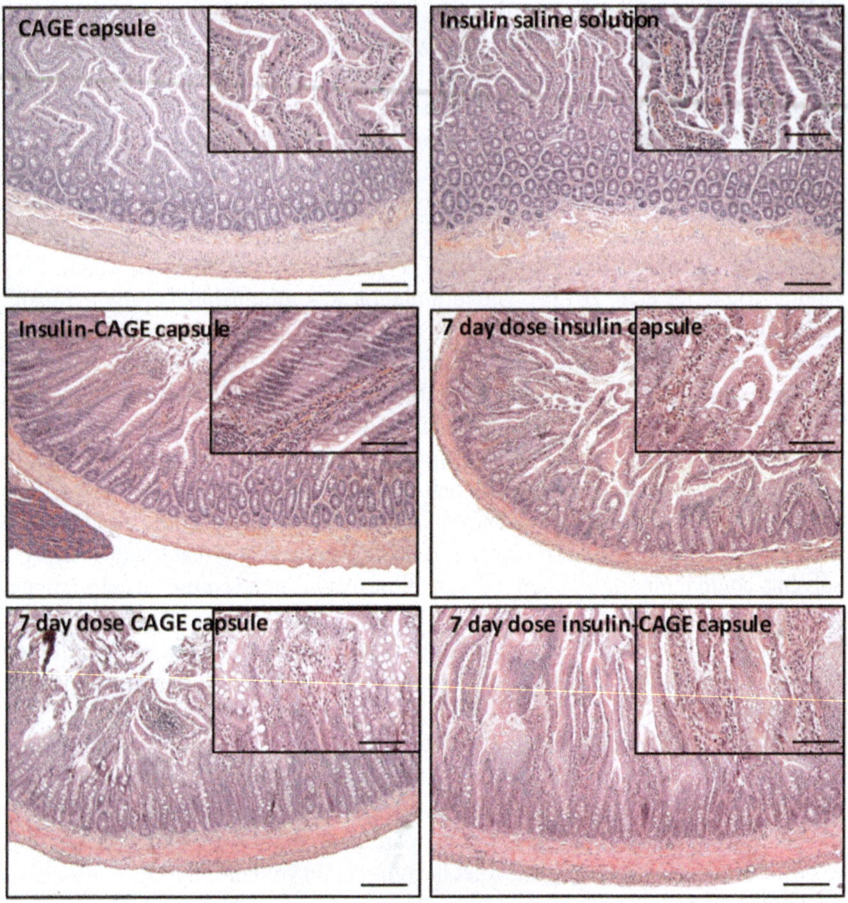

Fig. 2.7 Photomicrographs of hematoxylin and eosin-stained small intestine tissue sections (scale bar: 200 μm, inserts are intestinal sections with scale bar of 50 μm.). Sections represent oral administration of neat CAGE, insulin-saline, or insulin-CAGE capsules after a single dose or after daily dosing for 7 days of insulin, CAGE, or insulin-CAGE capsules. (Banerjee et al. 2018a, reprinted with permission of Proceedings of the National Academy of Sciences)

Chen et al. have determined the pharmacokinetics and acute toxicity of salvianolic acid B after oral administration within a choline chloride:glycerol (1:2) deep eutectic solvent (Chen et al. 2017). It was clear from the results that this solvent promoted the absorption of the active compound, with a higher maximum plasma concentration achieved early than from an aqueous solution, although the areas under the curve and the mean retention time were similar for the two formulations. The prolife of metabolites detected in the blood was also similar.

As a preliminary to its use as an oral formulation, Jeliński et al. tested the solubility in synthetic gastrointestinal medium of curcumin either added as an aqueous solution or in natural deep eutectic solvents formed from choline chloride and

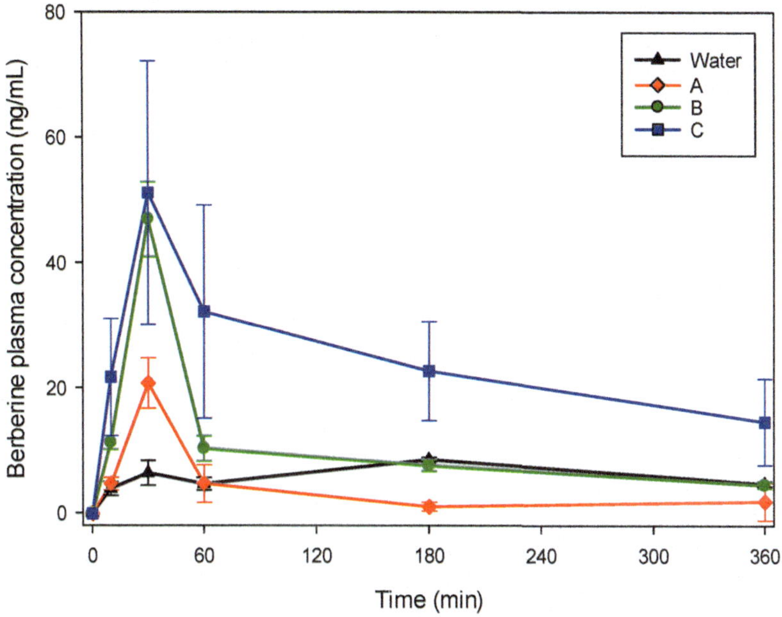

Fig. 2.8 Plasma pharmacokinetics (mean ± SD) of berberine in Balb/c mice following oral administration of 50 mg/kg in NADES solution (A = proline:malic acid (1:2), B = proline:urea (2:1), C = lactic acid:proline:malic acid:water (1:0.2:0.3:0.5)) or in suspension in water. (Sut et al. 2017)

Table 6 Oral and nasal administration of deep eutectic solvent formulations

Active agent	Deep eutectic solvent composition	Results in vivo	References
Oral administration			
Insulin	C:GE 1:2	Lowering of blood glucose in normal rats	Banerjee et al. (2018a)
None	C:GE 1:2	Reduced fat absorption; lower body weight gain on high-fat diet	Nurunnabi et al. (2019)
Rutin	P:GLA 2:1	Increased blood concentration; improved pharmacokinetics	Faggian et al. (2016)
Berberine	P:MA:LA:W 1:0.2:0.3:0.5	Increased blood concentration; improved pharmacokinetics	Sut et al. (2017)
Salvianolic acid B	CC:GL 1:2	Increased blood concentration; improved pharmacokinetics	Chen et al. (2017)
Curcumin	CC:GL 1:1	Increased solubility in gastric and intestinal media	Jeliński et al. (2019b)
Nasal administration			
Insulin	CC:MA 2:1	Lowering of blood glucose in normal rats	Li et al. (2019)

For abbreviations, see Table 1

various sugars or sugar alcohol (Jeliński et al. 2019b). Presentation of the drug in the form of a natural deep eutectic solvent greatly increased its solubility in both gastric and intestinal media, suggesting that its absorption would be facilitated by this formulation.

2.3.6 *Nasal Applications*

Nasal administration has also attracted attention as an alternative route for the administration of insulin and other proteins. As for transdermal and oral applications, this route requires a formulation that can promote absorption while remaining nontoxic. With this in mind, Li et al. investigated a formulation of insulin in a deep eutectic solvent composed of chlorine chloride and malic acid (2:1 molar ratio) containing porcine insulin for intranasal administration (Li et al. 2019). Circular dichroism studies showed that the structural conformation of the insulin was conserved within the solvent. The release of insulin from the solvent in a Franz diffusion cell was slower than that of the protein from a hydrogel. This could be explained by the limited interaction of the deep eutectic solvent, which had high viscosity, with water. The passage of insulin across ex vivo nasal mucosa was visualized after fluorescent labelling, and the deep eutectic solvent formulation was observed to promote penetration whereas the hydrogel did not. Furthermore, a dose-dependent hypoglycemic effect was obtained when the deep eutectic solvent was applied intranasally to nondiabetic rats. Histological studies of the nasal epithelium showed no evidence of nasal toxicity (Fig. 2.9; Li et al. 2019). Thus, as for oral and transdermal applications, deep eutectic solvents show promise as vehicles to facilitate protein absorption.

The results obtained with the oral and nasal administration of active molecules in deep eutectic solvents are summarized in Table 6.

2.3.7 *Applications in Formulation for Drug Delivery and Biotechnology*

As well as specific applications related to particular active molecules, some recent research has covered the contribution that deep eutectic solvents could bring to pharmaceutical technology as a whole. A recent review by Pedro et al. covers the role of deep eutectic solvents and therapeutic deep eutectic solvents containing active molecules in the development of drug delivery systems (Pedro et al. 2019). This report stresses the role of deep eutectic solvents in controlling drug polymorphism and also the ease with which deep eutectic solvent-solubilized molecules can be incorporated into biopolymer systems. Potticary et al. in a study deposited on arXiv.org in 2019, also discuss the role of deep eutectic solvents, or deep eutomic

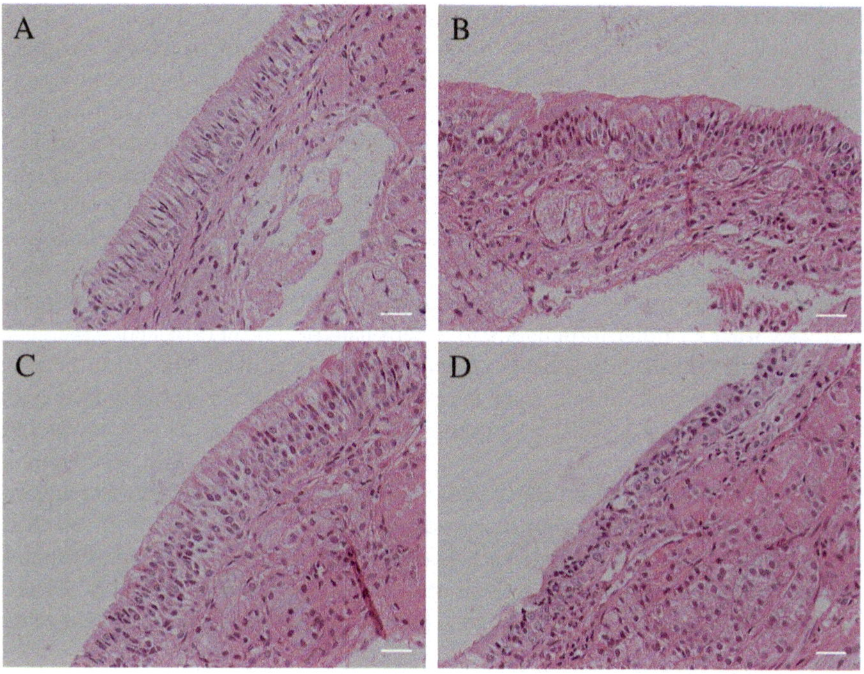

Fig. 2.9 Photomicrographs of hematoxylin and eosin-stained nasal epithelia sections (scale bar: 25 μm.). Sections represent nasal administration of (**a**) normal saline, (**b**) choline chloride:malic acid DES, (**c**) insulin-loaded DES, or (**d**) sodium deoxycholate once a day for 7 days. (Li et al. 2019, reprinted with permission of Elsevier)

solvents as they call them, with respect to drug polymorphism (Potticary et al. 2019). In particular, they investigated the interaction of different polymorphs of paracetamol that can act as a hydrogen bond acceptor with phenol as hydrogen bond donor. By selecting the right deep eutectic solvent composition, the most active form of the drug could be selected at room temperature.

An example of the incorporation of a deep eutectic solvent into a polymeric drug delivery system was given by Aroso et al. who formulated a therapeutic deep eutectic solvent with ibuprofen and menthol that they incorporated into a commercial preparation of cornstarch with poly-ε-caprolactone using supercritical carbon dioxide ($scCO_2$) technology to provide a controlled release system (Aroso et al. 2015). A review by Barros et al. describes the use of deep eutectic solvents or therapeutic deep eutectic solvents for impregnation of active molecules into polymeric devices using $scCO_2$ (Barros et al. 2017). The plasticizing properties of the deep eutectic solvents allow matrices of higher porosity to be produced.

In a more recent study carried out by Pradeepkumar et al. an anticancer drug, chlorambucil, was chemically modified to allow it to become one component of a deep eutectic solvent that was incorporated into a polymeric system modified with folic acid to target breast cancer cells (Pradeepkumar et al. 2018). Furthermore, a second anticancer drug, doxorubicin, was incorporated into the polymer micelles. Chlorambucil was conjugated with dimethylaminopropanol to form a hydrogen bond acceptor known as CABAL. This was mixed with 1,4-butanediol in various ratios to form deep eutectic solvent. A copolymer was synthesized from folic acid-modified β-alanine and poly-ε-caprolactone. When mixed with the optimized deep eutectic solvent, the polymer formed micelles that could then be loaded with doxorubicin. Drug-loaded micelles were more cytotoxic to the MDA-MB-231 human breast cancer cell line than to L929 mouse fibroblasts. Finally, tumor volume was reduced in an in vivo model of rats bearing DMBA-induced mammary tumors when the animals were treated with doxorubicin-loaded polymer micelles. In a later article, they used a similar formulation to encapsulate paclitaxel (Pradeepkumar et al. 2019a). After an extensive physicochemical characterization, the micelles were tested for cytotoxicity to HeLa human cervical cancer cells and normal blood cells. Activity against cancer cells was observed, while there was no toxicity toward the blood cells.

The same team has also produced a polymer-deep eutectic solvent formulation using a choline chloride:xylitol deep eutectic solvent as a medium in which to polymerize 2-hydroxylethyl methacrylate (HEMA) in the form of nanoparticles, which were then loaded with 5-fluorouracil (Pradeepkumar et al. 2019b). These nanoparticles were taken up by and showed antiproliferative activity against HeLa cells.

Florindo et al. have prepared a deep eutectic solvent composed of sodium dodecanoate and decanoic acid in the form of a hydrogel with rheological properties suitable for pharmaceutical applications (Florindo et al. 2018).

Deep eutectic solvents also have numerous applications in biotechnology that could lead to the discovery of new drugs or the preparation of new delivery systems. The use of deep eutectic solvents to extract natural products has already been referred to above, for example, the extraction of polyphenols from coffee beans (Benlebna et al. 2018), antioxidants from grape skin (Radošević et al. 2016), and flavonoids from plants (Tang et al. 2016). Furthermore, Fu et al. have used a deep eutectic solvent as a medium for producing a molecularly imprinted polymer in order to extract β-lactoglobulin (Fu et al. 2019). Also in the section on the incorporation of high-molecular-weight compounds into deep eutectic solvents, it has been shown that nanofibers can be produced from lysozyme (Silva et al. 2018a, b), and a pharmaceutically useful material can be created from gelatin in the "Glyceline" deep eutectic solvent (Qu et al. 2019).

It has also been demonstrated in work cited above that the activity of many enzymes can be conserved or even improved in deep eutectic solvents, opening up the way to their use as alternative media for biotransformations. Thus, the thermal stability of lysozyme can be increased (Su and Klibanov 2015); the catalytic activity of catalase (Harifi-Mood et al. 2017), laccase (Khodaverdian et al. 2018), and lipase (Nascimento et al. 2019) was conserved; and the biotransformation of steroids by bacteria was efficient (Mao et al. 2018). Finally, Sivapragasam et al. have shown that some choline-based ionic liquids promoted the growth of the yeast

Saccharomyces cerevisiae, suggesting that deep eutectic solvents could act as an alternative medium for microbial cultures (Sivapragasam et al. 2019).

2.4 Conclusions

This chapter shows that deep eutectic solvent-based systems are promising, versatile tools for many formulations destined for various routes of administration, including, but not limited to, oral, dermal, and nasal. They provide an alternative to traditional solvents for a wide range of known molecules and could also be applied to new chemical entities and natural products. This could also pave the way for the development of efficient antimicrobial formulations or novel delivery systems of high-molecular-weight molecules. However, most research up to this point has focused mainly on the "superiority" aspects of deep eutectic solvents but has only briefly considered their potential drawbacks. In fact, stability tests were generally short term, while the toxicity of deep eutectic solvents was investigated mostly on single cell lines using freshly prepared deep eutectic solvent samples. Therefore, it is essential to study the long-term stability of deep eutectic solvent-based formulations in more depth as well as the chronic and systemic toxicity of these systems before any attempt to commercialize them is made.

Acknowledgments The authors thank the French Embassy in Vietnam for the PhD fellowship of Canh-Hung Nguyen and the French National Research Agency (ANR) for the funding of the project ParasiDES (ANR-19-CE18-0027).

References

Abbott AP, Ahmed EI, Prasad K, Qader IB, Ryder KS (2017) Liquid pharmaceuticals formulation by eutectic formation. Fluid Phase Equilib 448:2–8. https://doi.org/10.1016/j.fluid.2017.05.009

Abranches DO, Larriba M, Silva LP, Melle-Franco M, Palomar JF, Pinho SP, Coutinho JAP (2019) Using COSMO-RS to design choline chloride pharmaceutical eutectic solvents. Fluid Phase Equilib 497:71–78. https://doi.org/10.1016/j.fluid.2019.06.005

Adawiyah N, Moniruzzaman M, Hawatulaila S, Goto M (2016) Ionic liquids as a potential tool for drug delivery systems. Med Chem Commun 7:1881–1189. https://doi.org/10.1039/C6MD00358C

Agatemor C, Ibsen KN, Tanner EE, Mitragotri S (2018) Ionic liquids for addressing unmet needs in healthcare. Bioeng Transl Med 3:7–25. https://doi.org/10.1002/btm2.10083

Ahmadi R, Hemmateenejad B, Safavi A, Shojaeifard Z, Mohabbati M, Firuzi O (2018) Assessment of cytotoxicity of choline chloride-based natural deep eutectic solvents against human HEK-293 cells: A QSAR analysis. Chemosphere 209:821–838. https://doi.org/10.1016/j.chemosphere.2018.06.103

Aroso IM, Craveiro R, Rocha Â, Dionísio M, Barreiros S, Reis RL, Paiva A, Duarte AR (2015) Design of controlled release systems for THEDES-therapeutic deep eutectic solvents, using supercritical fluid technology. Int J Pharm 492:73–79. https://doi.org/10.1016/j.ijpharm.2015.06.038

Aroso IM, Silva JC, Mano F, Ferreira AS, Dionísio M, Sá-Nogueira I, Barreiros S, Reis RL, Paiva A, Duarte AR (2016) Dissolution enhancement of active pharmaceutical ingredients by therapeutic deep eutectic systems. Eur J Pharm Biopharm 98:57–66. https://doi.org/10.1016/j.ejpb.2015.11.002

Banerjee A, Ibsen K, Iwao Y, Zakrewsky M, Mitragotri S (2017) Transdermal protein delivery using choline and geranate (CAGE) deep eutectic solvent. Adv Healthc Mater 6:1601411. https://doi.org/10.1002/adhm.201601411

Banerjee A, Ibsen K, Brown T, Chen R, Agatemor C, Mitragotri S (2018a) Ionic liquids for oral insulin delivery. Proc Natl Acad Sci U S A 115:7296–7301. https://doi.org/10.1073/pnas.1722338115

Banerjee A, Ibsen K, Brown T, Chen R, Agatemor C, Mitragotri S (2018b) Reply to Rogers and Gurau: definitions of ionic liquids and deep eutectic solvents. Proc Natl Acad Sci U S A 115:E11000–E11001. https://doi.org/10.1073/pnas.1815526115

Barros AA, Silva JM, Craveiro R, Paiva A, Reis RL, Duarte ARC (2017) Green solvents for enhanced impregnation processes in biomedicine. Curr Opin Green Sustain Chem 5:82–87. https://doi.org/10.1016/j.cogsc.2017.03.014

Benlebna M, Ruesgas-Ramón M, Bonafos B, Fouret G, Casas F, Coudray C, Durand E, Cruz Figueroa-Espinoza M, Feillet-Coudray C (2018) Toxicity of natural deep eutectic solvent betaine:glycerol in rats. J Agric Food Chem 66:6205–6212. https://doi.org/10.1021/acs.jafc.8b01746

Chen J, Li SF, Yao ZF, Yanga DW, Zhang LW (2016) Improved stability of salvianolic acid B from Radix Salviae miltiorrhizae in deep eutectic solvents. Ana Methods 8:2502–2509. https://doi.org/10.1039/C5AY03351A

Chen J, Wang Q, Liu M, Zhang L (2017) The effect of deep eutectic solvent on the pharmacokinetics of salvianolic acid B in rats and its acute toxicity test. J Chromatogr B Analyt Technol Biomed Life Sci 1063:60–66. https://doi.org/10.1016/j.jchromb.2017.08.016

Choi YH, van Spronsen J, Dai Y, Verberne M, Hollmann F, Arends IWCE, Witkamp GJ, Verpoorte R (2011) Are natural deep eutectic solvents the missing link in understanding cellular metabolism and physiology? Plant Physiol 156:1701–1705. https://doi.org/10.1104/pp.111.178426

Colombo Dugoni G, Di Pietro ME, Ferro M, Castiglione F, Ruellan S, Moufawad T, Moura L, Costa Gomes MF, Fourmentin S, Mele A (2019) Effect of water on deep eutectic solvent/β-cyclodextrin systems. ACS Sustain Chem Eng 7:7277–7285. https://doi.org/10.1021/acssuschemeng.9b00315

Dai Y, van Spronsen J, Witkamp GJ, Verpoorte R, Choi YH (2013) Natural deep eutectic solvents as new potential media for green technology. Anal Chim Acta 766:61–68. https://doi.org/10.1016/j.aca.2012.12.019

Dai Y, Witkamp GJ, Verpoorte R, Choi YH (2015) Tailoring properties of natural deep eutectic solvents with water to facilitate their applications. Food Chem 187:14–19. https://doi.org/10.1016/j.foodchem.2015.03.123

Duarte AR, Ferreira AS, Barreiros S, Cabrita E, Reis RL, Paiva A (2017) A comparison between pure active pharmaceutical ingredients and therapeutic deep eutectic solvents: solubility and permeability studies. Eur J Pharm Biopharm 114:296–304. https://doi.org/10.1016/j.ejpb.2017.02.003

Durand E, Lecomte J, Villeneuve P (2016) From green chemistry to nature: the versatile role of low transition temperature mixtures. Biochimie 120:119–123. https://doi.org/10.1016/j.biochi.2015.09.019

Durand E, Lecomte J, Upasani R, Chabi B, Bayrasy C, Baréa B, Jublanc E, Clarke MJ, Moore DJ, Crowther J, Wrutniak-Cabello C, Villeneuve P (2017) Evaluation of the ROS inhibiting activity and mitochondrial targeting of phenolic compounds in fibroblast cells model system and enhancement of efficiency by natural deep eutectic solvent (NADES) formulation. Pharm Res 34:1134–1146. https://doi.org/10.1007/s11095-017-2124-4

El Achkar T, Moufawad T, Ruellan S, Landy D, Greige-Gerges H, Fourmentin S (2020) Cyclodextrins: from solute to solvent. Chem Commun 56:3385–3388. https://doi.org/10.1039/D0CC00460J

Esquembre R, Sanz JM, Wall JG, del Monte F, Mateo CR, Ferrer ML (2013) Thermal unfolding and refolding of lysozyme in deep eutectic solvents and their aqueous dilutions. Phys Chem Chem Phys 15:11248–11256. https://doi.org/10.1039/c3cp44299c

Faggian M, Sut S, Perissutti B, Baldan V, Grabnar I, Dall'Acqua S (2016) Natural deep eutectic solvents (NADES) as a tool for bioavailability improvement: pharmacokinetics of rutin dissolved in proline/glycine after oral administration in rats: possible application in nutraceuticals. Molecules 21:E1531. https://doi.org/10.3390/molecules21111531

Florindo C, Celia-Silva LG, Martins LFG, Branco LC, Marrucho IM (2018) Supramolecular hydrogel based on a sodium deep eutectic solvent. Chem Commun 54:7527–7530. https://doi.org/10.1039/c8cc03266a

Fourmentin S, Landy D, Cunha Gomes de Moura LM, Tilloy S, Bricout H, Ferreira M (2018) Procédé d'épuration d'un effluent gazeux. FR3058905

Fu N, Li L, Liu K, Kim CK, Li J, Zhu T, Li J, Tang B (2019) A choline chloride-acrylic acid deep eutectic solvent polymer based on Fe3O4 particles and MoS2 sheets (poly(ChCl-AA DES)@Fe3O4@MoS2) with specific recognition and good antibacterial properties for β-lactoglobulin in milk. Talanta 197:567–577. https://doi.org/10.1016/j.talanta.2019.01.072

Gállego I, Grover MA, Hud NV (2015) Folding and imaging of DNA nanostructures in anhydrous and hydrated deep-eutectic solvents. Angew Chem Int Ed Engl 54:6765–6769. https://doi.org/10.1002/anie.201412354

Gautam RK, Ahmed SA, Seth D (2018) Photophysics of thioflavin T in deep eutectic solvents. J Lumin 198:508–516. https://doi.org/10.1016/j.jlumin.2018.02.055

Gutiérrez A, Atilhan M, Aparicio S (2018) A theoretical study on lidocaine solubility in deep eutectic solvents. Phys Chem Chem Phys 20:27464–27473. https://doi.org/10.1039/c8cp05641b

Gutiérrez A, Aparicio S, Atilhan M (2019) Design of arginine-based therapeutic deep eutectic solvents as drug solubilization vehicles for active pharmaceutical ingredients. Phys Chem Chem Phys 21:10621–10634. https://doi.org/10.1039/c9cp01408j

Halder AK, Natalia M, Cordeiro DS (2019) Probing the environmental toxicity of deep eutectic solvents and their components: an in silico modeling approach. ACS Sustain Chem Eng 7:10649–10660. https://doi.org/10.1021/acssuschemeng.9b01306

Haraźna K, Walas K, Urbańska P, Witko T, Snoch W, Siemek A, Jachimska B, Krzan M, Napruszewska BD, Witko M, Bednarz S, Guzik M (2019) Polyhydroxyalkanoate-derived hydrogen-bond donors for the synthesis of new deep eutectic solvents. Green Chem 21:3116–3126. https://doi.org/10.1039/C9GC00387H

Harifi-Mood AR, Ghobadi R, Divsalar A (2017) The effect of deep eutectic solvents on catalytic function and structure of bovine liver catalase. Int J Biol Macromol 95:115–120. https://doi.org/10.1016/j.ijbiomac.2016.11.043

Hattori T, Tagawa H, Inai M, Kan T, Kimura S, Itai S, Mitragotri S, Iwao Y (2019) Transdermal delivery of nobiletin using ionic liquids. Sci Rep 9:20191. https://doi.org/10.1038/s41598-019-56731-1

Hayyan M, Hashim MA, Hayyan A, Al-Saadi MA, AlNashef IM, Mirghani ME, Saheed OK (2013) Are deep eutectic solvents benign or toxic? Chemosphere 90:2193–2195. https://doi.org/10.1016/j.chemosphere.2012.11.004

Hayyan M, Looi CY, Hayyan A, Wong WF, Hashim MA (2015) In vitro and in vivo toxicity profiling of ammonium-based deep eutectic solvents. PLoS One 10:e0117934. https://doi.org/10.1371/journal.pone.0117934

Hayyan M, Mbous YP, Looi CY, Wong WF, Hayyan A, Salleh Z, Mohd-Ali O (2016) Natural deep eutectic solvents: cytotoxic profile. Springerplus 5:913. https://doi.org/10.1186/s40064-016-2575-9

Ibsen KN, Ma H, Banerjee A, Tanner EEL, Nangia S, Mitragotri S (2018) Mechanism of antibacterial activity of choline-based ionic liquids (CAGE). ACS Biomater Sci Eng 4:2370–2379. https://doi.org/10.1021/acsbiomaterials.8b00486

Jeliński T, Przybyłek M, Cysewski P (2019a) Solubility advantage of sulfanilamide and sulfacetamide in natural deep eutectic systems: experimental and theoretical investigations. Drug Dev Ind Pharm 45:1120–1129. https://doi.org/10.1080/03639045.2019.1597104

Jeliński T, Przybyłek M, Cysewski P (2019b) Natural deep eutectic solvents as agents for improving solubility, stability and delivery of curcumin. Pharm Res 36:116. https://doi.org/10.1007/s11095-019-2643-2
Juneidi I, Hayyan M, Mohd Ali O (2016) Toxicity profile of choline chloride-based deep eutectic solvents for fungi and Cyprinus carpio fish. Environ Sci Pollut Res Int 23:7648–7659. https://doi.org/10.1007/s11356-015-6003-4
Kellar R, Nieto NC, Koppisch A, Del Sesto R (2018) Ionic liquids that sterilize and prevent biofilm formation in skin wound healing devices. US20180093011
Khodaverdian S, Dabirmanesh B, Heydari A, Dashtban-Moghadam E, Khajeh K, Ghazi F (2018) Activity, stability and structure of laccase in betaine based natural deep eutectic solvents. Int J Biol Macromol 107:2574–2579. https://doi.org/10.1016/j.ijbiomac.2017.10.144
Lee MS, Lee K, Nam MW, Jeong KM, Lee JE, Kim NW, Yin Y, Lim SY, Yoo DE, Lee J, Jeong JH (2018) Natural deep eutectic solvents as a storage medium for human interferon-α2: a green and improved strategy for room-temperature biologics. J Ind Eng Chem 65:343–348. https://doi.org/10.1016/j.jiec.2018.05.005
Li Z, Lee PI (2016) Investigation on drug solubility enhancement using deep eutectic solvents and their derivatives. Int J Pharm 505:283–288. https://doi.org/10.1016/j.ijpharm.2016.04.018
Li Y, Wu X, Zhu Q, Chen Z, Lu Y, Qi J, Wu W (2019) Improving the hypoglycemic effect of insulin via the nasal administration of deep eutectic solvents. Int J Pharm 569:118584. https://doi.org/10.1016/j.ijpharm.2019.118584
Lim JH, Song SH, Park HS, Lee JR, Lee SM (2017) Spontaneous detachment of Streptococcus mutans biofilm by synergistic effect between zwitterion and sugar alcohol. Sci Rep 7:8107. https://doi.org/10.1038/s41598-017-08558-x
Lim JH, Jeong Y, Song SH, Ahn JH, Lee JR, Lee SM (2018) Penetration of an antimicrobial zinc-sugar alcohol complex into Streptococcus mutans biofilms. Sci Rep 8:16154. https://doi.org/10.1038/s41598-018-34366-y
Lu C, Cao J, Wang N, Su E (2016) Significantly improving the solubility of non-steroidal anti-inflammatory drugs in deep eutectic solvents for potential non-aqueous liquid administration. Med Chem Commun 7:955–959. https://doi.org/10.1039/C5MD00551E
Macário IPE, Jesus F, Pereira JL, Ventura SPM, Gonçalves AMM, Coutinho JAP, Gonçalves FJM (2018a) Unraveling the ecotoxicity of deep eutectic solvents using the mixture toxicity theory. Chemosphere 212:890–897. https://doi.org/10.1016/j.chemosphere.2018.08.153
Macário IPE, Ventura SPM, Pereira JL, Gonçalves AMM, Coutinho JAP, Gonçalves FJM (2018b) The antagonist and synergist potential of cholinium-based deep eutectic solvents. Ecotoxicol Environ Saf 165:597–602. https://doi.org/10.1016/j.ecoenv.2018.09.027
Macário IPE, Oliveira H, Menezes AC, Ventura SPM, Pereira JL, Gonçalves AMM, Coutinho JAP, Gonçalves FJM (2019) Cytotoxicity profiling of deep eutectic solvents to human skin cells. Sci Rep 9:3932. https://doi.org/10.1038/s41598-019-39910-y
Mamashli F, Badraghi J, Delavari B, Lanjanian H, Sabbaghian M, Hosseini M, Saboury AA (2018) Improvement of versatile peroxidase activity and stability by a cholinium-based ionic liquid. J Mol Liq 272:597–608. https://doi.org/10.1016/j.molliq.2018.09.128
Mao S, Li K, Hou Y, Liu Y, Ji S, Qin H, Lu F (2018) Synergistic effects of components in deep eutectic solvents relieve toxicity and improve the performance of steroid biotransformation catalyzed by Arthrobacter simplex. J Chem Technol Biotechnol 93:2729–2736. https://doi.org/10.1002/jctb.5629
Mbous YP, Hayyan M, Wong WF, Looi CY, Hashim MA (2017) Unraveling the cytotoxicity and metabolic pathways of binary natural deep eutectic solvent systems. Sci Rep 7:41257. https://doi.org/10.1038/srep41257
McCune JA, Kunz S, Olesińska M, Scherman OA (2017) DESolution of CD and CB macrocycles. Chem Eur J 23:8601–8604. https://doi.org/10.1002/chem.201701275
Mitar A, Panić M, Prlić Kardum J, Halambek J, Sander A, Zagajski Kučan K, Radojčić Redovniković I, Radošević K (2019) Physicochemical properties, cytotoxicity, and antioxidative activity of natural deep eutectic solvents containing organic acid. Chem Biochem Eng Q 33:1–18. https://doi.org/10.15255/CABEQ.2018.1454

Mokhtarpour M, Shekaari H, Martinez F, Zafarani-Moattar MT (2019a) Study of naproxen in some aqueous solutions of choline-based deep eutectic solvents: solubility measurements, volumetric and compressibility properties. Int J Pharm 564:197–206. https://doi.org/10.1016/j.ijpharm.2019.04.029

Mokhtarpour M, Shekaari H, Martinez F, Zafarani-Moattar MT (2019b) Effect of tetrabutylammonium bromide-based deep eutectic solvents on the aqueous solubility of indomethacin at various temperatures: measurement, modeling, and prediction with three-dimensional Hansen solubility parameters. AAPS PharmSciTech 20:204. https://doi.org/10.1208/s12249-019-1373-4

Morrison HG, Sun CC, Neervannan S (2009) Characterization of thermal behavior of deep eutectic solvents and their potential as drug solubilization vehicles. Int J Pharm 378:136–139. https://doi.org/10.1016/j.ijpharm.2009.05.039

Moufawad T, Moura L, Ferreira M, Bricout H, Tilloy S, Monflier E, Costa Gomes M, Landy D, Fourmentin S (2019) First evidence of cyclodextrin inclusion complexes in a deep eutectic solvent. ACS Sustain Chem Eng 7:6345–6351. https://doi.org/10.1021/acssuschemeng.9b00044

Moura L, Moufawad T, Ferreira M, Bricout H, Tilloy S, Monflier E, Costa Gomes MF, Landy D, Fourmentin S (2017) Deep eutectic solvents as green absorbents of volatile organic pollutants. Environ Chem Lett 15:747–753. https://doi.org/10.1007/s10311-017-0654-y

Nascimento PAM, Picheli FP, Lopes AM, Pereira JFB, Santos-Ebinuma VC (2019) Effects of cholinium-based ionic liquids on Aspergillus Niger lipase: stabilizers or inhibitors. Biotechnol Prog 35:e2838. https://doi.org/10.1002/btpr.2838

Niknaddaf F, Shahangian SS, Heydari A, Hosseinkhani S, Sajedi RH (2018) Deep eutectic solvents as a new generation of chemical chaperones. Chem Select 3:10603–10607. https://doi.org/10.1002/slct.201802235

Nurunnabi M, Ibsen KN, Tanner EEL, Mitragotri S (2019) Oral ionic liquid for the treatment of diet-induced obesity. Proc Natl Acad Sci U S A 116:25042–25047. https://doi.org/10.1073/pnas.1914426116

Olivares B, Martínez F, Rivas L, Calderón C, M Munita J, R Campodonico P (2018) A natural deep eutectic solvent formulated to stabilize β-lactam antibiotics. Sci Rep 8:14900. https://doi.org/10.1038/s41598-018-33148-w

Palmelund H, Andersson MP, Asgreen CJ, Boyd BJ, Rantanen J, Löbmann K (2019) Tailor-made solvents for pharmaceutical use? Experimental and computational approach for determining solubility in deep eutectic solvents (DES). Int J Pharm X 1:100034. https://doi.org/10.1016/j.ijpx.2019.100034

Pedro SN, Freire MG, Freire CSR, Silvestre AJD (2019) Deep eutectic solvents comprising active pharmaceutical ingredients in the development of drug delivery systems. Expert Opin Drug Deliv 16:497–506. https://doi.org/10.1080/17425247.2019.1604680

Pereira CV, Silva JM, Rodrigues L, Reis RL, Paiva A, Duarte ARC, Matias A (2019) Unveil the anticancer potential of limomene based therapeutic deep eutectic solvents. Sci Re 9:14926. https://doi.org/10.1038/s41598-019-51472-7

Potticary J, Hall C, Hamilton V, McCabe JF, Hall SR (2019) Deep eutomic solvents. https://arxiv.org/abs/1902.08376v2

Pradeepkumar P, Rajendran NK, Alarfaj AA, Munusamy MA, Rajan M (2018) Deep eutectic solvent-mediated FA-g-β-alanine-co-PCL drug carrier for sustainable and site-specific drug delivery. ACS Appl Bio Mater (6):2094–2109. https://doi.org/10.1021/acsabm.8b00554

Pradeepkumar P, Sangeetha R, Gunaseelan S, Varalakshmi P, A Chuturgoon A, Rajan M (2019a) Folic acid conjugated polyglutamic acid drug vehicle synthesis through deep eutectic solvent for targeted release of paclitaxel. Chem Select 4:10225–10235. https://doi.org/10.1002/slct.201902256

Pradeepkumar P, Subbiah A, Rajan M (2019b) Synthesis of bio-degradable poly(2-hydroxyethyl methacrylate) using natural deep eutectic solvents for sustainable cancer drug delivery. SN Appl Sci 1:568. https://doi.org/10.1007/s42452-019-0591-4

Qi QM, Mitragotri S (2019) Mechanistic study of transdermal delivery of macromolecules assisted by ionic liquids. J Control Release 311-312:162–169. https://doi.org/10.1016/j.jconrel.2019.08.029

Qu W, Häkkinen R, Allen J, D'Agostino C, Abbott AP (2019) Globular and fibrous proteins modified with deep eutectic solvents: materials for drug delivery. Molecules 24:3583. https://doi.org/10.3390/molecules24193583

Radošević K, Bubalo MC, Srček VG, Grgas D, Dragičević TL, Redovniković IR (2015) Evaluation of toxicity and biodegradability of choline chloride based deep eutectic solvents. Ecotoxicol Environ Saf 112:46–53. https://doi.org/10.1016/j.ecoenv.2014.09.034

Radošević K, Ćurko N, Srček VG, Bubalo MC, Tomašević M, Ganić KK, Redovniković IR (2016) Natural deep eutectic solvents as beneficial extractants for enhancement of plant extracts bioactivity. LWT Food Sci Technol 73:45–51. https://doi.org/10.1016/j.lwt.2016.05.037

Radošević K, Čanak I, Panić M, Markov K, Bubalo MC, Frece J, Srček VG, Redovniković IR (2018) Antimicrobial, cytotoxic and antioxidative evaluation of natural deep eutectic solvents. Environ Sci Pollut Res Int 25:14188–14196. https://doi.org/10.1007/s11356-018-1669-z

Rogers RD, Gurau G (2018) Is "choline and geranate" an ionic liquid or deep eutectic solvent system? Proc Natl Acad Sci U S A 115:E10999. https://doi.org/10.1073/pnas.1814976115

Rozema E, van Dam AD, Sips HCM, Verpoorte R, Meijer OC, Kooijman S, Choi YH (2015) Extending pharmacological dose-response curves for salsalate with natural deep eutectic solvents. RSC Adv 5:61398–61401. https://doi.org/10.1039/C5RA10196D

Sadaf A, Kumari A, Khare SK (2018) Potential of ionic liquids for inhibiting the growth and β-lactamase production by Bacillus cereus EMB20. Int J Biol Macromol 107:1915–1921. https://doi.org/10.1016/j.ijbiomac.2017.10.053

Sanchez-Fernandez A, Edler KJ, Arnold T, Alba Venero D, Jackson AJ (2017) Protein conformation in pure and hydrated deep eutectic solvents. Phys Chem Chem Phys 19:8667–8670. https://doi.org/10.1039/c7cp00459a

Santos de Almeida T, Júlio A, Saraiva N, Fernandes AS, Araújo MEM, Baby AR, Rosado C, Mota JP (2017) Choline- versus imidazole-based ionic liquids as functional ingredients in topical delivery systems: cytotoxicity, solubility, and skin permeation studies. Drug Dev Ind Pharm 43:1858–1865. https://doi.org/10.1080/03639045.2017.1349788

Santos F, Leitão MIPS, Duarte ARC (2019) Properties of therapeutic deep eutectic solvents of L-arginine and ethambutol for tuberculosis treatment. Molecules 24:55. https://doi.org/10.3390/molecules24010055

Scherman OA, McCune JA (2018) Deep eutectic solvent compositions. WO2018167315

Shamseddin A, Crauste C, Durand E, Villeneuve P, Dubois G, Durand T, Vercauteren J, Veas F (2017) Resveratrol formulated with a natural deep eutectic solvent inhibits active matrix metalloprotease-9 in hormetic conditions. Eur J Lipid Sci Technol 119:1700171. https://doi.org/10.1002/ejlt.201700171

Shekaari H, Zafarani-Moattar MT, Mokhtarpour M (2017) Solubility, volumetric and compressibility properties of acetaminophen in some aqueous solutions of choline based deep eutectic solvents at T=(288.15 to 318.15) K. Eur J Pharm Sci 109:121–130. https://doi.org/10.1016/j.ejps.2017.07.021

Shekaari H, Zafarani-Moattar MT, Mokhtarpour M (2018a) Experimental determination and correlation of acetaminophen solubility in aqueous solutions of choline chloride based deep eutectic solvents at various temperatures. Fluid Phase Equilib 462:100–110. https://doi.org/10.1016/j.fluid.2018.01.017

Shekaari H, Zafarani-Moattar MT, Shayanfar A, Mokhtarpour M (2018b) Effect of choline chloride/ethylene glycol or glycerol as deep eutectic solvents on the solubility and thermodynamic properties of acetaminophen. J Mol Liq 249:1222–1235. https://doi.org/10.1016/j.molliq.2017.11.057

Shekaari H, Zafarani-Moattar MT, Mokhtarpour M, Faraji S (2019) Exploring cytotoxicity of some choline-based deep eutectic solvents and their effect on the solubility of lamotrigine in aqueous media. J Mol Liq 283:834–842. https://doi.org/10.1016/j.molliq.2019.03.079

Sidat Z, Marimuthu T, Kumar P, du Toit LC, Kondiah PPD, Choonara YE, Pillay V (2019) Ionic liquids as potential and synergistic permeation enhancers for transdermal drug delivery. Pharmaceutics 11:96. https://doi.org/10.3390/pharmaceutics11020096

Silva NHCS, Pinto RJB, Martins MA, Ferreira R, Correia I, Freire CSR, Marrucho IM (2018a) Ionic liquids as promoters of fast lysozyme fibrillation. J Mol Liq 272:456–467. https://doi.org/10.1016/j.molliq.2018.08.064
Silva NHCS, Vilela C, Pinto RJB, Martins MA, Marrucho IM, Freire CSR (2018b) Tuning lysozyme nanofibers dimensions using deep eutectic solvents for improved reinforcement ability. Int J Biol Macromol 115:518–527. https://doi.org/10.1016/j.ijbiomac.2018.03.150
Silva JM, Silva E, Reis RL, Duarte ARC (2019) A closer look in the antimicrobial properties of deep eutectic solvents based on fatty acids. Sustain Chem Pharm 14:100192. https://doi.org/10.1016/j.scp.2019.100192
Sivapragasam M, Wilfred CD, Jaganathan JR, Krishnan S, Ghani WAWAK (2019) Choline-based ionic liquids as media for the growth of Saccharomyces cerevisiae. PRO 7:471. https://doi.org/10.3390/pr7070471
Stott PW, Williams AC, Barry BW (1998) Transdermal delivery from eutectic systems: enhanced permeation of a model drug, ibuprofen. J Control Release 50:297–308. https://doi.org/10.1016/S0168-3659(97)00153-3
Su E, Klibanov AM (2015) Low-transition-temperature mixtures (LTTMs) for dissolving proteins and for drug formulation. Appl Biochem Biotechnol 177:753–758. https://doi.org/10.1007/s12010-015-1777-x
Sut S, Faggian M, Baldan V, Poloniato G, Castagliuolo I, Grabnar I, Perissutti B, Brun P, Maggi F, Voinovich D, Peron G, Dall'Acqua S (2017) Natural deep eutectic solvents (NADES) to enhance Berberine absorption: an in vivo pharmacokinetic study. Molecules 22:E1921. https://doi.org/10.3390/molecules22111921
Tang N, Zhong J, Yan W (2016) Solubilities of three flavonoids in different natural deep eutectic solvents at T = (288.15 to 328.15) K. J Chem Eng Data 61:4203–4208. https://doi.org/10.1021/acs.jced.6b00552
Tanner EEL, Ibsen KN, Mitragotri S (2018) Transdermal insulin delivery using choline-based ionic liquids (CAGE). J Control Release 286:137–144. https://doi.org/10.1016/j.jconrel.2018.07.029
Tanner EEL, Curreri AM, JP B, Selig-Wober NC, Yang AB, Kendig C, Fluhr MP, Kim N, Mitragotri S (2019) Design principles of ionic liquids for transdermal drug delivery. Adv Mater e1901103:31. https://doi.org/10.1002/adma.201901103
Tarate B, Bansal AK (2015) Characterization of CoQ 10-lauric acid eutectic system. Thermochim Acta 605:100–106. https://doi.org/10.1016/j.tca.2015.01.018
Tateishi-Karimata H, Sugimoto N (2014) Structure, stability and behaviour of nucleic acids in ionic liquids. Nucleic Acids Res 42:8831–8844. https://doi.org/10.1093/nar/gku499
Tønnesen HH, Wikene KO (2016) Eutectic solvents and uses thereof. WO2016108083
Torregrosa-Crespo J, Marset X, Guillena G, Ramón DJ, María Martínez-Espinosa R (2020) New guidelines for testing "Deep eutectic solvents" toxicity and their effects on the environment and living beings. Sci Total Environ 704:135382. https://doi.org/10.1016/j.scitotenv.2019.135382
Vanda H, Dai Y, Wilson EG, Verpoorte R, Choi YH (2018) Green solvents from ionic liquids and deep eutectic solvents to natural deep eutectic solvents. CR Chimie 21:628–638. https://doi.org/10.1016/j.crci.2018.04.002
Wang J, Dong X, Yu Q, Baker SN, Li H, Larm NE, Baker GA, Chen L, Tan J, Chen M (2017) Incorporation of antibacterial agent derived deep eutectic solvent into an active dental composite. Dent Mater 33:1445–1455. https://doi.org/10.1016/j.dental.2017.09.014
Wen Q, Chen JX, Tang YL, Wang J, Yang Z (2015) Assessing the toxicity and biodegradability of deep eutectic solvents. Chemosphere 132:63–69. https://doi.org/10.1016/j.chemosphere.2015.02.061
Wikene KO, Bruzell E, Tønnesen HH (2015a) Characterization and antimicrobial phototoxicity of curcumin dissolved in natural deep eutectic solvents. Eur J Pharm Sci 80:26–32. https://doi.org/10.1016/j.ejps.2015.09.013
Wikene KO, Bruzell E, Tønnesen HH (2015b) Improved antibacterial phototoxicity of a neutral porphyrin in natural deep eutectic solvents. J Photochem Photobiol B 148:188–196. https://doi.org/10.1016/j.jphotobiol.2015.04.022

Wikene KO, Rukke HV, Bruzell E, Tønnesen HH (2016) Physicochemical characterisation and antimicrobial phototoxicity of an anionic porphyrin in natural deep eutectic solvents. Eur J Pharm Biopharm 105:75–84. https://doi.org/10.1016/j.ejpb.2016.06.001

Wikene KO, Rukke HV, Bruzell E, Tønnesen HH (2017) Investigation of the antimicrobial effect of natural deep eutectic solvents (NADES) as solvents in antimicrobial photodynamic therapy. J Photochem Photobiol B 171:27–33. https://doi.org/10.1016/j.jphotobiol.2017.04.030

Wojnarowska Z, Smolka W, Zotova J, Knapik-Kowalczuk J, Sherif A, Tajber L, Paluch M (2018) The effect of electrostatic interactions on the formation of pharmaceutical eutectics. Phys Chem Chem Phys 20:27361–27367. https://doi.org/10.1039/c8cp05905e

Wolbert F, Brandenbusch C, Sadowski G (2019) Selecting excipients forming therapeutic deep eutectic systems – a mechanistic approach. Mol Pharm 16:3091–3099. https://doi.org/10.1021/acs.molpharmaceut.9b00336

Zakrewsky M, Lovejoy KS, Kern TL, Miller TE, Le V, Nagy A, Goumas AM, Iyer RS, Del Sesto RE, Koppisch AT, Fox DT, Mitragotri S (2014) Ionic liquids as a class of materials for transdermal delivery and pathogen neutralization. Proc Natl Acad Sci U S A 111:13313–13318. https://doi.org/10.1073/pnas.1403995111

Zakrewsky M, Mitragotri S, Fox DT, Koppisch A, Del Sesto R, Lovejoy K (2015) Ionic liquids for transdermal drug delivery. WO2015066647

Zakrewsky M, Banerjee A, Apte S, Kern TL, Jones MR, Sesto RE, Koppisch AT, Fox DT, Mitragotri S (2016a) Choline and geranate deep eutectic solvent as a broad-spectrum antiseptic agent for preventive and therapeutic applications. Adv Healthc Mater 5:1282–1289. https://doi.org/10.1002/adhm.201600086

Zakrewsky M, Mitragotri S, Fox DT, Koppisch A, Del Sesto R, Lovejoy K (2016b) Ionic liquids for transdermal drug delivery. US20160263225

Zhang Y, Du C, Cong Y, Xue Y, Qiao B, Ye T, Wang M (2019) Solubility increment and thermodynamic analysis of bioactive antofloxacin hydrochloride in aqueous ChCl/PTS deep eutectic solvent and cosolvent mixtures. J Chem Eng Data 64:5748–5754. https://doi.org/10.1021/acs.jced.9b00717

Chapter 3
Therapeutic Deep Eutectic Systems for the Enhancement of Drug Bioavailability

Filipa Santos and Ana Rita C. Duarte

Contents

Abstract Nowadays, green and sustainable approaches are one of the principles that companies have presented in their products and processes, and the pharmaceutical industry is not an exception. A constant search for new developments that allow the use of methods with minor risks to the environment is a transversal concern throughout the industry. For pharmaceutical and biomedical companies, green chemistry is a principle to have in mind, particularly in their search for new substances or matrices to deliver drugs or solve problems of solubility, polymorphism, toxicity, bioavailability, and pharmacokinetic. Currently, despite the investment on research and development, it is increasingly difficult to find new molecules to substitute the "old" ones with enhanced therapeutic efficacy. Because of that, the

F. Santos · A. R. C. Duarte (✉)
LAQV, REQUIMTE, Departamento de Química da Faculdade de Ciências e Tecnologia, Universidade Nova de Lisboa, Caparica, Portugal
e-mail: aduarte@fct.unl.pt

S. Fourmentin et al. (eds.), *Deep Eutectic Solvents for Medicine, Gas Solubilization and Extraction of Natural Substances*, Environmental Chemistry for a Sustainable World 56, https://doi.org/10.1007/978-3-030-53069-3_3

investigation in pharmaceutical field and in drug development have been very focused on the improvement of the characteristics of existing drugs.

The major points of this review focus on the development of deep eutectic systems for therapeutic applications, namely, therapeutic deep eutectic systems (THEDES), that emerged as a good alternative to overcome the drawbacks of some of the existing drugs, mainly in terms of bioavailability. These systems are presented as alternative solvents that could be prepared with an active pharmaceutical ingredient incorporated in the eutectic mixture or could act as a delivery vehicle, dissolving the drug in it. Recent developments on these systems show that THEDES could also be impregnated in matrices for the development of medical devices or controlled drug delivery systems, for example, using supercritical CO_2.

Keywords Deep eutectic systems · Pharmacy · Controlled drug delivery · Alternative solvents · Green chemistry · Bioavailability

Abbreviations

THEDES	Therapeutic deep eutectic systems
API	Active pharmaceutical ingredient
ILs	Ionic liquids
VOCs	Volatile organic compounds
DES	Deep eutectic solvent
LTTM	Low transition temperature mixtures
NADES	Natural deep eutectic solvents
DSC	Differential scanning calorimetry
TGA	Thermogravimetric analysis
T_g	Glass transition temperature
NMR	Nuclear magnetic resonance
IR	Infrared spectroscopy
PFG-NMR	Pulsed-field gradient of nuclear magnetic resonance
NOESY	2D nuclear Overhauser spectroscopy
FTIR	Fourier-transform infrared spectroscopy
ATR	Attenuated total reflection
BCS	Biopharmaceutics classification system
FDA	Food and Drug Administration
MTS	3-(4,5-Dimethylthiazol-2-yl)-5-(3-carboxymethoxyphenyl)-2-(4-sulfophenyl)-2*H*-tetrazolium (dye for cell viability assay)
MTT	3-(4,5-Dimethylthiazol-2-yl)-2,5-diphenyltetrazolium bromide (dye for cell viability assay)
IC_{50}	Half-maximal inhibitory concentration
POM	Polarized optical microscopy
CAGE	Deep eutectic solvent choline and geranate

HaCatT	Cell line of human keratinocytes
HT29	Cell line of human colorectal adenocarcinoma
FDDS	Fast-dissolving delivery systems
RESS	Rapid expansion of supercritical solution
PGSS	Particle from gas saturated solution
SAS	Supercritical antisolvent
NSAID	Nonsteroidal anti-inflammatory drug
VLE	Vapor liquid equilibria
TEER	Transepithelial/transendothelial electrical resistance

3.1 Introduction

The development of green chemistry methods was initially reported in the 1990s and has a major importance in the development of more sustainable processes and less hazardous substances (Anastas and Eghbali 2010; Dunn 2012; Anastas and Kirchhoff 2002). The green chemistry concept was defined by the Environmental Protection Agency as the "design of chemical products and processes to reduce or eliminate the use and generation of hazardous substances" (Anastas and Eghbali 2010; Anastas and Warner 1998). This concept is diffused throughout chemical industry and academia in a search for the development of sustainable processes and new substances to be applied in different fields, such as healthcare and pharmaceuticals, cosmetics, agriculture, energy, advanced materials, and many other areas of research (Dunn 2012; Warner et al. 2004).

The green chemistry approaches in pharmaceutical field are frequently applied to solve problems concerning toxicity and bioavailability of the active pharmaceutical ingredients (APIs). Nonetheless, it is also crucial to design new sustainable processes for the synthesis of APIs. The sustainability in pharmaceutical industry is nowadays an essential point to take into consideration, due to the possibility of reducing costs of production and produce safer products that contribute to decrease their toxicity in human health, but that also reduce the risk of contamination of the environment (Kümmerer 2007, 2010; Cizmas et al. 2015; Ali and Khan 2017; Blasco and DelValls 2008). It was reported early in the 1990s that the amount of waste produced for the synthesis of 1 kg of an API was around 50–100 kg. These findings triggered pharmaceutical industries to find alternatives for the reduction and/or elimination of the waste generated, implement the use of green solvents, and develop new synthesis processes (Kümmerer 2007, 2010).

The investigation and application of green solvents has been one of the most active areas in the research of green chemistry, because solvents currently used contribute with a mass waste in synthesis processes and usually present high toxicity and flammability and could be corrosive (Anastas and Eghbali 2010). In this sense, the implementation of processes that avoid the use of solvents or use alternative solvents such as water, supercritical fluids, ionic liquids, and more recently deep eutectic systems is taken into consideration for improvement of industrial processes and research methods (Anastas and Eghbali 2010; Welton 2015). In this chapter, the focus will be on the application and development of deep eutectic systems in pharmaceutical field and their importance for the improvement of the API properties, namely, the increase of their bioavailability.

3.2 History of Green Solvents and Evolution Until Therapeutic Deep Eutectic Systems

The concept of green solvents was introduced by John Warner and Paul Anastas, in 1998, when they listed the 12 principles of green chemistry. One of the green chemistry principles assumes that the use of solvents should be as an auxiliary substances, in the synthesis and preparation processes, and emphases that they should be made unnecessary whenever possible, and if used they should be innocuous (Anastas and Warner 1998; Warner et al. 2004). The sustainability of a process or a product is, hence, a result of the complex interaction between the product or process implemented and environment, technology, and economic factors (Welton 2015).

In the last decades, the research on green or safer solvents, recognized as environmentally benign, has been growing exponentially. However, it is necessary to consider the source and synthesis of the solvent, its properties, and the disposal. A variety of compounds have been considered as green solvents such as water, supercritical and subcritical fluids, ionic liquids, solvents derived from biomass, and more recently deep eutectic systems (Bubalo et al. 2015; Francisco et al. 2013; Hayyan et al. 2015; Welton 2015). Strategies for the development of green solvents that lead to the substitution of petroleum-based solvents by solvents from renewable resources have been explored. Furthermore, the substitution of hazardous solvents by solvents that present better environmental, health, and safety properties is also a topic of intense research (Bubalo et al. 2015; Welton 2015).

Ionic liquids (ILs) appeared for the first time in 1914, described by Paul Walden in the synthesis of ethylammonium nitrate in a reaction that involves nitric acid (Domínguez de María and Maugeri 2011; Kudlak et al. 2015). An IL is defined as an organic salt composed by ions that result in a single salt with a melting point below 100 °C; some examples of cations and anions that are commonly used for the synthesis of ionic liquids are presented in Fig. 3.1. In the last years, they have been frequently used and investigated as green solvents (Ferraz et al. 2011; Hough et al. 2007; Liu et al. 2018a, b; Pena-Pereira and Namieśnik 2014). One of the applications of ionic liquids which was foreseen as a major breakthrough in industry was to replace volatile organic compounds (VOCs), presenting a safer alternative and exhibiting high solvency and a negligible vapor pressure (Brennecke and Maginn 2001; Heckenbach et al. 2016; Hough and Rogers 2007).

Ionic liquids have been developed in sequent generations. The first generation of ILs appeared in the 1960s and describes compounds that are designed in accordance with desired physical properties, such as hydrophobicity, viscosity, density, thermal

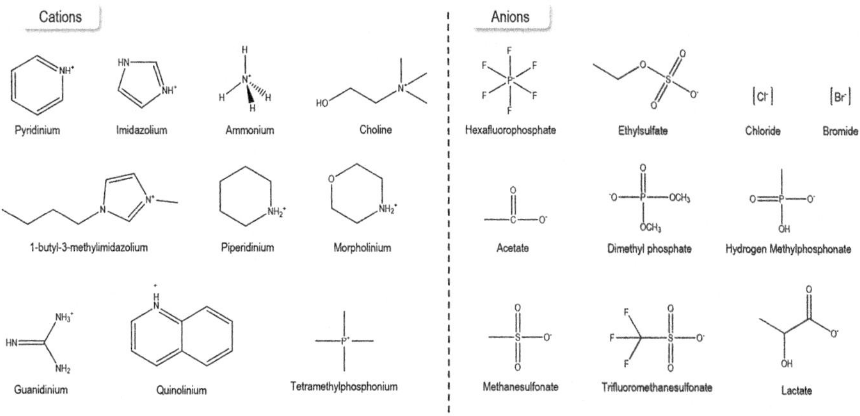

Fig. 3.1 Examples of some cations and anions that are commonly used in the synthesis of ionic liquids for different applications (Sivapragasam et al. 2016; Kudlak et al. 2015)

stability, conductivity, and melting point. However, these generation has the disadvantage to be sensitive to the presence of water and air (Domínguez de María and Maugeri 2011). In the early 1990s, the second generation of ILs emerged, which are considered advanced materials, in which the design of compounds combines both physical and chemical properties, such as chemical reactivity, energy density, oxygen balance, flammability, and others (Kudlak et al. 2015). The third generation of ILs appeared in the 2000s and consists in the synthesis of ionic liquids that have biological properties combined with physicochemical properties. In this group, the ILs that are used in pharmaceutical and biomedicine applications are included. In this case, a combination of cations and anions could lead to specific biological activity, for example, in drug delivery and production processes (Dias et al. 2017; Domínguez de María and Maugeri 2011; Ferraz et al. 2011; Kudlak et al. 2015).

Although ionic liquids are considered green solvents and have demonstrated good results for some applications, they are expensive and sometimes difficult to prepare and some of them have high toxicity (Kudlak et al. 2015; Paiva et al. 2014). In 2003, Abbott and coworkers reported a new type of green solvents that could act as an alternative to ionic liquids (Abbott et al. 2003, 2004, 2006; Francisco et al. 2013; Kudlak et al. 2015; Ruß and König 2012). When they experimented mixing urea and choline chloride, in which the initial components are in solid state, they verified that they formed a eutectic mixture, liquid at room temperature and presenting interesting properties to be used as a solvent (Abbott et al. 2004; Francisco et al. 2013; Kudlak et al. 2015; Ruß and König 2012). Despite eutectic mixtures have been described decades ago for many applications, it was only a few years ago that they started to be studied as solvents, as illustrated in Fig. 3.3. Abbott and coworkers introduced the term "deep eutectic solvent" (DES) to refer to a mixture of two or more starting materials with high melting points that often form a liquid at room temperature with lower melting temperature than the initial compounds (Abbott et al. 2017; Aroso et al. 2016; Dai et al. 2013; Francisco et al. 2013). Different theories were reported to explain the formation of eutectic mixtures as stable liquids, namely, the idea of a cluster formation or a mechanical mixture of the components. Abbott and coworkers suggested hydrogen bond interactions, in which the mixture results from the interactions between a hydrogen bond donor and a hydrogen bond acceptor, which leads to lowering the entropic differences of the phase transitions and allows decreasing the melting point (Francisco et al. 2013; Pedro et al. 2019). The nature of the interactions between the components could affect the capacity of the initial compounds to interact, and the charge delocalization could modulate the different physicochemical properties of DES, when compared to initial components, since the interactions could be hydrogen bonds, van der Waals, and/or electrostatic forces (Francisco et al. 2013; Santana et al. 2019). The van der Waals interactions in liquids could be observed by molecular dynamics using the empirical Lennard-Jones (LJ) equation that includes repulsive and attractive forces (Wojnarowska et al. 2018). The higher capacity of the components to establish hydrogen bonds is related with the phase-transition temperature and stability of the components (Francisco et al. 2013).

DES represent a group of systems that can be prepared from a variety of compounds, which may lead to thousands of different combinations (up to 10^6) (Barros et al. 2017; Francisco et al. 2013). These systems emerged as alternative candidates to ionic liquids and have also been described as low transition temperature mixtures (LTTM) (Durand et al. 2016). In 2011, Choi and coworkers reported 30 different combinations with choline chloride, natural carboxylic acids, sugars, and water that form viscous liquids and were named as natural deep eutectic solvents (NADES). NADES could be eutectic mixtures that we observe in our daily life; as an example, honey and syrup are eutectic mixtures of sugars at room temperature, and they could be used in food, as dietary supplement, and for medical formulations, because they are easily biodegradable and often have low toxicity (Fig. 3.2) (Choi et al. 2011; Dai et al. 2015; Liu et al. 2018a, b; Kudlak et al. 2015).

The use of eutectic mixtures for therapeutic applications was reported decades ago for transdermal delivery of anesthetic and anti-inflammatory drugs (Evers et al. 1985). Later, Stott and coworkers observed that ibuprofen could form eutectic mixtures with different terpenes and enhance skin permeation, dissolving API and increasing its solubility, permeability, and absorption (Aroso et al. 2015; Stott et al. 1998). In 2015, Aroso and coworkers named these solvents as therapeutic deep eutectic systems (THEDES), and they include deep eutectic systems that have an API incorporated in the mixture or a DES that can dissolve an API and improve their characteristics in terms of bioavailability and toxicity (Aroso et al. 2015; Gala et al. 2014; Wojnarowska et al. 2018) (Fig. 3.3).

3.2.1 Eutectics in Pharmacy

The term eutectic is derived from the Greek word *eutectos*, which means easily fused, and it was used for the first time by Frederik Guthrie, in 1884, to describe "bodies made up of two or more constituents, which constituents are in such

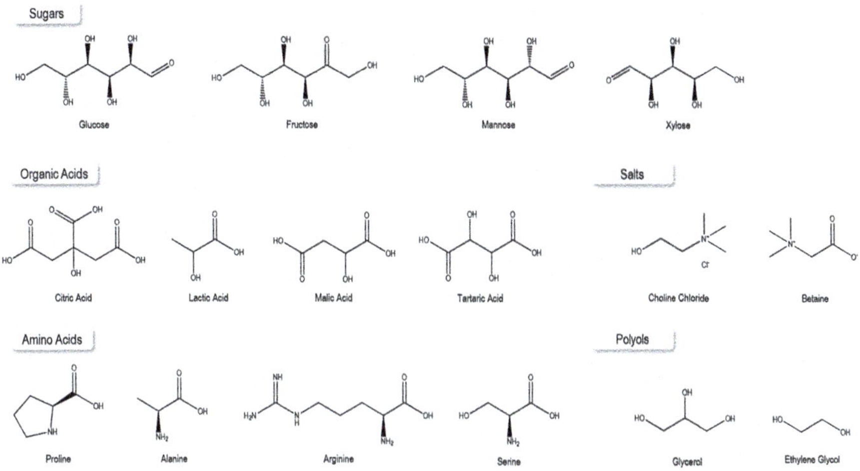

Fig. 3.2 Structures of molecules that could be used to prepare deep eutectic systems as sugars, organic acids, salts, amino acids, and polyols (Yang 2018)

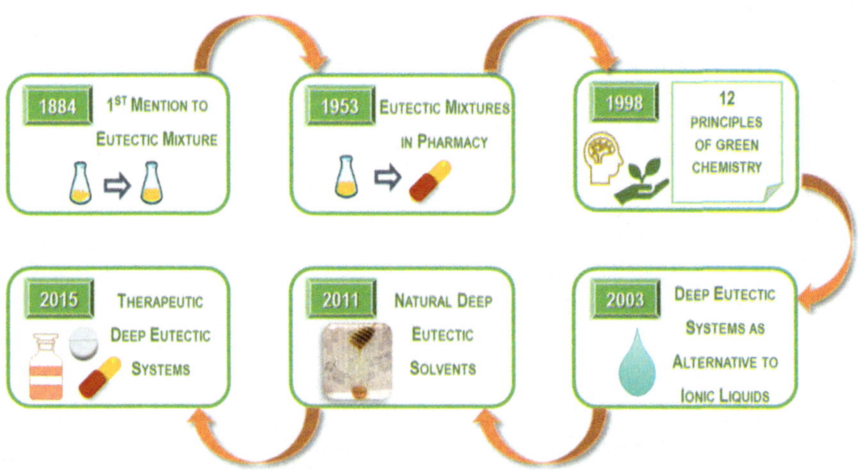

Fig. 3.3 Main developments on eutectic mixtures since they were referred for the first time until they started to be used as therapeutic deep eutectic systems. Deep eutectic solvents (DES) have been considered green and alternative solvents since they comply the principles of green chemistry, allow to reduce waste, develop safer products, minimize the energy used, develop degradable products, and reduce the use of hazardous substances

proportion to one another as to give to the resultant compound body a minimum temperature of liquefaction, that is, a lower temperature of liquefaction than that given by any other proportion" (Guthrie 1884; Martins et al. 2018). Cherukuvada and Nangia characterize eutectics as a discontinuous solid solution having a heterogeneous structural organization in the crystal lattice (Cherukuvada and Nangia 2014).

The eutectic mixtures are used since the beginning of the century, and until now many researchers investigated these mixtures for therapeutic applications. Bellafiore reported the use of eutectic mixtures in pharmacy; however, the aim of this study was to prevent the liquefaction and incompatibilities between the components of the eutectic mixtures (camphor and salol), incorporating powders with absorbent characteristics as magnesium carbonate, kaolin, and magnesium oxide (Bellafiore 1953; Prista et al. 2008). Another study performed with camphor and salol showed that these components form a eutectic mixture with a decrease in the melting point from 43 °C of salol and 179 °C of camphor to 6 °C of the mixture salol-camphor (Prista et al. 2008). Sekiguchi and coworkers studied eutectic mixtures and admitted that eutectic compounds are composed by two components. Being one of them water soluble and when exposed to gastrointestinal environment, this soluble compound dissolves fast, keeping the insoluble part with a large superficial area and more susceptible to absorption. It was reported that for binary combinations, the eutectic mixtures usually present a characteristic "V" type phase diagram, while the cocrystals exhibit a characteristic "W" type phase diagram (Prista et al. 2008; Cherukuvada 2016).

Eutectic mixtures sometimes present high viscosity or cannot be formed without the presence of water, and it was reported that adding water to a eutectic mixture in a certain molar ratio, being water part of the mixture, could help in the preparation of eutectic mixtures and decreases their viscosity, without compromising the system (Dai et al. 2015). Dai and coworkers studied the effects of water incorporated in natural deep eutectic solvents and observed that small amounts of water resulted in mixtures with low viscosity and reduced preparation time and in general the stability and solubility are increased. Nevertheless, adding up to 50% of water to the system could lead to the break of hydrogen bonds and dilution of the components, making a solution and not a DES (Craveiro et al. 2016; Dai et al. 2015).

3.2.2 *Advantages of Using Eutectics in Pharmacy*

The use of eutectic mixtures for therapeutic applications could represent an advantage to improve formulations and avoid the drawbacks of polymorphic drugs, since the mixtures, frequently, are in liquid form at room temperature, and could therefore provide a better solvent for several low soluble or insoluble drugs. The intrinsic characteristics of eutectic mixtures such as low melting point, liquid form, 100% yield, high solvent stability, and low toxicity confer to these systems an opportunity for the search of mixtures that could improve the drug bioavailability and pharmacokinetics, by adding characteristics to the API that lead to an efficient absorption, high biocompatibility, high solubility and permeability, and low toxicity. These parameters are essential for determining the bioavailability of an API, and the use of eutectic mixtures could enable to have new formulations or improved formulations without modifying the API (Álvarez and Zhang 2019).

The pharmaceutical industry has designed mostly crystalline APIs for formulation; however, many of these drugs fail in testing, due to issues with delivery mechanisms, like dissolution, transport, bioavailability, and polymorphism that could change the properties of the drugs. The polymorphism is a problem of crystalline APIs and is defined as the ability of a substance to exist in two or more crystalline forms with different arrangements and/or conformations of the molecules in the crystalline matrix. Several classes of drugs exhibit polymorphism, and the API could crystallize as a solvate that can be stoichiometric or nonstoichiometric. The presence of polymorphs or solvates can affect the mechanism of action of an API; subsequently, the solid-state structure determines its properties as dissolution rate, solubility, permeability, bioavailability, and others. In the production processes, polymorphism could also occur and lead to formulations with an API with ineffective doses (Domingos et al. 2015; Hough and Rogers 2007; Zainal-Abidin et al. 2019). Frequently, polymorphism is a solid-state phenomenon and is more prominent in compounds that contain more than one functional group promoting the interactions and form multiple supramolecular synthons (Domingos et al. 2015). One example of polymorphism that occurs in drugs was reported with risperidone, an antipsychotic drug commonly used in schizophrenia. This drug presented three forms, and just one of them could be used safely in therapeutics as it does not suffer polymorphic transformation during the manufacturing or storage process (Domingos et al. 2015).

Towards avoid unstable polymorphic forms of the APIs and enhance the solubility and/or permeability, it has been studied the development of metastable polymorphs, amorphous structures, salts or cocrystals formation and the use of eutectic mixtures as therapeutic deep eutectic systems, as represented in Table 3.1 (Domingos et al. 2015; Duarte et al. 2017). The therapeutic deep eutectic systems are frequently mistaken with unstable cocrystals; however, they present different types of components and interactions that allow the development of new drug delivery systems or improve the properties of the existing drugs (Cherukuvada and Nangia 2014; Duarte et al. 2017). The choice of right combinations for preparing THEDES potentiates the modulation of their characteristics and the tuning of the properties of a specific

Table 3.1 Biopharmaceutics classification system (BCS) based on problems and solutions presented by the pharmaceutical industry in the last years

Biopharmaceutics classification system (BCS)					
CLASS I	CLASS II		CLASS III		CLASS IV
Improved bioavailability ↑ **permeability** ↓ **solubility**	↑ **permeability** ↓ **solubility**	Solubility enhancement -Particle size reduction -Solid dispersions -Nanoparticles -Soluble salts -Cosolvents -Surfactants -pH adjustment -SMEDDS/ SEDDS -DES	↓ **permeability** ↑ **solubility**	Permeability enhancement -Use absorption enhancers -Efflux inhibitors -Motility modifiers -Prodrugs -DES	Combine approaches for enhancing solubility and permeability ↓ **permeability** ↓ **solubility**

API, since their features could offer solubility and dissolution enhancement of poorly soluble drugs (Aroso et al. 2016; Cherukuvada and Nangia 2014).

Goldberg and coworkers studied eutectic mixtures with urea and paracetamol, concluding that the mixture leads to a significant increase in the solubility and gastrointestinal absorption of paracetamol (Prista et al. 2008; Goldberg et al. 1966). Later on, Dichi and coworkers reinvestigate eutectic mixtures formed with paracetamol, acetylsalicylic acid, and caffeine and observed eutectic and metatectic points close to each other in the mixture paracetamol-caffeine (138.6 and 139.5 °C), and then they determined the eutectic composition of the mixture by Tamman's triangle for each phase diagram (Dichi et al. 2018).

In another study, Stott and coworkers observed that deep eutectic systems could be formed with the incorporation of an API in the DES, creating, hence, the possibility of develop controlled drug delivery devices, and improving the characteristics of the APIs themselves (Aroso et al. 2016; Stott et al. 1998), also showed an enhancement of solubility of drugs in different eutectic mixtures with choline chloride (Morrison et al. 2009).

Goud and colleagues investigated the formation of eutectic mixtures with curcumin that is an active compound and a hydrophobic polyphenol that could present diverse therapeutic activities like anti-inflammatory, antioxidant, anticancer and potentially used for Alzheimer's disease (Goud et al. 2012). Curcumin has problems of solubility and, consequently, bioavailability, and because of that, many strategies have been used for enhancing the stability and bioavailability of curcumin, such as use of nanoparticles and micelles, polymorphs, cocrystals, and then eutectic compositions with nicotinamide, ferulic acid, hydroquinone, *p*-hydroxybenzoic acid, and L-tartaric acid. These eutectic compositions with therapeutic components have shown an enhancement of the solubility and were prepared by mechanochemical grinding to provide stabilized mixtures by weak and short-range interactions (Goud et al. 2012).

3.2.3 Different Methods for Preparation of Therapeutic Deep Eutectic Systems

In terms of preparation of therapeutic deep eutectic systems (THEDES), there are a few methods described in the literature for the preparation of these mixtures (Meneses et al. 2019). The most commonly used procedure is to mix the components at a certain molar ratio, stirring and heating (with temperatures that usually vary between 40 and 80 °C), until a clear liquid is formed. The mixture could also be prepared in a mortar and a pestle, placed in stirring or vortexing. The temperature used is always dependent on the components present in the mixture, since sugars, for example, are not stable at high temperatures for long periods of time (Aroso et al. 2016; García-Argüelles et al. 2013; Serrano et al. 2012; Meneses et al. 2019). Another method described for the preparation of eutectic systems is through

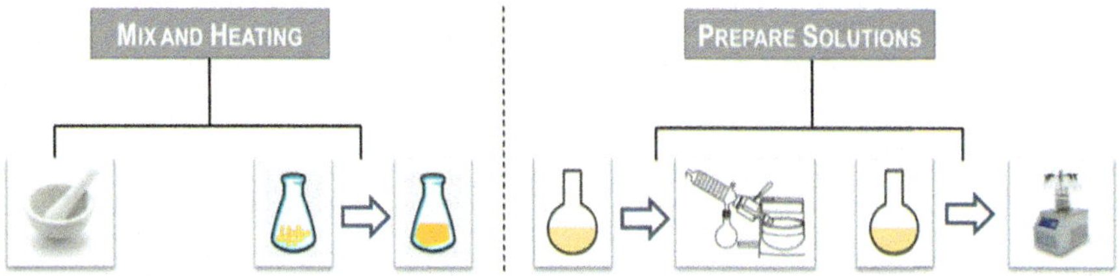

Fig. 3.4 Different methods described for the preparation of therapeutic deep eutectic systems (THEDES). In the mix and heating method, it is described that a mortar and a pestle could be used or otherwise it is possible to mix the compounds and by heating and stirring form a deep eutectic system. Other methods that have been reported are the preparation of solutions of different components and then evaporating the solvent or freeze-dried and a deep eutectic solvent (DES) is obtained

evaporation of water from an aqueous solution. The components are dissolved in an excess of water and then evaporated at approximately 50 °C in a rotary evaporator; the obtained liquid is then placed in a desiccator to dry, with silica gel until constant weight is reached (Dai et al. 2013; Liu et al. 2018a, b; Owczarek et al. 2016; Santana et al. 2019; Meneses et sal. 2019). Other method used is freeze-drying and consists in the preparation of an aqueous solution of each component as in evaporation method, but in this method, the resulting solution is freeze-dried to form a clear viscous liquid (Gutiérrez et al. 2009; Owczarek et al. 2016; Santana et al. 2019; Meneses et al. 2019) (Fig. 3.4).

Recently, new methods were reported for preparation of deep eutectic solvents. In particular, the microwave-assisted synthesis of deep eutectic solvents, where the components are irradiated, in a closed system, with microwave radiation and at controlled temperature, which cause dipole rotation and the molecules collide resulting in a dielectric heating and a reduced time of preparation. Another method that has been used for preparation of DES is the use of ultrasounds that leads to the formation and collapse of bubbles, which when they reach to the critical size release the energy that promotes the interactions between hydrogen bond donors and hydrogen bond acceptors (Santana et al. 2019).

3.2.4 Characterization of Therapeutic Deep Eutectic Systems

The characterization of deep eutectic systems is different depending on the application; however, the physicochemical characterization is always necessary. For pharmaceutical and biomedical applications, more information in terms of biocompatibility and formulation is needed. Regarding this, several techniques for the physicochemical and biological characterization of THEDES could be used, and they are presented in Table 3.2 and described in more detail in the next subsections.

Table 3.2 Summary of potential methods of characterization of therapeutic deep eutectic systems (THEDES)

Characterization method	Identification and morphology	Molecular structure	Hydrogen bonding interactions	Thermal stability	Phase transitions	Melting point, glass transition	Bioavailability	In vitro cytotoxicity
Polarized optical microscopy	√							
^{1}H nuclear magnetic resonance (2D techniques)	√	√	√					
Differential scanning calorimetry	√			√	√	√		
Thermogravimetric analysis	√			√	√			
Fourier-transform infrared spectroscopy/attenuated total reflection	√	√	√					
Water content	√							
Solubility							√	
Permeability							√	
Viscosity	√							
Density	√							
Polarity	√							
MTS/MTT/Alamar Blue (cell viability assays)								√

Physicochemical Characterization

- *Thermal characterization* – this analysis is important to create a phase diagram of the eutectic system, determining the eutectic point that corresponds to the ratio in which a single melting point exists, corresponding to the point of the eutectic liquid (Phaechamud et al. 2015). The degradation temperature and other variations of temperature that could be important in these compounds also can be observed with this analysis. In general, the thermal characterization can be performed with techniques such as differential scanning calorimetry (DSC) and thermogravimetric analysis (TGA) that are complementary techniques that provide information about different phase transitions in different endothermic and exothermic stages, glass transition temperature (T_g), and degradation temperature (Morrison et al. 2009; Phaechamud et al. 2015; Aroso et al. 2016).
- *Spectroscopic characterization* – this provides information about the structure of the compounds and the possible interactions that exist between the different components of the mixtures. Some examples of techniques used to observe hydrogen bonds and intermolecular interactions are nuclear magnetic resonance (NMR) and infrared spectroscopy (IR). The structure of THEDES was, for example, investigated by pulsed-field gradient (PFG) NMR (Duarte et al. 2017; Mann et al. 2019) and ^{1}H-^{1}H nuclear Overhauser spectroscopy (NOESY) (Dai et al. 2013; Aroso et al. 2016; Mann et al. 2019), which allowed the exploration of which interactions exist in the mixture and the determination of the most probable ratios between the components. The infrared spectroscopy involves the use of infrared radiation and is performed usually by Fourier-transform infrared spectroscopy (FTIR) or attenuated total reflection (ATR). These techniques show the spectral bands of known groups, and the delocalization of the shifts of these groups could indicate the presence of hydrogen interactions between the components in the mixture (Aroso et al. 2016; Dai et al. 2013, Dai et al. 2015; Xin et al. 2017).
- *Physicochemical properties* – the eutectic mixtures, usually, are liquid forms, and for further applications, properties such as viscosity, density, polarity, and water content are relevant for application and formulation. THEDES, generally, are viscous fluids and their properties can be seen as an advantage or disadvantage, depending on the application in different formulations for drug delivery. Viscosity and rheological measurements are performed using a rheometer or viscosimeter. Density can be measured using a densimeter and polarity can be determined in a relative scale through colorimetric assays (Craveiro et al. 2016). The water content that can be measured by Karl Fischer titration is an important characteristic to measure, as water content values above 50% compromise the structure and stability of the hydrogen interactions established among the components of the mixture. Further than that, it is the water content that influences directly the viscosity, density, and polarity of the mixtures (Craveiro et al. 2016; Dai et al. 2015).

Additional techniques could be used to expand or complement the characterization of THEDES, like polarized optical microscopy (POM) that shows information about the morphology of the eutectic mixture. The mixture is observed under polarized light, and the presence or absence of crystals is evaluated as an evidence of the formation of a homogeneous and amorphous mixture (Silva et al. 2018; Aroso et al. 2016).

Biological Characterization

- *Biological characterization* – for pharmaceutical applications, specific characterization methodologies should be performed, in particular the evaluation of the stability of the THEDES for long periods of time, its response to adverse environments, the evaluation of dissolution rates (Aroso et al. 2015), their bioavailability through biopharmaceutics classification system (BCS) and pharmacokinetic studies, and their toxicity.

For pharmaceutical substances, the Food and Drug Administration (FDA) provides a classification system named biopharmaceutics classification system (BCS) that serves as a guideline to predict intestinal drug absorption; however, this system restricts the simulations to the solubility and intestinal permeability (Duarte et al. 2017; Savjani et al. 2012). The parameters established by BCS described in the literature classifies an API in four different classes (Table 3.1), in which class I is high soluble and high permeable, class II is low soluble and high permeable, class III is high soluble and low permeable, and class IV is low soluble and low permeable (Chavda et al. 2010; Duarte et al. 2017; Silva et al. 2018; Varma et al. 2004). Solubility measurements can be performed in a buffer solution which is most adequate to the administration route of the THEDES. The maximum solubility is determined by saturating solutions under the same conditions. In the case of the permeability, in studies performed with therapeutic deep eutectic systems, it is possible to use synthetic membranes to mimic the effect on tissue, for example, membranes of polyethersulfone (Duarte et al. 2017). These studies allow the relative comparison of the permeability between the API and the THEDES. Also, permeability studies could be made by transwell assays that allow to generate cell layers in vitro and reconstitute their microenvironment and measure the permeability of a compound, by measuring the barrier integrity of the cell layer through transepithelial/transendothelial electrical resistance (TEER).

In vitro cytotoxicity tests are necessary mainly for applications for human consumption. This screening is based on assays that test cell viability (MTS, MTT, Alamar Blue) and measure the IC_{50} for THEDES and APIs, in different cell lines depending on the application (Faggian et al. 2016; Hayyan et al. 2016; Mano et al. 2016). Other in vitro assays could be performed regarding the future application of THEDES, such as antioxidant activity, apoptosis, and reactive oxygen species, among others (Hayyan et al. 2015; Mbous et al. 2017a, b).

3.3 Applications of Therapeutic Deep Eutectic Systems in Research and Industry

The therapeutic deep eutectic systems represent an attractive and sustainable method to explore an infinite number of possible combinations with different components that incorporate or dissolve a drug. For industry and research, it is a very interesting area that provides the possibility of improving the characteristics of the APIs (solubility and permeability, for example). However, other properties could be enhanced depending on the compounds that are part of the mixture. Incorporating natural compounds with very low toxicity can, for instance, reduce the toxicity of a drug, while the incorporation of an antioxidant can tune the antioxidant properties of the API (Zainal-Abidin et al. 2019; Álvarez and Zhang 2019).

In the last decade, the interest and knowledge for providing solutions for old biomedical and pharmaceutical problems has increased. The objective remains to find alternative and safe biomaterials, active compounds, and routes of administration of "old" drugs that improve their characteristics, turn them less toxic, and facilitate the therapeutics (Aroso et al. 2016).

3.3.1 Therapeutic Deep Eutectic Systems for Improvement of the Bioavailability of Drugs

The most common approach to form eutectic mixtures and use them in pharmacy is trying to incorporate one or more APIs as components of the eutectic mixture. Bonain discovered a eutectic mixture for topical anesthesia that includes cocaine hydrochloride, phenol, or menthol to form a homogeneous liquid at room temperature. Nevertheless, it was observed that cocaine has toxic effects and phenol has caustic properties (Gala et al. 2014). Evers and coworkers, in 1981, reported the use of a eutectic mixture of lidocaine and prilocaine (1:1) in an oil-in-water emulsion to increase the analgesic effect through the skin. With this mixture, they described an increase of the concentration of local anesthetic in the emulsion of approximately 20% to 80% of lidocaine-prilocaine, maintaining the total concentration of the local anesthetic low (5%) (Buckley and Benfield (1993); Evers et al. (1985); Lowrie et al. 1989). The transdermal permeation enhancement of this cream (trade name EMLA) is possible through the combination of the components that decreases the melting point of the mixture to approximately 22 °C, while the compounds individually have low permeation through the skin, due to their melting points higher than the body temperature (37 °C) (Cherukuvada and Nangia (2014); Yin and Jiang (2018); Woolfson et al. 2000). It was though observed that prilocaine has high tendency to cause methemoglobinemia, and this discovery led Gala and coworkers to study other eutectic mixtures for lidocaine, such as lidocaine-tetracaine and lidocaine-camphor (Gala et al. 2014).

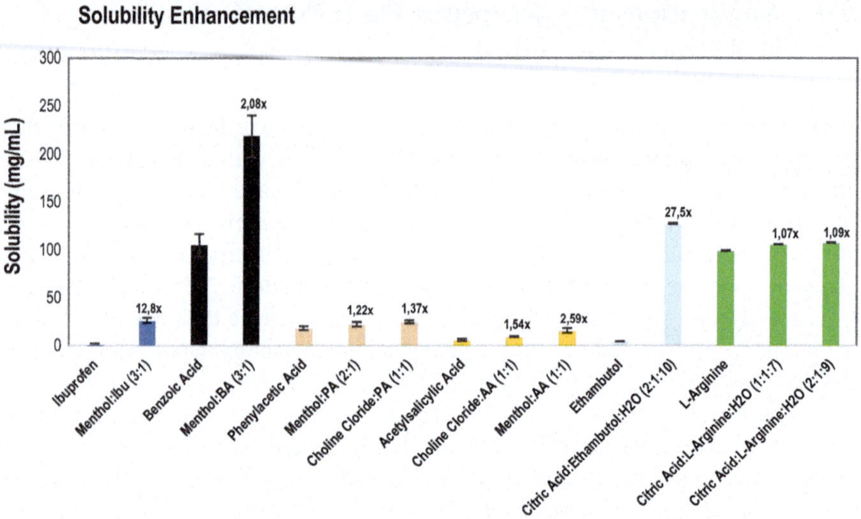

Fig. 3.5 Representative results presented in the literature with enhancement of solubility of APIs when present in the eutectic mixture as part of therapeutic deep eutectic systems (THEDES) (Aroso et al. (2016); Duarte et al. (2017); Aroso et al. (2015); Barros et al. (2017); Santos et al. 2019)

These studies that include the formation of eutectic mixtures with therapeutic agents demonstrated an improvement of the characteristics of the therapeutic agent. These promising results lead to investigation and development of studies of solubility with therapeutic eutectic mixtures and using the eutectic mixtures as a vehicle to solubilize poorly soluble drugs, to improve their solubility, as we can observe in Fig. 3.5.

Shen and coworkers have studied the intestinal absorption of daidzein in the presence of a eutectic mixture of borneol:menthol and with a microemulsion to improve the bioavailability of daidzein. The results indicate that the eutectic mixture increased the solubility of daidzein, increasing the membrane fluidity and permeability through the intestinal mucosa. The use of daidzein borneol:menthol microemulsion with a surfactant to improve the intestinal absorption could result in a synergistic effect between the eutectic mixture and microemulsion with surfactant (Shen et al. 2011).

Morrison et al. tested potential solubilization of drugs such as benzoic acid, griseofulvin, danazol, itraconazole, and AMG517 in NADES. In this study, was measured the solubility with pure NADES and mixtures of NADES and water in different proportions, the results showed an increase of 5- to 22,000-fold in the solubility of the drugs comparing with water. In addition, NADES could represent a promising vehicle for oral drug delivery, since their components are, in general, pharmaceutically tolerable (Morrison et al. (2009); Vanda et al. 2018); Álvarez and Zhang 2019).

In another study was formulated an eutectic mixture with choline bicarbonate and geranic acid in the ratio 1:2, that is called CAGE, for enhancing topical drug delivery of proteins like bovine serum albumin, ovalbumin, and insulin. The proteins were dispersed in the DES to increase their penetration through epidermis and dermis, and it was observed that insulin, for example, when dispersed in CAGE presents a decrease of blood glucose levels 4 h later. And it was proven that these CAGE eutectic mixtures could potentiate the transdermal delivery of therapeutic peptides and proteins and increase the permeability of macromolecules through the skin (Banerjee et al. 2017). Zakrewsky et al. reported the use of this eutectic solvent CAGE as antimicrobial agent against some of the drug-resistant bacterial strains, like *Mycobacterium tuberculosis*, *Staphylococcus aureus*, *Candida albicans*, and herpes simplex virus (Zakrewsky et al. 2016; Zainal-Abidin et al. 2019).

Li et al. have investigated the solubility enhancement of some poorly water-soluble drugs, such as itraconazole, piroxicam, lidocaine, and posaconazole. The DES prepared were based on choline chloride and carboxylic acids; they showed an improvement of the solubility of 6700-fold for itraconazole, 430-fold for piroxicam, 28-fold for lidocaine, and 6400-fold for posaconazole, when compared to water solubility. They also prepared a ternary system with choline chloride:glycolic acid:oxalic acid (1:1.6:0.4) that increases the solubility of itraconazole to 53,600-fold (Li and Lee (2016); Álvarez and Zhang 2019). In another study, Shekaari et al. tested the solubility enhancement of acetaminophen in DES of choline chloride and glycerol or ethylene glycol. The solubility of acetaminophen was higher in DES with ethylene glycol than with glycerol, and complementary studies indicate that the interactions established between the DES and acetaminophen are stronger with increasing concentration of the cosolvent (Shekaari et al. 2018).

Recently, Silva and coworkers reported the use of fatty acids to formulate DES with antimicrobial properties, preparing DES with capric acid, myristic acid, stearic acid, and lauric acid. The system that shows better antimicrobial activity for gram-positive bacteria and *C. albicans* was capric acid:lauric acid (2:1). This system seems to be promising for detachment of biofilms of several types of microorganisms including gram-positive and gram-negative bacteria (Silva et al. 2019a, b, c). The same group has investigated the preparation and application of THEDES with fatty acids for wound healing. They prepared systems with menthol and lauric acid, myristic acid, and stearic acid in different molar ratios. All the systems were characterized, and it was observed that the hydrophobic THEDES with menthol:stearic acid exhibit important properties as wound healing agents, since it was noticed a recover of the cell layer with HaCaT after 24 hour, which could be important for the development of new formulations (Silva et al. 2019a, b, c).

The antitumor potential of DES was studied by Mbous and coworkers; they prepare natural DES and studied their anticancer activity in different cell lines and observe that NADES present a higher cellular tolerance and potential for the development of anticancer agents. Despite this, the mixtures studied by Mbous et al. do not have incorporated active pharmaceutical ingredients or compounds that have anticancer activity reported (Mbous et al. 2017a, b). Later, Pereira et al. studied the antiproliferative properties of THEDES prepared with limonene that present antitumor activity itself but present some cytotoxicity. The systems prepared were myristic acid:limonene (1:1, 1:2, and 2:1), menthol:limonene (1:1, 1:2, and 2:1), capric acid:limonene (1:1, 1:2, and 2:1), and ibuprofen:limonene (1:1, 1:2, 2:1, 1:4, and 1:8); all of them present antitumor activity; however, only ibuprofen:limonene (1:4) was able to inhibit HT29 without compromising cell viability and also improve the anti-inflammatory activity of ibuprofen (Pereira et al. 2019). The studies of these therapeutic mixtures in vitro and in vivo are essential for understanding the mechanisms that govern the therapeutic effect and the possible interactions that these mixtures have with different cells and tissues. The implementation of these biological studies accompanied with modeling of the interaction and permeation through cellular membranes could be very promising and help formulate mixtures with specific targets and low toxicity.

3.3.2 *Biomedical Formulation with Therapeutic Deep Eutectic Systems*

The design and formulation of suitable drug delivery carriers is another major concern of the pharmaceutical industry, and different examples of formulations are reported in the literature. Tuntarawongsa and Phaechamud have previously prepared eutectic mixtures with menthol and camphor, borneol, and WS-3 and observe an increase of drug bioavailability. Later, they incorporate a polymer in the eutectic mixture menthol and camphor to prolong the drug release and have a slow drug diffusion. The polymeric eutectic system with Eudragit® was used as a vehicle for ibuprofen, due to its higher solubility in the system menthol:camphor, and the hydrophobic part turns this mixture suitable for controlled release of ibuprofen, for example, in periodontitis (Tuntarawongsa and Phaechamud (2012); Mbous et al. 2017a, b). Mano et al. prepared THEDES with choline chloride and mandelic acid (1:2) which were encapsulated in gelatin fibers by electrospinning, producing fibers with a smooth surface and that can adopt many conformations. These fibers were designed to obtain fast-dissolving delivery systems (FDDS) with simple techniques (Mano et al. 2016; Roda et al. 2019).

Zainal-Abidin et al. explored the use of DES to functionalize the surface of nanodrug carriers of graphene. In spite of these, nanodrug carriers of graphene per se represent an alternative way for increasing the efficiency of drug delivery, due to their high surface area, intrinsic mobility, thermal stability, and high loading capacity. But these nanocarriers of graphene could present some toxicity for humans and for the environment, and the functionalization of these carriers with DES allows surface modifications and introduction of functional groups that increase the biocompatibility of graphene (Zainal-Abidin et al. 2019).

In order to observe an improvement of the characteristics of a drug in an eutectic mixture, Patel et al. studied the eutectic mixture nimesulide:nicotinamide (1:2) and produce a powder through spray dried and verify that the solubility of this mixture was enhanced 14-fold and the dissolution in water enhanced 2-fold, when compared with pure drug. In spite of eutectic mixtures are most of the times in a liquid form, they can be used directly for processing and manufacturing of solid forms and at the same time improve the characteristics of the drugs (Patel et al. 2019; Álvarez and Zhang 2019).

Several papers in the literature report the use of supercritical fluid technology as a versatile technology for the preparation of different formulations. Using different particle formation techniques such as rapid expansion of supercritical solutions (RESS), particle from gas saturated solution (PGSS), supercritical antisolvent (SAS), or GAS for coprecipitation of the drug and the polymer, it is possible to produce particles, for example, encapsulating drug in a polymer matrix (Guney and Akgerman 2002; Reverchon et al. 2009). Silva et al. explored the loading of gauzes with a eutectic blend of lauric acid:myristic acid by supercritical CO_2. With these novel approaches, it was possible to obtain homogeneous eutectic blends and improve their antibacterial properties, which could be explained by the increase of the hydrophobicity of the blend formulated that may improve their ability to interact with the membrane of the bacteria (Silva et al. 2019a, b, c). In an attempt to find and effective therapy for tuberculosis treatment, Roda and coworkers investigated for the first time the encapsulation of THEDES with anti-tuberculosis drugs, as a component of the mixture, through PGSS and evaluate the influence of different water ratios present on the mixtures on the PGSS process (Roda et al. 2020).

Supercritical carbon dioxide technology can be used for the incorporation of a drug in a polymeric matrix rendering a controlled release systems basis on a slight plasticization of the polymeric particles that fused together and produce a 3D structure. Silva et al. reported the possibility of using the THEDES choline chloride:ascorbic acid and solubilized dexamethasone in the eutectic mixture, which was then impregnated in a polymeric matrix by supercritical CO_2. This system could be important in studies of bone tissue engineering, because its components could assist osteogenic differentiation from stem cells (Barros et al. 2017; Silva et al. 2018).

Aroso and coworkers studied the development of controlled drug delivery systems of anti-inflammatory drugs, namely, using the THEDES menthol:ibuprofen and a biodegradable polymer, composed by a blend of starch, and obtained three-dimensional porous materials with supercritical fluid sintering (Aroso et al. 2015; Roda et al. 2019). In this work, the release of ibuprofen from THEDES form has shown significant differences, being the ibuprofen dissolved in THEDES released much faster than API itself.

The impregnation of THEDES with supercritical CO_2 is still a research area that remains relatively unexplored, particularly due to the lack of measurements on the binary systems of THEDES and CO_2. The vapor liquid equilibria (VLE) experiments are relevant to optimize the operating conditions for impregnation, such that the amount of THEDES impregnated is within its therapeutic window. However, Barros et al. explore the impregnation of a THEDES system (menthol:ibuprofen) in alginate sponges that were prepared by freeze-drying, supercritical CO_2 was used for the impregnation of THEDES, and the solubility of these binary systems (THEDES + CO_2) was studied (Barros et al. 2017).

3.3.3 Future Perspectives

The THEDES could be used as a single mixture, providing higher dissolution rates of the API and enhancement of solubility, permeability, and absorption through tissues, or they can be incorporated in suitable polymer to potentiate the efficiency of biomedical devices (Aroso et al. 2016, 2015; Roda et al. 2019). The different applications that THEDES have been used and reported are represented in Table 3.3.

The use of DES as therapeutic agents reveals to be very promising to enhance the characteristics of existing drugs and optimize or develop new formulations that could be easily administrated and more effective. However, the formulation of eutectic mixtures is still based on trial and error experiments to obtain a liquid and stable mixture at room temperature that can be used in different applications, including pharmaceutical research. To facilitate these studies, it is important to understand how molecular interactions are established between the components of the mixture and what molecules could be mixed, leading to enhanced properties, avoiding trial and error experiments. To better understand this type of mixtures and their characteristics, how they can be modulated for different applications will represent a huge step on the research of these mixtures and in their development. The computational modeling and molecular dynamics studies could represent one part of the investigation of these mixtures and could help researchers understand better their advantages as well as provide tools for a more systematic preparation of these formulations.

Table 3.3 Different systems reported in the literature for different applications of therapeutic deep eutectic systems (THEDES)

THEDES	Application	References
Ibuprofen:terpenes (menthol, thymol, menthone, 1,8-cineole, *d*-limonene, *p*-cymene)	Preparation and characterization of eutectic mixtures with ibuprofen and several terpenes. Evaluate them as transdermal permeation enhancers	Stott et al. (1998)
Lidocaine:prilocaine (3:7); (4:6); (5:5); (6:4); (7:3)	Evaluation of the enhancement of transmembrane drug transport	Fiala et al. (2010)
Borneol:menthol (25:75)	Enhance the intestinal absorption of daidzein	Shen et al. (2011)
Lidocaine:tetracaine (1:1) Lidocaine:camphor (1:1)	Preparation of eutectic mixtures for anesthetic applications	Gala et al. (2014)
Choline chloride:glycolic acid (1:2) Choline chloride:glycolic acid:oxalic acid (1:1.6:0.4)	Enhance the solubility of poorly soluble drugs	Li and Lee (2016)
Choline chloride:acetylsalicylic acid (1:1) Choline chloride:phenylacetic acid (1:1) Menthol:benzoic acid (1:1); (2:1); (3:1) Menthol:acetylsalicylic acid (1:1); (2:1); (3:1) Menthol:phenylacetic acid (1:1); (2:1); (3:1)	Preparation of THEDES with API incorporated in the system, without solvent addition	Aroso et al. (2016), and Duarte et al. (2017)

(continued)

Table 3.3 (continued)

THEDES	Application	References
Menthol:ibuprofen (3:1)	Preparation of THEDES with API (NSAID) incorporated in the system for impregnation in polymer matrices with supercritical CO_2	Aroso et al. (2015), Barros et al. (2017), and Duarte et al. (2017)
Choline chloride:mandelic acid (1:2)	Preparation of THEDES and encapsulate them in gelatin fiber membranes	Mano et al. (2016)
Choline chloride:ascorbic acid (1:1), (1:2), (2:1)	Preparation of THEDES to dissolve an API (dexamethasone), with possible applications in tissue engineering	Silva et al. (2018)
Choline chloride:lactic acid (1:1) β-Alanine:lactic acid (1:1)	Solubility enhancement of lidocaine in DES (theoretical study)	Gutiérrez et al. (2018)
Choline chloride:glycerol (1:2) Choline chloride:ethylene glycol (1:2)	Solubility enhancement of acetaminophen with DES	Shekaari et al. (2018)
Citric acid:ethambutol:H_2O (2:1:10) Citric acid:L-arginine:H_2O (1:1:6) Citric acid:L-arginine:H_2O (1:1:7) Citric acid:L-arginine:H_2O (2:1:7) Citric acid:L-arginine:H_2O (2:1:8) Citric acid:L-arginine:H_2O (2:1:9)	Preparation of THEDES with API and water incorporated in the system	Santos et al. (2019)
Choline chloride:sucrose:H2O (4:1:4) Choline chloride:glycerol:H2O (1:2:1)	Preparation of DES for functionalization of drug nanocarriers of graphene	Zainal-Abidin et al. (2019)
Menthol:lauric acid (1:1); (1:2); (2:1); (4:1); (8:1) Menthol:myristic acid (1:1); (2:1); (4:1); (8:1); (10:1) Menthol:stearic acid (4:1); (8:1); (20:1)	Preparation and characterization of THEDES with terpenes and fatty acids with wound healing properties	Silva et al. (2019a, b, c)
Lauric acid:myristic acid (1:1)	Preparation of eutectic blends and loading onto gauzes	Silva et al. (2019a, b, c)
Capric acid:lauric acid (2:1) Capric acid:stearic acid (4:1) Capric acid:myristic acid (3:1)	Preparation of different DES with fatty acids with antimicrobial activity and biofilm detachment	Silva et al. (2019a, b, c)
Myristic acid:limonene (1:1); (1:2); (2:1) Capric acid:limonene (1:1); (1:2); (2:1) Menthol:limonene (1:1); (1:2); (2:1) Ibuprofen:limonene (1:1); (1:2); (2:1); (4:1); (8:1)	Preparation of THEDES with limonene and evaluate their antitumor activity	Pereira et al. (2019)
Nimesulide:nicotinamide (1:2)	Direct preparation of compressible dosage forms for drug delivery	Patel et al. (2019)

3.4 Conclusion

The present review emphasizes the importance and evolution of therapeutic deep eutectic solvents. THEDES are systems that have an infinite number of possible combinations of its components, which allows the modulation of their features and the application in different pharmaceutical formulations. For pharmaceutical research, it is important to have this possibility that involves less toxicity, improves the properties of the drugs, is cost-effective, and is an ecological alternative to the preparation of enhanced biopharmaceuticals.

Acknowledgments The research leading to these results has received funding from the European Union Horizon 2020 Programme under the agreement number ERC-2016-CoG 725034 (ERC Consolidator Grant Des.solve). This work was supported also by the Associate Laboratory for Green Chemistry-LAQV which is financed by the ERDF under the PT2020 Partnership Agreement (POCI-01-0145-FEDER-007265).

References

Abbott AP, Capper G, Davies DL, Rasheed RK, Tambyrajah V (2003) Novel solvent properties of choline chloride/urea mixtures. Chem Commun:70–71. https://doi.org/10.1039/B210714G

Abbott AP, Boothby D, Capper G, Davies DL, Rasheed R (2004) Deep eutectic solvents formed between choline chloride and carboxylic acids. J Am Chem Soc 126:9142–9147. https://doi.org/10.1021/ja048266j

Abbott AP, Capper G, Gray S (2006) Design of improved deep eutectic solvents using hole theory. ChemPhysChem 7:803–806. https://doi.org/10.1002/cphc.200500489

Abbott AP, Ahmed EI, Prasad K, Qader IB, Ryder KS (2017) Liquid pharmaceuticals formulation by eutectic formation. Fluid Phase Equilib 448:2–8. https://doi.org/10.1016/j.fluid.2017.05.009

Ali H, Khan E (2017) Environmental chemistry in the twenty-first century. Environ Chem Lett 15:329–346. https://doi.org/10.1007/s10311-016-0601-3

Álvarez MS, Zhang Y (2019) Sketching neoteric solvents for boosting drugs bioavailability. J Control Release 311–312:225–232. https://doi.org/10.1016/j.jconrel.2019.09.008

Anastas P, Eghbali N (2010) Green chemistry: principles and practice. Chem Soc Rev 39:301–312. https://doi.org/10.1039/b918763b

Anastas PT, Kirchhoff MM (2002) Origins, current status, and future challenges of green chemistry. Acc Chem Res 35:686–694. https://doi.org/10.1021/ar010065m

Anastas PT, Warner JC (1998) Green chemistry: theory and practice. Oxford University Press. ISBN: 9780198506980

Aroso IM, Craveiro R, Rocha A, Dionísio M, Barreiros S, Reis RL, Paiva A, Duarte ARC (2015) Design of controlled release systems for THEDES — therapeutic deep eutectic solvents, using supercritical fluid technology. Int J Pharm 492:73–979. https://doi.org/10.1016/j.ijpharm.2015.06.038

Aroso IM, Silva JC, Mano F, Ferreira ASD, Dionísio M, Sá-nogueira I, Barreiros S, Reis RL, Paiva A, Duarte ARC (2016) Dissolution enhancement of active pharmaceutical ingredients by therapeutic deep eutectic systems. Eur J Pharm Biopharm 98:57–66. https://doi.org/10.1016/j.ejpb.2015.11.002

Banerjee A, Ibsen K, Iwao Y, Zakrewsky M, Mitragotri S (2017) Transdermal protein delivery using choline and geranate (CAGE) deep eutectic solvent. Adv Healthc Mater 6:1–11. https://doi.org/10.1002/adhm.201601411

Barros AA, Silva JM, Craveiro R, Paiva A, Reis RL, Duarte ARC (2017) Green solvents for enhanced impregnation processes in biomedicine. Curr Opin Green Sustain Chem 5:82–87. https://doi.org/10.1016/j.cogsc.2017.03.014

Bellafiore IJ (1953) Stabilization of capsules of eutectic mixtures against liquefaction. J Am Pharm Assoc 14:580–582. https://doi.org/10.1016/S0095-9561(16)33172-3

Blasco J, DelValls A (2008) Impact of emergent contaminants in the environment: environmental risk assessment. Hdb Env Chem 5(S/1):169–188. https://doi.org/10.1007/698_5_107

Brennecke JF, Maginn EJ (2001) Ionic liquids: innovative fluids for chemical processing. AICHE J 47:2384–2389. https://doi.org/10.1002/aic.690471102

Bubalo MC, Vidović S, Redovniković IR, Jokić S (2015) Green solvents for green technologies. J Chem Technol Biotechnol 90:1631–1639. https://doi.org/10.1002/jctb.4668

Buckley MM, Benfield P (1993) Eutectic lidocaine/prilocaine cream: a review of the topical anaesthetic/analgesic efficacy of a eutectic mixture of local anaesthetics (EMLA). Drugs 46:126–151. https://doi.org/10.2165/00003495-199346010-00008

Chavda HP, Patel CN, Anand IS (2010) Biopharmaceutics classification system. Sys Rev Pharm (1):62–69

Cherukuvada S (2016) On the issues of resolving a low melting combination as a definite eutectic or an elusive cocrystal: a critical evaluation. J Chem Sci 128:487–499. https://doi.org/10.1007/s12039-016-1055-7

Cherukuvada S, Nangia A (2014) Eutectics as improved pharmaceutical materials: design, properties and characterization. Chem Commun 50:906–923. https://doi.org/10.1039/c3cc47521b

Choi YH, Spronsen JV, Dai Y, Verberne M, Hollmann F, Arends IWCE, Witkamp GJ, Verpoorte R (2011) Are natural deep eutectic solvents the missing link in understanding cellular metabolism and physiology? Plant Physiol 156:1701–1705. https://doi.org/10.1104/pp.111.178426

Cizmas L, Sharma VK, Gray CM, McDonald TJ (2015) Pharmaceuticals and personal care products in waters: occurrence, toxicity, and risk. Environ Chem Lett 13:381–394. https://doi.org/10.1007/s10311-015-0524-4

Craveiro R, Aroso I, Flammia V, Carvalho T, Viciosa MT, Dionísio M, Barreiros S, Reis RL, Duarte ARC, Paiva A (2016) Properties and thermal behavior of natural deep eutectic solvents. J Mol Liq 215:534–540. https://doi.org/10.1016/j.molliq.2016.01.038

Dai Y, Spronsen JV, Witkamp GJ, Verpoorte R, Choi YH (2013) Natural deep eutectic solvents as new potential media for green technology. Anal Chim Acta 766:61–68. https://doi.org/10.1016/j.aca.2012.12.019

Dai Y, Witkamp GJ, Verpoorte R, Choi YH (2015) Tailoring properties of natural deep eutectic solvents with water to facilitate their applications. Food Chem 187:14–19. https://doi.org/10.1016/j.foodchem.2015.03.123

Dias AR, Costa-Rodrigues J, Fernandes MH, Ferraz R, Prudêncio C (2017) The anticancer potential of ionic liquids. Chem Med Chem 12:11–18. https://doi.org/10.1002/cmdc.201600480

Dichi E, Sghaier M, Guiblin N (2018) Reinvestigation of the paracetamol-caffeine, aspirin-caffeine and paracetamol-aspirin phase equilibria diagrams. J Therm Anal Calorim 131:2141–2155. https://doi.org/10.1007/s10973-017-6855-6

Domingos S, André V, Quaresma S, Martins ICB, Piedade MFM, Duarte MT (2015) New forms of old drugs: improving without changing. J Pharm Pharmacol 67:830–846. https://doi.org/10.1111/jphp.12384

Domínguez de María P, Maugeri Z (2011) Ionic liquids in biotransformations: from proof-of-concept to emerging deep-eutectic-solvents. Curr Opin Chem Biol 15:220–225. https://doi.org/10.1016/j.cbpa.2010.11.008

Duarte ARC, Ferreira ASD, Barreiros S, Cabrita E, Reis RL, Paiva A (2017) A comparison between pure active pharmaceutical ingredients and therapeutic deep eutectic solvents: solubility and permeability studies. Eur J Pharm Biopharm 114:296–304. https://doi.org/10.1016/j.ejpb.2017.02.003

Dunn PJ (2012) The importance of green chemistry in process research and development. Chem Soc Rev 41:1452–1461. https://doi.org/10.1039/c1cs15041c

Durand E, Lecomte J, Villeneuve P (2016) From green chemistry to nature: the versatile role of low transition temperature mixtures. Biochimie 120:119–123. https://doi.org/10.1016/j.biochi.2015.09.019

Evers H, Dardel OV, Juhlin L, Ohlsen L, Vinnars E (1985) Dermal effects of compositions based on the eutectic mixture of lignocaine and prilocaine (EMLA). Br J Anaesth 57:997–1005. https://doi.org/10.1093/bja/57.10.997

Faggian M, Sut S, Perissutti B, Baldan V, Grabnar I, Dall'Acqua S (2016) Natural deep eutectic solvents (NADES) as a tool for bioavailability improvement: pharmacokinetics of rutin dissolved in proline/glycine after oral administration in rats: possible application in nutraceuticals. Molecules 21:1531–1542. https://doi.org/10.3390/molecules21111531

Ferraz R, Branco LC, Prudêncio C, Noronha JP, Petrovski Z (2011) Ionic liquids as active pharmaceutical ingredients. Chem Med Chem 6:975–985. https://doi.org/10.1002/cmdc.201100082

Fiala S, Jones SA, Brown MB (2010) A fundamental investigation into the effects of eutectic formation on transmembrane transport. Int J Pharm 393:68–73. https://doi.org/10.1016/j.ijpharm.2010.04.001

Francisco M, Bruinhorst AVD, Kroon MC (2013) Low-transition-temperature mixtures (LTTMs): a new generation of designer solvents. Angew Chem Int Ed 52:3074–3085. https://doi.org/10.1002/anie.201207548

Gala U, Chuong MC, Varanasi R, Chauhan H (2014) Characterization and comparison of lidocaine-tetracaine and lidocaine-camphor eutectic mixtures based on their crystallization and hydrogen-bonding abilities. AAPS Pharm Sci Tech 16:528–536. https://doi.org/10.1208/s12249-014-0242-4

García-Argüelles S, Serrano MC, Gutiérrez MC, Ferrer ML, Yuste L, Rojo F, Del Monte F (2013) Deep eutectic solvent-assisted synthesis of biodegradable polyesters with antibacterial properties. Langmuir 29:9525–9534. https://doi.org/10.1021/la401353r

Goldberg AH, Gibaldi M, Kanig JL (1966) Increasing dissolution rates and gastrointestinal absorption of drugs via solid solutions and eutectic mixtures II: experimental evaluation of a eutectic mixture: urea-acetaminophen systems. J Pharm Sci 55:482–487. https://doi.org/10.1002/jps.2600550507

Goud NR, Suresh K, Sanphui P, Nangia A (2012) Fast dissolving eutectic compositions of curcumin. Int J Pharm 439:63–72. https://doi.org/10.1016/j.ijpharm.2012.09.045

Gutiérrez MC, Ferrer ML, Mateo CR, Monte F (2009) Freeze-drying of aqueous solutions of deep eutectic solvents: a suitable approach to deep eutectic suspensions of self-assembled structures. Langmuir 25:5509–5515. https://doi.org/10.1021/la900552b

Guney O, Akgerman A (2002) Synthesis of controlled-release products in supercritical medium. AICHE J 48:856–866. https://doi.org/10.1002/aic.690480419

Guthrie F (1884) On *eutexia. The London, Edinburgh, and Dublin Philosophical Magazine and.* J Sci 17:462–482. https://doi.org/10.1080/14786448408627543

Gutiérrez A, Atilhan M, Aparicio S (2018) A theoretical study on lidocaine solubility in deep eutectic solvents. Phys Chem Chem Phys 20:27464–27473. https://doi.org/10.1039/c8cp05641b

Hayyan M, Looi CY, Hayyan A, Wong WF, Hashim MA (2015) *In vitro* and *in vivo* toxicity profiling of ammonium-based deep eutectic solvents. PLoS One 10:1–18. https://doi.org/10.1371/journal.pone.0117934

Hayyan M, Mbous YP, Looi CY, Wong WF, Hayyan A, Salleh Z, Mohd-Ali O (2016) Natural deep eutectic solvents: cytotoxic profile. Springer Plus 5:913–925. https://doi.org/10.1186/s40064-016-2575-9

Heckenbach ME, Romero FN, Green MD, Halden RU (2016) Meta-analysis of ionic liquid literature and toxicology. Chemosphere 150:266–274. https://doi.org/10.1016/j.chemosphere.2016.02.029

Hough WL, Rogers RD (2007) Ionic liquids then and now: from solvents to materials to active pharmaceutical ingredients. Bull Chem Soc Jpn 80:2262–2269. https://doi.org/10.1246/bcsj.80.2262

Hough WL, Smiglak M, Rodríguez H, Swatloski RP, Spear SK, Daly DT, Pernak J, Grisel JE, Carliss RD, Soutullo MD, Davis JH, Rogers RD (2007) The third evolution of ionic liquids: active pharmaceutical ingredients. New J Chem 31:1429–1436. https://doi.org/10.1039/b706677p

Kudlak B, Owczarek K, Namiesnik J (2015) Selected issues related to the toxicity of ionic liquids and deep eutectic solvents — a review. Environ Sci Pollut Res 22:11975–11992. https://doi.org/10.1007/s11356-015-4794-y

Kümmerer K (2007) Sustainable from the very beginning: rational design of molecules by life cycle engineering as an important approach for green pharmacy and green chemistry. Green Chem 9:899–907. https://doi.org/10.1039/b618298b

Kümmerer K (2010) Pharmaceuticals in the environment. Annu Rev Environ Resour 35:57–75. https://doi.org/10.1146/annurev-environ-052809-161223

Li Z, Lee PI (2016) Investigation on drug solubility enhancement using deep eutectic solvents and their derivatives. Int J Pharm 505:283–288. https://doi.org/10.1016/j.ijpharm.2016.04.018

Liu Y, Friesen JB, McAlpine JB, Lankin DC, Chen SN, Pauli GF (2018a) Natural deep eutectic solvents: properties, applications, and perspectives. J Nat Prod 81:679–690. https://doi.org/10.1021/acs.jnatprod.7b00945

Liu Y, Zhang Y, Chen SN, Friesen JB, Nikolić D, Choules MP, McAlpine JB, Lankin DC, Gemeinhart RA, Pauli GF (2018b) The influence of natural deep eutectic solvents on bioactive natural products: studying interactions between a hydrogel model and *Schisandra chinensis* metabolites. Fitoterapia 127:212–219. https://doi.org/10.1016/j.fitote.2018.02.024

Lowrie A, Jones MJ, Eastley RJ (1989) Effect of a eutectic mixture of local anaesthetic (EMLA) on tourniquet pain in volunteers. Br J Anaesth 63:751–753. https://doi.org/10.1093/bja/63.6.751

Mann SK, Pham TN, Mcqueen LL, Lewandowski JR, Brown SP (2019) Revealing intermolecular hydrogen bonding structure and dynamics in a deep eutectic pharmaceutical by magic-angle spinning NMR spectroscopy. Mol Pharm. https://doi.org/10.1021/acs.molpharmaceut.9b01075

Mano F, Martins M, Sá-Nogueira I, Barreiros S, Borges JP, Reis RL, Duarte ARC, Paiva A (2016) Production of electrospun fast-dissolving drug delivery systems with therapeutic eutectic systems encapsulated in gelatin. AAPS Pharm Sci Tech 18:2579–2585. https://doi.org/10.1208/s12249-016-0703-z

Martins MAR, Pinho SP, Coutinho JAP (2018) Insights into the nature of eutectic and deep eutectic mixtures. J Solut Chem. https://doi.org/10.1007/s10953-018-0793-1

Mbous YP, Hayyan M, Hayyan A, Wong WF, Hashim MA, Looi CY (2017a) Applications of deep eutectic solvents in biotechnology and bioengineering - promises and challenges. Biotechnol Adv 35:105–134. https://doi.org/10.1016/j.biotechadv.2016.11.006

Mbous YP, Hayyan M, Wong WF, Looi CY, Hashim MA (2017b) Unraveling the cytotoxicity and metabolic pathways of binary natural deep eutectic solvent systems. Sci Rep 7:41257–41271. https://doi.org/10.1038/srep41257

Meneses L, Santos F, Gameiro AR, Paiva A, Duarte ARC (2019) Preparation of binary and ternary deep eutectic systems. JoVE 152:1–5. https://doi.org/10.3791/60326

Morrison HG, Sun CC, Neervannan S (2009) Characterization of thermal behavior of deep eutectic solvents and their potential as drug solubilization vehicles. Int J Pharm 378:136–139. https://doi.org/10.1016/j.ijpharm.2009.05.039

Owczarek K, Szczepanska N, Plotka-Wasylka J, Rutkowska M, Shyshchak O, Bratychak M, Namiesnik J (2016) Natural deep eutectic solvents in extraction process. Ch&ChT 10:601–606. https://doi.org/10.23939/chcht10.04si.601

Paiva A, Craveiro R, Aroso I, Martins M, Reis RL, Duarte ARC (2014) Natural deep eutectic solvents–solvents for the 21st century. ACS Sustain Chem Eng 2:1063–1071. https://doi.org/10.1021/sc500096j

Patel RD, Raval MK, Bagathariya AA, Sheth NR (2019) Functionality improvement of nimesulide by eutectic formation with nicotinamide: exploration using temperature-composition phase diagram. Adv Powder Technol 30:961–973. https://doi.org/10.1016/j.apt.2019.02.010

Pedro SN, Freire MG, Freire CSR, Silvestre AJD (2019) Deep eutectic solvents comprising active pharmaceutical ingredients in the development of drug delivery systems. Expert Opin Drug Deliv 16:407–506. https://doi.org/10.1080/17425247.2019.1604680

Pena-Pereira F, Namieśnik J (2014) Ionic liquids and deep eutectic mixtures: sustainable solvents for extraction processes. ChemSusChem 7:1784–1800. https://doi.org/10.1002/cssc.201301192

Pereira CV, Silva JM, Rodrigues L, Reis RL, Paiva A, Duarte ARC, Matias A (2019) Unveil the anticancer potential of limomene based therapeutic deep eutectic solvents. Sci Rep 9:1–11. https://doi.org/10.1038/s41598-019-51472-7

Phaechamud T, Tuntarawongsa S, Charoensuksai P (2015) Evaporation behavior and characterization of eutectic solvent and ibuprofen eutectic solution. AAPS Pharm Sci Tech 17:1213–1220. https://doi.org/10.1208/s12249-015-0459-x

Prista LN, Alves AC, Morgado R, Lobo JS (2008) Tecnologia Farmacêutica – volume I. In: *Fundação Calouste Gulbenkian*, 7th edn ISBN: 9789723109757

Reverchon E, Adami R, Cardea S, Porta GD (2009) Supercritical fluids processing of polymers for pharmaceutical and medical applications. J Supercrit Fluids 47:484–492. https://doi.org/10.1016/j.supflu.2008.10.001

Roda A, Matias AA, Paiva A, Duarte ARC (2019) Polymer science and engineering using deep eutectic solvents. Polymers 11:1–22. https://doi.org/10.3390/polym11050912

Roda A, Santos F, Matias AA, Paiva A, Duarte ARC (2020) Design and processing of drug delivery formulations of therapeutic deep eutectic systems for tuberculosis. J Supercrit Fluids 161:104826. https://doi.org/10.1016/j.supflu.2020.104826

Ruß C, König B (2012) Low melting mixtures in organic synthesis - an alternative to ionic liquids? Green Chem 14:2969–2982. https://doi.org/10.1039/c2gc36005e

Santana APR, Mora-Vargas JA, Guimarães TGS, Amaral CDB, Oliveira A, Gonzalez MH (2019) Sustainable synthesis of natural deep eutectic solvents (NADES) by different methods. J Mol Liq 293:1–6. https://doi.org/10.1016/j.molliq.2019.111452

Santos F, Leitão MIPS, Duarte ARC (2019) Properties of therapeutic deep eutectic solvents of L-arginine and ethambutol for tuberculosis treatment. Molecules 24:1–12. https://doi.org/10.3390/molecules24010055

Savjani KT, Gajjar AK, Savjani JK (2012) Drug solubility: importance and enhancement techniques. ISRN Pharm 2012:1–10. https://doi.org/10.5402/2012/195727

Serrano MC, Gutiérrez MC, Jiménez R, Ferrer ML, Del Monte F (2012) Synthesis of novel lidocaine-releasing poly(diol-co-citrate) elastomers by using deep eutectic solvents. Chem Commun 48:579–581. https://doi.org/10.1039/c1cc15284j

Shekaari H, Zafarani-Moattar MT, Shayanfar A, Mokhtarpour M (2018) Effect of choline chloride/ethylene glycol or glycerol as deep eutectic solvents on the solubility and thermodynamic properties of acetaminophen. J Mol Liq 249:1222–1235. https://doi.org/10.1016/j.molliq.2017.11.057

Shen Q, Li X, Li W, Zhao X (2011) Enhanced intestinal absorption of daidzein by borneol/menthol eutectic mixture and microemulsion. AAPS Pharm Sci Tech 12:1044–1049. https://doi.org/10.1208/s12249-011-9672-4

Silva JM, Reis RL, Paiva A, Duarte ARC (2018) Design of functional therapeutic deep eutectic solvents based on choline chloride and ascorbic acid. ACSS Sustain Chem Eng 6:10355–10363. https://doi.org/10.1021/acssuschemeng.8b01687

Silva JM, Akkache S, Araújo AC, Masmoudi Y, Reis RL, Badens E, Duarte ARC (2019a) Development of innovative medical devices by dispersing fatty acid eutectic blend on gauzes using supercritical particle generation processes. Mater Sci Eng C 99:599–610. https://doi.org/10.1016/j.msec.2019.02.012

Silva JM, Pereira CV, Mano F, Silva E, Castro VIB, Sá-Nogueira I, Reis RL, Paiva A, Matias A, Duarte ARC (2019b) Therapeutic role of deep eutectic solvents based on menthol and saturated fatty acids on wound healing. ACS Appl Bio Mater 2:4346–4355. https://doi.org/10.1021/acsabm.9b00598

Silva JM, Silva E, Reis RL, Duarte ARC (2019c) A closer look in the antimicrobial properties of deep eutectic solvents based on fatty acids. Sustain Chem Pharm 14:1–9. https://doi.org/10.1016/j.scp.2019.100192

Sivapragasam M, Moniruzzaman M, Goto M (2016) Recent advances in exploiting ionic liquids for biomolecules: solubility, stability and applications. Biotechnol J 11:1000–1013. https://doi.org/10.1002/biot.201500603

Stott PW, Williams AC, Barry BW (1998) Transdermal delivery from eutectic systems: enhanced permeation of a model drug, ibuprofen. J Control Release 50:297–308. https://doi.org/10.1016/S0168-3659(97)00153-3

Tuntarawongsa S, Phaechamud T (2012) Polymeric eutectic system. Adv Mater Res 528:180–183. https://doi.org/10.4028/www.scientific.net/AMR.528.180

Vanda H, Dai Y, Wilson EG, Verpoorte R, Choi YH (2018) Green solvents from ionic liquids and deep eutectic solvents to natural deep eutectic solvents. C R Chim 21:628–638. https://doi.org/10.1016/j.crci.2018.04.002

Varma MVS, Khandavilli S, Ashokraj Y, Jain A, Dhanikula A, Sood A, Thomas NS, Pillai O, Sharma P, Gandhi R, Agrawal S, Nair V, Panchagnula R (2004) Biopharmaceutic classification system: a scientific framework for pharmacokinetic optimization in drug research. Curr Drug Metab 5:375–388. https://doi.org/10.2174/1389200043335423

Warner JC, Cannon AS, Dye KM (2004) Green chemistry. Environ Impact Assess Rev 24:775–799. https://doi.org/10.1016/j.eiar.2004.06.006

Welton T (2015) Solvents and sustainable chemistry. Proc R Soc Lond A 471:20150502. https://doi.org/10.1098/rspa.2015.0502

Wojnarowska Z, Smolka W, Zotova J, Knapik-Kowalczuk J, Sherif A, Tajber L, Paluch M (2018) The effect of electrostatic interactions on the formation of pharmaceutical eutectics. Phys Chem Chem Phys 20:27361–27367. https://doi.org/10.1039/c8cp05905e

Woolfson AD, Malcolm RK, Campbell K, Jones DS, Russell JA (2000) Rheological, mechanical and membrane penetration properties of novel dual drug systems for percutaneous delivery. J Control Release 67:395–408. https://doi.org/10.1016/S0168-3659(00)00230-3

Xin R, Qi S, Zeng C, Khan FI, Yang B, Wang Y (2017) A functional natural deep eutectic solvent based on trehalose: structural and physicochemical properties. Food Chem 217:560–567. https://doi.org/10.1016/j.foodchem.2016.09.012

Yang Z (2018) Natural deep eutectic solvents and their applications in biotechnology. Adv Biochem Eng Biotechnol. https://doi.org/10.1007/10_2018_67

Yin L, Jiang S (2018) Evaluation of EMLA cream for relieving pain during needle insertion on totally implantable venous access device. J Vascular Access 19:634–638. https://doi.org/10.1177/1129729818768182

Zainal-Abidin MH, Hayyan M, Ngoh GC, Wong WF, Looi CY (2019) Emerging frontiers of deep eutectic solvents in drug discovery and drug delivery systems. J Control Release 316:168–195. https://doi.org/10.1016/j.jconrel.2019.09.019

Zainal-Abidin MH, Hayyan M, Ngoh GC, Wong WF (2019) Doxorubicin loading on functional graphene as a promising nanocarrier using ternary deep eutectic solvent systems. ACS Omega 5:1656–1668. https://doi.org/10.1021/acsomega.9b03709

Zakrewsky M, Banerjee A, Apte S, Kern TL, Jones MR, Sesto RED, Koppisch AT, Fox DT, Mitragotri S (2016) Choline and geranate deep eutectic solvent as a broad-spectrum antiseptic agent for preventive and therapeutic applications. Adv Healthc Mater 5:1282–1289. https://doi.org/10.1002/adhm.201600086

Chapter 4
Solubility of Gases in Deep Eutectic Solvents

Leila Moura, Laura Kollau, and Margarida Costa Gomes

Contents

Abstract Deep eutectic solvents are promising alternative media for more sustainable chemistry and chemical engineering. The way they dissolve gases provides valuable information about solvation mechanisms but also points the way to new applications of these interesting liquid mixtures. Here we review the solubility of 10 permanent gases in 87 deep eutectic solvents: carbon dioxide (CO_2), carbon monoxide (CO), nitric oxide (NO), nitrogen dioxide (NO_2), hydrogen (H_2), hydrogen sulfide (H_2S), sulfur dioxide (SO_2), nitrogen (N_2), ammonia (NH_3), and methane (CH_4). The gas solubility studies were performed using either saturation or gravimetric techniques. Carbon dioxide is the most studied gas so far, followed by SO_2 and NH_3. Although for some gases the absorption process is only physical, e.g., N_2, CO, and CH_4, for others it was reported to be both physical and chemical, e.g., SO_2, NO, and CO_2. The solubility of the most commonly studied gases varies in the order $SO_2 > NO > CO_2 > H_2S$, the larger absorption for SO_2 being reported for the common choline chloride:urea mixtures with Henry's law constant lower than 0.18 bar at 293 K. The larger solubilities for NO, H_2S, and CO_2 were measured for mixtures based in tetraalkylphosphonium salts, with Henry's law constants typically below

L. Moura
School of Chemistry and Chemical Engineering, QUILL, Queen's University Belfast, Belfast, UK
e-mail: L.Moura@qub.ac.uk

L. Kollau · M. C. Gomes (✉)
Chemistry Laboratory ENS Lyon, CNRS & Lyon University, Lyon, France
e-mail: laura.kollau@ens-lyon.fr; margarida.costa-gomes@ens-lyon.fr

S. Fourmentin et al. (eds.), *Deep Eutectic Solvents for Medicine, Gas Solubilization and Extraction of Natural Substances*, Environmental Chemistry for a Sustainable World 56, https://doi.org/10.1007/978-3-030-53069-3_4

0.20 bar at 303 K. Eutectic mixtures containing amines are also reported as excellent absorbers of CO_2, this behavior being probably explained by a chemical reaction between the gas and these components of the eutectic mixture.

Keywords Deep eutectic solvent · Gas solubility · Carbon dioxide · Sulfur dioxide · Ammonia · Carbon monoxide · Nitric oxide · Nitrogen dioxide · Hydrogen · Hydrogen sulfide · Nitrogen · Methane

Abbreviations

[AcCh]Cl	Acetylcholine chloride
[ATPP]Br	Allyltriphenylphosphonium bromide
[BDHA]Cl	Benzyldimethyl(2-hydroxyethyl)ammonium chloride
[BMIm][$MeSO_4$]	1-Butyl-3-methylimidazolium methylsulfate
[BMIm]Cl	1-Butyl-3-methylimidazolium chloride
[BTEA]Cl	Benzyltriethylammonium chloride
[BTMA]Cl	Benzyltrimethylammonium chloride
[BTPP]Cl	Benzyltriphenylphosphonium chloride
[BuTPP]Br	N-Butyltriphenylphosphonium bromide
[Ch]Cl	Choline chloride
[EMIm]Cl	1-Ethyl-3-methylimidazolium chloride
[HMIm]Cl	1-Hexyl-3-methylimidazolium chloride
[MEA]Cl	Monoethanolammonium chloride
[MIm]Cl	1-Methylimidazolium chloride
[MTPP]Br	Methyltriphenylphosphonium bromide
[N_{1111}]Cl	Tetramethylammonium chloride
[N_{1888}]Br	Methyltrioctylammonium bromide
[N_{1888}]Cl	Trioctylmethylammonium chloride
[N_{2221}]Cl	Triethylmethylammonium chloride
[N_{2222}][Im]	Tetraethylammonium imidazolate
[N_{2222}][TrZ]	Tetraethylammonium 1,2,4-triazolate
[N_{2222}]Br	Tetraethylammonium bromide
[N_{2222}]Cl	Tetraethylammonium chloride
[N_{4444}]Br	Tetrabutylammonium bromide
[N_{4444}]Cl	Tetrabutylammonium chloride
[N_{8888}]Br	Tetraoctylammonium bromide
[N_{8888}]Cl	Tetraoctylammonium chloride
[NH_4][SCN]	Ammonium thiocyanate
[P_{2222}][Im]	Tetraethylphosphonium imidazolate
[P_{2222}][TrZ]	Tetraethylphosphonium 1,2,4-triazolate
[P_{4444}]Br	Tetrabutylphosphonium bromide
[P_{4444}]Cl	Tetrabutylphosphonium chloride

$[P_{4444}]F$	Tetrabutylammonium fluoride
[Pip]Br	1-Hydroxyethyl-1,4-dimethyl-piperazinium bromide
1,2-PrOH	1,2-Propanediol
1,4-BuOH	1,4-Butanediol
2,3-BuOH	2,3-Butanediol
Ac	Acetamide
AcH	Acetic acid
Al	Alanine
AMP	Aminomethyl propanol
AMPD	2-Amino-2-methyl-1,3-propanediol
AP	3-Amino-1-propanol
Arg	L-Arginine
Bet	Betaine
BzH	Benzoic acid
BzIm	Benzimidazole
Cap	Caprolactam
Car	Cardanol
Cat	Catechol
DBN	1,5-Diazabicyclo[4.3.0]-non-5-ene
DBU	1,8-Diazabicyclo[5.4.0]undec-7-ene
DEA	Diethanolamine
DEA-HCl	Diethylamine hydrochloride
DecH	Decanoic acid
DEG	Diethylene glycol
DETA	Diethylenetriamine
EA	Ethanolamine
EA-HCl	Ethylamine hydrochloride
EDA	Ethylenediamine
EG	Ethylene glycol
FAc	Trifluoroacetamide
Form	Formic acid
Fru	d-Fructose
FurH	Furoic acid
FurOH	Furfuryl alcohol
Gly	Glycerol
Gua	Guaiacol
Guan-CO_3	Guanidine carbonate
Guan-HCl	Guanidinium hydrochloride
Im	Imidazole
K[SCN]	Potassium thiocyanate
K_2CO_3	Potassium carbonate
Lac	Lactic acid
Lev	Levulinic acid
Mal	dl-Malic acid
Mal	Malic acid

Malo	Malonic acid
MDEA	Methyldiethanolamine
Me2U	1,3-Dimethylthiourea
MeIm	4-Methylimidazole
MeU	N-Methylurea
MTBD	7-Methyl-1,5,7-triazabicyclo[4.4.0]dec-5-ene
OcH	Octanoic acid
oTolH	o-Toluic acid
Ox	Oxalic acid
PEG200	Polyethylene glycol 200
PEHA	Pentaethylenehexamine
Phe	Phenol
Pheac	Phenylacetic acid
Pro	Proline
ProH	Propionic acid
Pyr	Pyrazole
Res	Resorcinol
Suc	Succinonitrile
TAE-HCl	Thioacetamide hydrochloride
TBD	1,5,7-Triazabicyclo[4.4.0]-dec-5-ene
TEA-HCl	Triethanolamine hydrochloride
TEG	Triethylene glycol
TEPA	Tetraethylenepentamine
Tet	Tetrazole
TETA-HCl	Triethylenetetramine hydrochloride
Thio2U	1,3-Dimethylurea
ThioU	Thiourea
TMGua	Tetramethylguanidine
Tri	1,2,4-Triazole
TriCN	3-Cyano-1,2,4-triazole
$TriNH_2$	3-Amino-1,2,4-triazole
TrZ	Triazole
U	Urea
U-HCl	Urea hydrochloride

4.1 Introduction

Studying solubility is a useful way to assess molecular interactions in solution and to obtain significant information on the microscopic structure of the liquid solvents. In the case of new, complex solvents, like mixtures showing deep eutectics, the study of the solubility of simple solutes, that are gases at room conditions, can constitute a starting point to the understanding of solvation in these media.

Besides, the knowledge of solubility data is essential to calculate phase equilibria for the design of chemical engineering processes and of course to determine the fate of substances in the environment. The scientific and technological aspects are often associated with each other, for example, in the search for new strategies to improve the choice of novel solvents or separation media for chemical reactions or industrial processes. In a society demanding less hazardous and more efficient chemistry, the choice of alternative solvents (acceptable both from an economical and environmental point of view) for reactions or separations is regarded as one of the promising ways for sustainable development.

Mixtures formed by salts and molecular compounds that can act as hydrogen bond donors may often lead to non-ideal binary mixtures that form eutectics, with melting temperatures significantly lower than that of the predictions for their ideal eutectic mixture. These mixtures are often designated, after the pioneer work by Abbott et al. (2003, 2004), as deep eutectic solvents, or DES, and are considered as promising liquids with low environmental impact for different applications. Amongst these applications, electroplating, biomass conversion, atmospheric air remediation, liquid-liquid extraction, and gas separation using deep eutectic solvents are considered as promising alternatives to traditional approaches based on conventional solvents.

In this chapter, we report the solubility of ten solutes, which are gaseous at room temperature, in different deep eutectic solvents, considered as such by authors of the publications mentioned. The gases reported include carbon dioxide (CO_2), carbon monoxide (CO), nitric oxide (NO), nitrogen dioxide (NO_2), hydrogen (H_2), hydrogen sulfide (H_2S), sulfur dioxide (SO_2), nitrogen (N_2), ammonia (NH_3), and methane (CH_4). Although some studies report the use of deep eutectic solvents for olefin/paraffin gas separations (Jiang et al. 2017a, b, c; Deng et al. 2017b; Dou et al. 2018), to our knowledge, the solubility of hydrocarbon gases other than methane has not been studied in eutectic mixtures. Eighty-seven different eutectic mixtures were studied with the ten different gases as reported in Tables 4.1 and 4.2. Carbon dioxide is, by far, the most studied gas in deep eutectic solvents with more than 50 published papers dealing with the experimental determination of the solubility at different conditions of temperature and partial pressure of gas. Sulfur dioxide and ammonia were also studied in 40 and 30 eutectic mixtures, respectively. All the other seven gases are much less studied in deep eutectic solvents.

The majority of the studies so far consider that the gases NH_3, H_2S, CO, CH_4, NO_2, and N_2 dissolve in the eutectic mixtures. The absorption capacity of deep eutectic solvents for the different gases varies greatly, and values of K_H varying between 367.7×10^5 Pa for CH_4 in [Ch]Cl:U (1:1.5) at 353.2 K (Liu et al. 2019) and 0.0167×10^5 Pa for NO in [P_{4444}]Br:EG (1:50) at 323.15 K (Dou et al. 2019) were found. One of the reasons for this major variety of Henry's law constants is the fact that the absorption process for some gases can be only a physical process (e.g., N_2 (Wu et al. 2015), CO (Wu et al. 2015), CH_4 (Wu et al. 2015; Liu et al. 2019)) while for others it might be both physical and chemical (as reported for SO_2 (Zhao et al. 2018; Chen et al. 2018; Cui et al. 2019a) in [BMIm]Cl:Ac (1:1) (Zhao et al. 2018) and Pip:Gly (1:6) (Cui et al. 2019a), NO (Sun et al. 2017a; Dou et al. 2019) in

Table 4.1 CO_2, CO, NO, NO_2, and H_2 gas solubility in each deep eutectic solvent

	CO_2	CO	NO	NO_2	H_2
[ATPP]Br:DEG	Ghaedi et al. (2017a)				
[ATPP]Br:TEG	Ghaedi et al. (2017a)				
[BMIm][$MeSO_4$]:U	Akhmetshina et al. (2018)				
[MEA]Cl:EDA	Trivedi et al. (2016) and Shukla and Mikkola (2018, 2019)				
[MEA]Cl:DETA	Shukla and Mikkola (2018)				
[MEA]Cl:TEPA	Shukla and Mikkola (2018)				
[MEA]Cl:PEHA	Shukla and Mikkola (2018)				
[MIm]Cl:EDA	Shukla and Mikkola (2018)				
[MIm]Cl:DETA	Shukla and Mikkola (2018)				
[MIm]Cl:TEPA	Shukla and Mikkola (2018)				
[MIm]Cl:PEHA	Shukla and Mikkola (2018)				
[MIm]Cl:AP	Shukla and Mikkola (2018)				
[N_{4444}]Br:AP	Shukla and Mikkola (2018)				
[N_{4444}]Br:AMP	Shukla and Mikkola (2018)				
[MIm]Cl:EDA	Shukla and Mikkola (2019)				
[N_{2222}]Br:Lev	Deng et al. (2015, 2016)				
[N_{2222}]Cl:Lev	Deng et al. (2015, 2016)				
[N_{4444}]Br:Lev	Deng et al. (2015, 2016)				
[N_{4444}]Br:$Thio_2$U			Sn et al. (2017b)		
[N_{4444}]Cl:Lev	Deng et al. (2015, 2016)				
[N_{4444}]Cl:$Thio_2$U			Sun et al. (2017b)		
[N_{4444}]Br:AcH	Sarmad et al. (2016) and Ma et al. (2017)				
[N_{4444}]Br:EA	Sarmad et al. (2016), Ali et al. (2016), Ma et al. (2017) and Li et al. (2019)				
[N_{4444}]Cl:EA	Li et al. (2019)				
[N_{4444}]Br:EG	Haider et al. (2018)				
[N_{4444}]Br:DEG	Haider et al. (2018)				
[N_{4444}]Br:DEA	Ali et al. (2016), Ma et al. (2017) and Haider et al. (2018)				
[N_{4444}]Br:MDEA	Haider et al. (2018)				
[N_{4444}]Br:TEA	Ali et al. (2016) and Li et al. (2019)				
[N_{2222}]Br:TEA	Li et al. (2019)				
[N_{2222}]Br:OcH	Sarmad et al. (2016)				
[N_{2222}]Cl:AcH	Sarmad et al. (2016)				
[N_{4444}]Cl:EA	Ma et al. (2017)				
[N_{2221}]Cl:AcH	Ma et al. (2017)				
[N_{1111}]Cl:AcH	Sarmad et al. (2016) and Ma et al. (2017)				
[N_{1111}]Cl:Phe	Ji et al. (2016)				

(continued)

Table 4.1 (continued)

	CO_2	CO	NO	NO_2	H_2
[N_{2222}]Cl:Phe	Ji et al. (2016)				
[N_{3333}]Cl:AcH	Sarmad et al. (2016) and Ma et al. (2017)				
[N_{3333}]Cl:EA	Sarmad et al. (2016) and Ma et al. (2017)				
[N_{2221}]Cl:AcH	Sarmad et al. (2016)				
[N_{1111}]Cl:Lac	Zubeir et al. (2014)				
[N_{2222}]Cl:Lac	Zubeir et al. (2014)				
[N_{4444}]Cl:Lac	Zubeir et al. (2014)				
[N_{2221}]Cl:EG	Sarmad et al. (2016)				
[N_{2221}]Cl:Lac	Sarmad et al. (2016)				
[N_{2221}]Cl:Lev	Sarmad et al. (2016)				
[N_{2221}]Cl:Gly	Sarmad et al. (2016)				
[N_{1888}]Br:DecH	Zubeir et al. (2018)				
[N_{1888}]Cl:DecH	Zubeir et al. (2018)				
[N_{4444}]Cl:DecH	Zubeir et al. (2018)				
[N_{8888}]Cl:DecH	Zubeir et al. (2018)				
[N_{8888}]Br:DecH	Zubeir et al. (2018)				
[N_{8888}]Cl:DecH	Zubeir et al. (2018)				
[P_{2222}][Im]:EG	Cui et al. (2019b)				
[P_{2222}][TrZ]:EG	Cui et al. (2019b)				
[N_{2222}][Im]:EG	Cui et al. (2019b)				
[N_{2222}][TrZ]:EG	Cui et al. (2019b)				
[N_{4444}] Cl:Gly:TMGua	Huang et al. (2017)				
[N_{4444}] Cl:DecH:TMGua	Huang et al. (2017)				
[P_{4444}]Br:EG			Dou et al. (2019)		
[P_{4444}]Br:Me_2U			Jiang et al. (2016)		
[P_{4444}]Br:$Thio_2U$			Sun et al. (2017b)		
[P_{4444}]Cl:$Thio_2U$			Sun et al. (2017b)		
[P_{4444}]F:Cap			Duan et al. (2011)		
[AcCh]Cl:Im	Li et al. (2018)				
[AcCh]Cl:Lev	Deng et al. (2015, 2016)				
[AcCh]Cl:Tri	Li et al. (2018)				
[AcCh]Cl:Gua	Liu et al. (2017)				

(continued)

Table 4.1 (continued)

	CO_2	CO	NO	NO_2	H_2
Al:Lac	Altamash et al. (2018)				
Al:Mal	Altamash et al. (2018)				
[ATPP]Br:DEG	Ghaedi et al. et al. (2017b)				
[ATPP]Br:TEG	Ghaedi et al. (2017b, c)				
Arg:Gly	Ren et al. (2018)				
[BDHA]Cl:AcH	Sarmad et al. (2016)				
[BDHA]Cl:Lac	Sarmad et al. (2016)				
[BDHA]Cl:Gly	Sarmad et al. (2016)				
Bet:Lac	Altamash et al. (2018)				
Bet:Mal	Altamash et al. (2018)				
[BTEA]Cl:AcH	Sarmad et al. (2016) and Ma et al. (2017)				
[BTMA]Cl:AcH	Sarmad et al. (2016) and Ma et al. (2017)				
[BTMA]Cl:Gly	Sarmad et al. (2016)				
[BTPP]Cl:Gly	Ali et al. (2016)				
[BuTPP]Br:EG	Ali et al. (2016)				
[MTPP]Br:EA	Ali et al. (2016)				
Guan-HCl:EA	Ma et al. (2017)				
[Ch]Cl:1,4-BuOH	Chen et al. (2014)				
[Ch]Cl:2,3-BuOH	Chen et al. (2014)				
[Ch]Cl:1,2-PrOH	Chen et al. (2014)				
[Ch]Cl:Lev	Ullah et al. (2015), Deng et al. (2015, 2016) and Lu et al. (2015)				
[Ch]Cl:Car	Liu et al. (2018)				
[Ch]Cl:EG	Ali et al. (2014), Mirza et al. (2015), Bhawna and Pandey (2017), Ma et al. (2017) and Haider et al. (2018)				
[Ch]Cl:FurH	Lu et al. (2015)				
[Ch]Cl:Gly	Ali et al. (2014), Ma et al. (2017)				
[Ch]Cl:Gly:AcH	Sarmad et al. (2016)				
[Ch]Cl:Gly:Arg	Chemat et al. (2016)				
[Ch]Cl:Gua	Liu et al. (2017, 2018)				
[Ch]Cl:Lac	Francisco et al. (2013) and Ma et al. (2017))				
[Ch]Cl:Mal	Mirza et al. (2015)				
[Ch]Cl:EA	Adeyemi et al. (2017a, b), Li et al. (2019), Ali et al. (2014), Sarmad et al. (2016), Bhawna and Pandey (2017) and Ma et al. (2017)				
[Ch] Cl:Pro:PEG200	Li et al. (2008b)				

(continued)

Table 4.1 (continued)

	CO_2	CO	NO	NO_2	H_2
[Ch]Cl:DEA	Ali et al. (2014), Adeyemi et al. (2017b), Ma et al. (2017), Haider et al. (2018) and Li et al. (2019))				
[Ch]Cl:TEA	Ali et al. (2014) and Li et al. (2019)				
[Ch]Cl:DEG	Chen et al. (2014), Ali et al. (2014) and Haider et al. (2018)				
[Ch]Cl:TEG	Chen et al. (2014)				
[Ch]Cl:MDEA	Adeyemi et al. (2017b), Haider et al. (2018) and Li et al. (2019)				
[Ch]Cl:Phe	Ji et al. (2016)				
[Ch]Cl:Pheac	Altamash et al. (2017)				
[Ch]Cl:Res:Gly	Li et al. (2017)				
[Ch]Cl:U	Li et al. (2008a), Su et al. (2009), Xie et al. (2013, 2014, 2016), Ali et al. (2014), Mirza et al. (2015), Bhawna and Pandey (2017), Ma et al. (2017) and Liu et al. (2019)	Xie et al. (2016)			Xie et al. (2016)
[Ch]Cl:U:TBD	Bhawna and Pandey (2017)				
[Ch]Cl:U:DBU	Bhawna and Pandey (2017)				
[Ch]Cl:U:DBN	Bhawna and Pandey (2017)				
[Ch]Cl:U:TBD:Gly	Bhawna and Pandey (2017)				
[Ch]Cl:U:DBU:Gly	Bhawna and Pandey (2017)				
[Ch]Cl:U:DBN:Gly	Bhawna and Pandey (2017)				
[Ch]Cl:EG:TBD	Bhawna and Pandey (2017)				
[Ch]Cl:EG:DBU	Bhawna and Pandey (2017)				
[Ch]Cl:EG:DBN	Bhawna and Pandey (2017)				
[Ch]Cl:EG:TBD:Gly	Bhawna and Pandey (2017)				
[Ch]Cl:EG:DBU:Gly	Bhawna and Pandey (2017)				
[Ch]Cl:EG:DBN:Gly	Bhawna and Pandey (2017)				
[Ch]Cl:EA:TBD	Bhawna and Pandey (2017)				
[Ch]Cl:EA:DBU	Bhawna and Pandey (2017)				
[Ch]Cl:EA:DBN	Bhawna and Pandey (2017)				
[Ch]Cl:EA:TBD:Gly	Bhawna and Pandey (2017)				
[Ch]Cl:EA:DBU:Gly	Bhawna and Pandey (2017)				
[Ch]Cl:EA:DBN:Gly	Bhawna and Pandey (2017)				
[Ch]Cl:Gly:MTBD	Sze et al. (2014)				
[Ch]Cl:Gly:DBU	Sze et al. (2014)				
[Ch]Cl:Gly:DBN	Sze et al. (2014)				

(continued)

Table 4.1 (continued)

	CO_2	CO	NO	NO_2	H_2
[Ch]Cl:Gly:TMGua	Huang et al. (2017)				
EH:Gly	Jiang et al. (2019)				
DEA-HCl:Gua	Liu et al. (2017)				
Gly:TMGua	Huang et al. (2017)				
Gly:DBU	Huang et al. (2017)				
Gly:DBN	Huang et al. (2017)				
Gly:EA-HCl	Jiang et al. (2019)				
Guan-HCL:EA	Sarmad et al. (2016)				
Guan-CO_3:Mal:EG	Mirza et al. (2017a, b)				
Guan-CO_3:Mal:EG:Ar	Mirza et al. (2017a, b)				
K_2CO_3:Gly	Ghaedi et al. (2018)				
K_2CO_3:EG	Ghaedi et al. (2018)				
K_2CO_3:Gly:AMPD	Ghaedi et al. (2018)				
K_2CO_3:EG:AMPD	Ghaedi et al. (2018)				
[MTPP]Br:AcH	Sarmad et al. (2016) and Ma et al. (2017)				
[MTPP]Br:1,2-PrOH	Sarmad et al. (2016)				
[MTPP]Br:EG	Sarmad et al. (2016)				
[MTPP]Br:Gly	Sarmad et al. (2016)				
[MTPP]Br:Lev	Sarmad et al. (2016)				
[MTPP]Br:AcH:Lev	Sarmad et al. (2016) and Ma et al. (2017)				
[MTPP]Br:EA	Ma et al. (2017)				
[MTPP]Br:DEA	Ma et al. (2017)				
TAE-HCl:EDA	Trivedi et al. (2016)				
TEA-HCl:EDA	Trivedi et al. (2016)				
TETA-HCl:EG	Zhang et al. (2018)				
TETA-HCl:TEG	Zhang et al. (2018)				
U-HCl:EDA	Trivedi et al. (2016)				

$[P_{4444}]$Cl:Me_2U (1:1) (Sun et al. 2017b) and $[N_{4444}]$Br:EG (1:50) (Dou et al. 2019), and CO_2 through carbamate formation (Ali et al. 2014; Adeyemi et al. 2017a, b) or calcium carbonate production (Karimi et al. 2018)). Some of the studies concerning SO_2 were carried out in simulated flue gas mixtures where the presence of N_2 is not explicitly taken into account as it is just considered as a filler gas (Deng et al. 2019a). Although in the cases where the gas reacts with the liquid absorber a value for Henry's law constant cannot be calculated, we have included in this report the values of K_H even for reactive gases when published as such by the different authors.

In Fig. 4.1, Henry's law constants, K_H, of different gases in [Ch]Cl:U (1:2) are presented. Hydrogen exhibits the lowest solubility (represented by the largest Henry's law constant) followed by carbon monoxide, nitrogen, and methane.

Table 4.2 H_2S, SO_2, N_2, NH_3, and CH_4 gas solubility in each deep eutectic solvent

	H_2S	SO_2	N_2	NH_3	CH_4
[BMIm][MeSO$_4$]:U	Akhmetshina et al. (2018)				
[BMIm]Cl:Ac		Liu et al. (2013a)			
[BMIm]Cl:Im		Chen et al. (2018)			
[BMIm]Cl:MeIm		Chen et al. (2018)			
[EMIm]Cl:Ac		Liu et al. (2013a)			
[EMIm]Cl:EG		Yang et al. (2017)			
[EMIm]Cl:Suc		Yang et al. (2019)			
[EMIm]Cl:TEG		Yang et al. (2018)			
[HMIm]Cl:Ac		Liu et al. (2013a)			
[N_{2222}]Br:Lev		Deng et al. (2015)			
[N_{2222}]Cl:Lev		Deng et al. (2015)			
[N_{4444}]Br:Im		Chen et al. (2018)			
[N_{4444}]Br:Lev		Deng et al. (2015)			
[N_{4444}]Cl:BzIm		Chen et al. (2018)			
[N_{4444}]Cl:Im		Chen et al. (2018)			
[N_{4444}]Cl:Lev		Deng et al. (2015)			
[N_{4444}]Cl:MeIm		Chen et al. (2018)			
[N_{4444}]Cl:Pyr		Chen et al. (2018)			
[N_{4444}]Cl:Tet		Chen et al. (2018)			
[P_{4444}]Br:Form	Wu et al. (2019)				
[P_{4444}]Br:ProH	Wu et al. (2019)				
[AcCh]Cl:Im		Deng et al. (2017a)			

(continued)

Table 4.2 (continued)

	H_2S	SO_2	N_2	NH_3	CH_4
[AcCh]Cl:Lev		Deng et al. (2015)			
[AcCh]Cl:Tri		Deng et al. (2017a,b)			
Al:Lac			Altamash et al. (2018)		Altamash et al. (2019)
Al:Mal			Altamash et al. (2018)		
Bet:Lac			Altamash et al. (2018)		Altamash et al. (2019)
Cap:Ac		Liu et al. (2013b)			
Cap:BzH		Liu et al. (2013b)			
Cap:FurH		Liu et al. (2013b)			
Cap:Im		Liu et al. (2013b)			
Cap:oTolH		Liu et al. (2013b)			
[Ch]Cl:Lev		Deng et al. (2015)			
[Ch]Cl:AcH	Wu et al. (2019)				
[Ch]Cl:Car		Liu et al. (2018)			
[Ch] Cl:Cat:Gly				Li et al. (2017)	
[Ch]Cl:EG		Sun et al. (2015)		Zhong et al. (2019c) and Duan et al. (2019)	
[Ch]Cl:FAc				Duan et al. (2019)	
[Ch]Cl:Form	Wu et al. (2019)				
[Ch] Cl:Fru:Gly				Li et al. (2017)	
[Ch]Cl:Gly		Yang et al. (2013)		Duan et al. (2019)	
[Ch] Cl:Gly:AcH					
[Ch] Cl:Gly:Arg					
[Ch]Cl:Gua		Liu et al. (2018)			

(continued)

Table 4.2 (continued)

	H_2S	SO_2	N_2	NH_3	CH_4
[Ch]Cl:Lac					Altamash et al. (2019)
[Ch]Cl:Lev:Gly				Li et al. (2017)	
[Ch]Cl:Mal		Sun et al. (2015)			Altamash et al. (2019)
[Ch]Cl:Mal:Gly				Li et al. (2017)	
[Ch]Cl:Malo:Gly				Li et al. (2017)	
[Ch]Cl:MeU				Duan et al. (2019)	
[Ch]Cl:Ox:Gly				Li et al. (2017)	
[Ch]Cl:Phe:EG				Li et al. (2017) and Zhong et al. (2019b)	
[Ch]Cl:Phe:Gly				Li et al. (2017)	
[Ch]Cl:Pheac					Altamash et al. (2019)
[Ch]Cl:Pheac:Gly				Li et al. (2017)	
[Ch]Cl:ProH	Wu et al. (2019)				
[Ch]Cl:Res:Gly				Li et al. (2017)	
[Ch]Cl:Tet:EG				Zhong et al. (2019c)	
[Ch]Cl:ThioU		Sun et al. (2015)			
[Ch]Cl:U	Liu et al. (2019)	Sun et al. (2015)	Xie et al. (2016)	Zhong et al. (2019a)	Xie et al. (2016) and Liu et al. (2019)
EH:Gly				Jiang et al. (2019)	
Im:Gly				Deng et al. (2019a)	
K[SCN]:Ac		Liu et al. (2013a)			
K[SCN]:Cap		Liu et al. (2013a)			
K[SCN]:Gly				Deng et al. (2019b)	

(continued)

Table 4.2 (continued)

	H_2S	SO_2	N_2	NH_3	CH_4
TriCN:Gly				Deng et al. (2019a)	
TriNH$_2$:Gly				Deng et al. (2019a)	

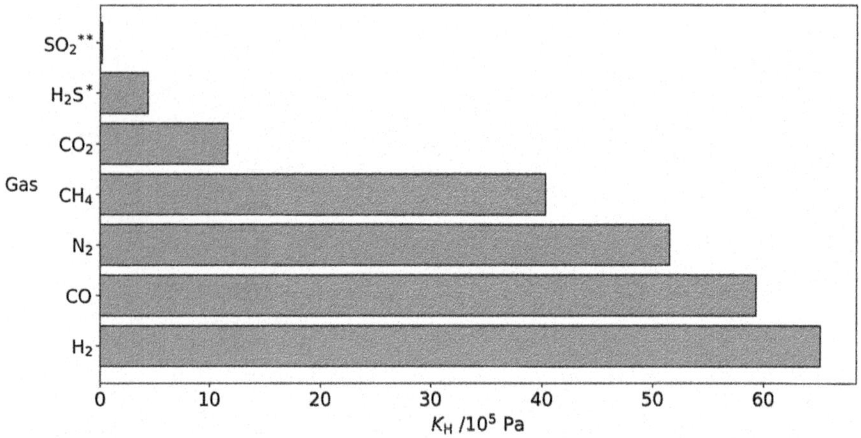

Fig. 4.1 Henry's law constant, K_H, for several gases in [Ch]Cl:U (1:2) (Wu et al. 2015; Sun et al. 2015; Liu et al. 2019) or at 308.2 K. * for values at 313.2 K; ** for values at 293 K. Hydrogen exhibits the lowest solubility followed by carbon monoxide, nitrogen, and methane. Carbon dioxide, hydrogen sulfide, and sulfur dioxide are the most soluble gases. These trends are still poorly understood

Carbon dioxide, hydrogen sulfide, and sulfur dioxide are the most soluble gases. The reasons for the observed trends are still poorly understood as they do not always follow the polarity or the size of the solute gases. Their understanding will surely contribute to a more generalized use of eutectic mixtures as solvents as well as to the design of task-specific deep eutectic solvents for a given application.

4.2 Experimental Methods

Both saturation and gravimetric techniques have been reported in the literature to study the gas absorption by deep eutectic solvents. In saturation methods, a precise quantity of gas is put in contact with a known amount of ionic liquid at a controlled temperature in a constant volume or a constant pressure cell. The gas solubility can be calculated at equilibrium conditions. When performed isochorically, the measurement method consists of accurately determining the volume of the cell and the volume of the vapor and liquid phases to calculate the quantity of gas remaining in the gas phase and, by difference, the amount of gas dissolved in the liquid. When measured isobarically, gas solubility is directly calculated by the difference of volume of the vapor phase before and after the gas absorption.

The mass of the solution in equilibrium with a given partial pressure of gas can also be used to determine the solubility in gravimetric methods. To measure the solubility of gases in deep eutectic solvents, two different gravimetric methods have been reported. One, somewhat less accurate, consists on the low pressure bubbling of gas in the liquid mixture that is weighed after equilibrium is reached. Other methodologies involve the use of gravimetric microbalances where a few milliliters of liquid are put in contact with the gas and the change in mass of the liquid phase is monitored as a function of the gas pressure at different temperatures. The amount of required sample is, in this case, much lower and the measurements are precise with relatively short equilibrium times. The disadvantages are linked with the necessity of having a very precise balance in order to be able to measure light or scarcely soluble gases. Furthermore, the accuracy of the gas absorption data obtained using a microbalance is considerably lower when the liquid samples have a non-negligible vapor pressure at the temperature of the measurements.

4.3 Gas Solubilities

Although the gas solubility data are reported in different units – mass ratio, molar ratio, molality, or mole fraction – here it was decided to represent only the gas solubility data reported as Henry's law constant, K_H, which is defined as:

$$K_{H,i} \equiv \lim_{x_i \to 0} \left(f_i / x_i \right),$$

where f_i is the fugacity of the gas and x_i its molar fraction concentration in the solution at a given pressure and temperature. Henry's law coefficients allow an easier comparison between different gases and absorbents, making it easier to evaluate the published data on gas absorption by different deep eutectic solvents in this chapter. For comparison purposes we considered the solubility data from Duan et al. (2019) although Henry's law constants in this case are based on molality instead of molar fraction.

In Fig. 4.2, Henry's law constants are presented for CO_2, H_2S, SO_2, and NO – the gases with larger solubilities in deep eutectic solvents. Solvents with different compositions but based on the same salts and hydrogen bond donors were studied by different authors. Figure 4.3 represents Henry's law constants for CO_2 in mixtures of choline chloride with urea at different compositions and as a function of temperature. The differences in K_H are significant (from *ca.* 18 bar to 30 bar in [Ch]Cl:U (1:2.5) and [Ch]Cl:U (1:1.5), respectively) but correspond to mole fraction concentrations of CO_2 of the same order of magnitude – 5.5×10^{-2} and 3.3×10^{-2} at a partial pressure of 1 bar of gas in [Ch]Cl:U (1:2.5) and [Ch]Cl:U (1:1.5), respectively. We have represented in Fig. 4.2 only one of the compositions studied for each deep eutectic solvent.

CO_2

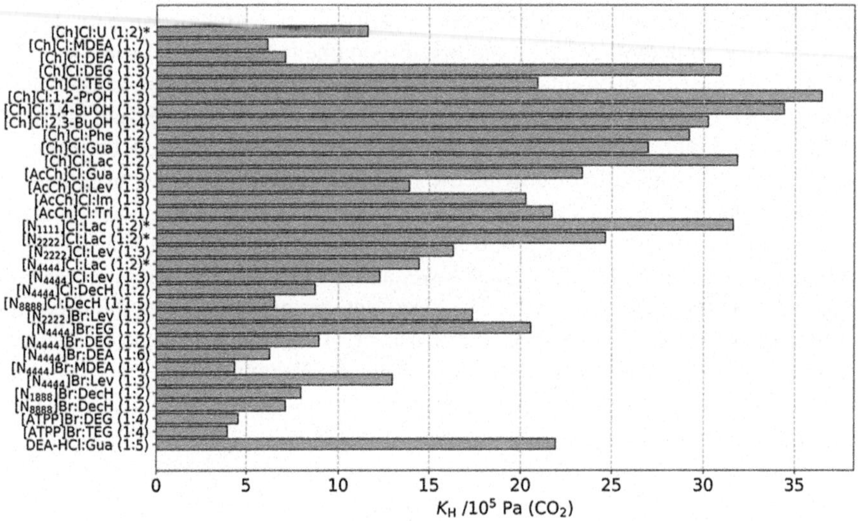

H_2S

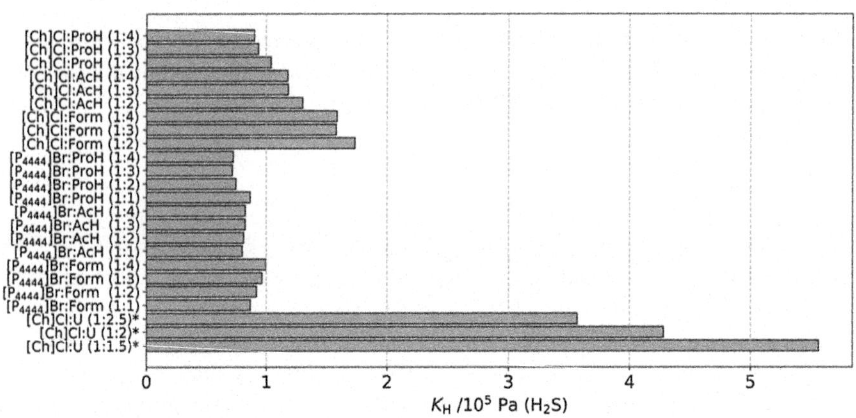

NO

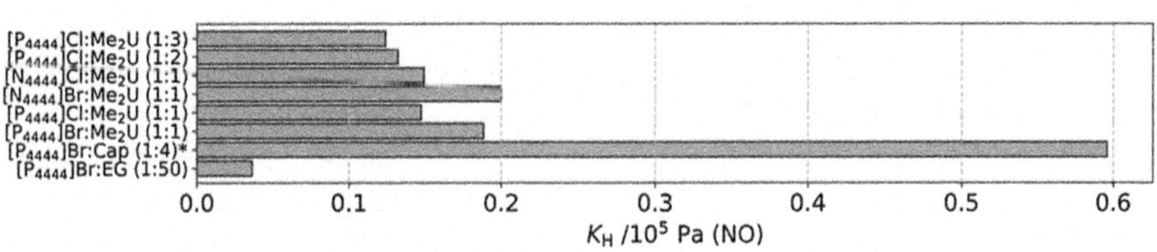

(continued)

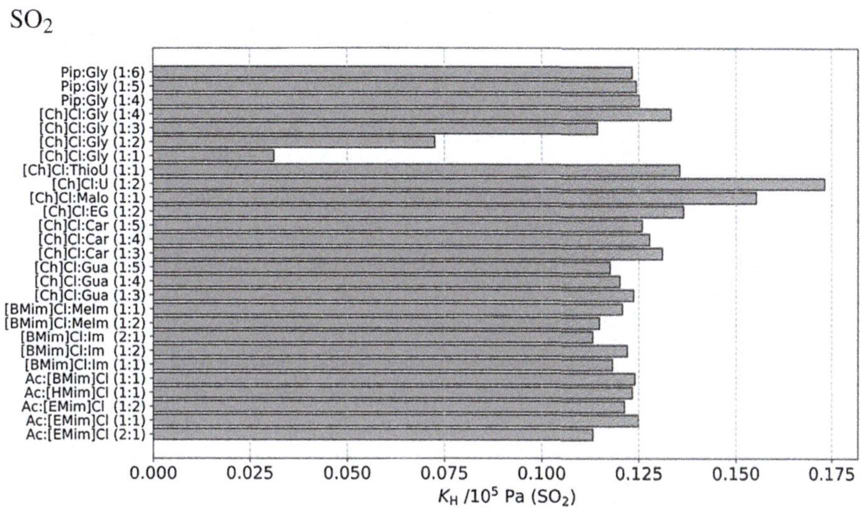

Fig. 4.2 Henry's law constant, K_H, for CO_2 at 303.15 K (*at 308.15 K), H_2S at 298.15 K (* at 313.2 K), NO at 303.15 K (* at 353.15 K), and SO_2 at 293.15 K. Sulfur dioxide, SO_2, is the most soluble of the four gases with reported K_H at 293 K lower than 0.18 bar. Notice the large variation in the range of solubilities for these gases. We observe that the solubility of the most commonly studied gases varies in the order SO_2 > NO > CO_2 > H_2S, the larger absorption for SO_2 being reported for the common choline chloride:urea mixture. The larger solubilities for NO, H_2S, and CO_2 were measured for mixtures based in tetraalkylphosphonium salts. Large solubility of CO_2 in eutectic mixtures containing amines is likely due to a chemical reaction between these components and the gas

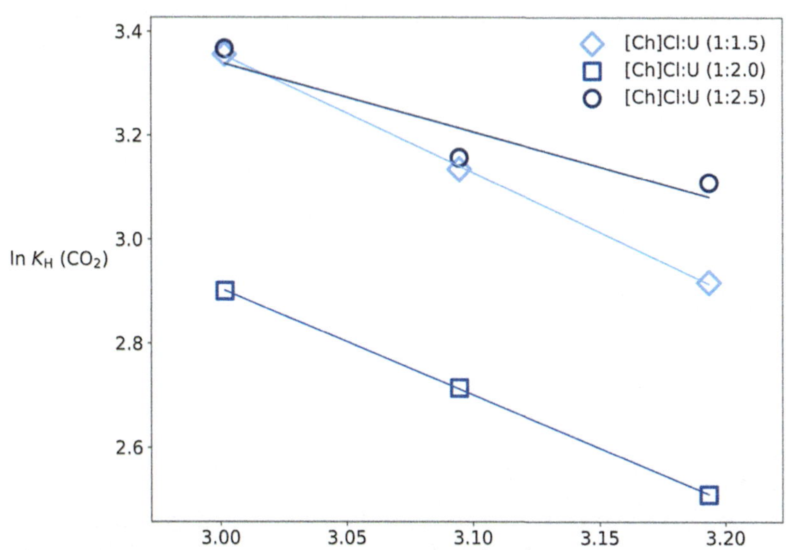

Fig. 4.3 Henry's law constant, K_H, for CO_2 in mixtures of [Ch]Cl:U with different molar ratio compositions (◇ [Ch]Cl:U (1:1.5), □ [Ch]Cl:U (1:2), ○ [Ch]Cl:U (1:2.5))

Sulfur dioxide, SO_2, is the most soluble of the four gases with reported K_H at 293 K lower than 0.18 bar. Many of the deep eutectic solvents studied also dissolve large quantities of NO, several of these mixtures, containing tetraalkylphosphonium and tetraalkylammonium salts, having K_H values at 303 K lower than 0.20 bar. H_2S and CO_2 are less soluble gases in the deep eutectic solvents studied so far with K_H values at 298 K and 303 K lower than 6 bar and 40 bar, respectively.

The mixtures that dissolve the largest quantities of SO_2 ($K_H < 0.12$ bar) are based on imidazolium or choline chloride salts with imidazole, methylimidazole, or glycerol as H-bond donors. For NO, these deep eutectic solvents were not studied; therefore, amongst the reported data, the eutectic mixtures based on tetraalkylphosphonium salts and ethylene glycol or 1,3-dimethylthiourea are the ones that absorb larger quantities of gas with reported K_H lower than 0.15 bar. Eutectic mixtures of tetraalkylphosphonium salts are also the ones that absorb larger quantities of H_2S with Henry's law constants at 298 K lower than 1 bar. For CO_2, tetrabutylammonium salts mixed with diethanolamine and allyltriphenylphosphonium salts mixed with triethylene glycol are the deep eutectic solvents with larger capacities with K_H values at 303 K lower than 5 bar, probably ought to a chemical reaction between the gas and one of the components of the liquid mixture. Even if no evident trends could be found, tetraalkylphosphonium or tetraalkylammonium salts seem to be the constituents of the deep eutectic solvents that are capable of absorbing larger quantities of gas.

Polarity seems to be at first sight a good descriptor for the gas absorption in deep eutectic solvents as polar gases, like SO_2 or H_2S, seem to be more soluble in deep eutectic solvents. Nevertheless, non-polar gases like NO are also very soluble in deep eutectic solvents (although a relatively small number of solvents have been studied) and quadrupolar gases as CO_2 have a much lower solubility.

The analysis of the behavior of the gas absorption with temperature is important as it allows to ascertain the relative importance of the gas-solvent interactions and of the structural organization of the solution to the solvation process. The Gibbs energy of solvation, defined as the difference in chemical potential when the solute is transferred from an ideal gas standard pressure into the reference state in the solution at infinite dilution, is related with Henry's law coefficient through

$$\Delta_{\text{solv}} G_i = RT \ln\left(\frac{K_{\text{H},i}}{p^0}\right)$$

its variation with temperature being directly related with the enthalpy of solvation

$$\Delta_{\text{solv}} H_i = -T^2 \frac{\partial}{\partial T}\left(\frac{\Delta_{\text{solv}} G_i}{T}\right)_p$$

Figure 4.4 represents Henry's law constants as a function of temperature for seven different gases in the deep eutectic solvent [Ch]Cl:U (1:2). Three groups of solubility data can be identified with the lowest values of K_H corresponding to SO_2, followed by a second group of gases including H_2S and CO_2 and a third group of less

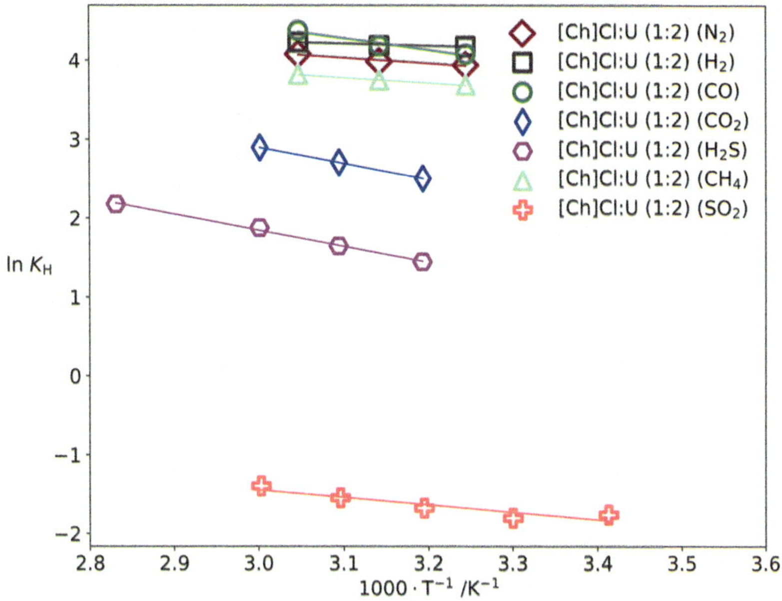

Fig. 4.4 Henry's law constant, K_H, for seven gases in [Ch]Cl:U (1:2) as a function of temperature in the range 290 K to 333 K (◇ N_2, □ H_2, ○ CO, ◇CO_2, ⬡ H_2S, △ CH_4, + SO_2)

soluble gases – CH_4, CO, H_2, and N_2. The natural logarithm of Henry's law constants decreases with decreasing temperature (meaning that the solubility of the different gases decreases with increasing temperature), thus corresponding to exothermic dissolutions. The larger the variation of Henry's law constant with temperature, the more negative the enthalpy of solvation is, which means that the gas dissolution is controlled by favorable interactions between the liquid and the gas in solution. As observed in Fig. 4.4, the slopes are indeed steeper for the three most soluble gases in [Ch]Cl:U when compared with the four less soluble gases reported for this deep eutectic solvent.

In Fig. 4.5, we represent the variation with temperature of Henry's law constants of CO_2 in several deep eutectic solvents. Here again three groups of deep eutectic solvents can be identified. The first corresponds to deep eutectic solvents formed by mixtures of salts with alcohols that dissolve relatively low quantities of carbon dioxide as [Ch]Cl:1,2-PrOH (1:3) or [Ch]Cl:DEG (1:3). The second group corresponds to deep eutectic solvents formed by quaternary ammonium salts with levulinic acid (with a behavior as a CO_2 solvent close to that of the deep eutectic solvents [Ch]Cl:U (1:2)), and the third includes deep eutectic solvents formed by mixtures of the same family of salts but with decanoic acid. This last group is the one that dissolves larger quantities of carbon dioxide but with a lower variation with temperature, meaning that the higher gas absorption cannot be attributed to a more favorable enthalpy of dissolution.

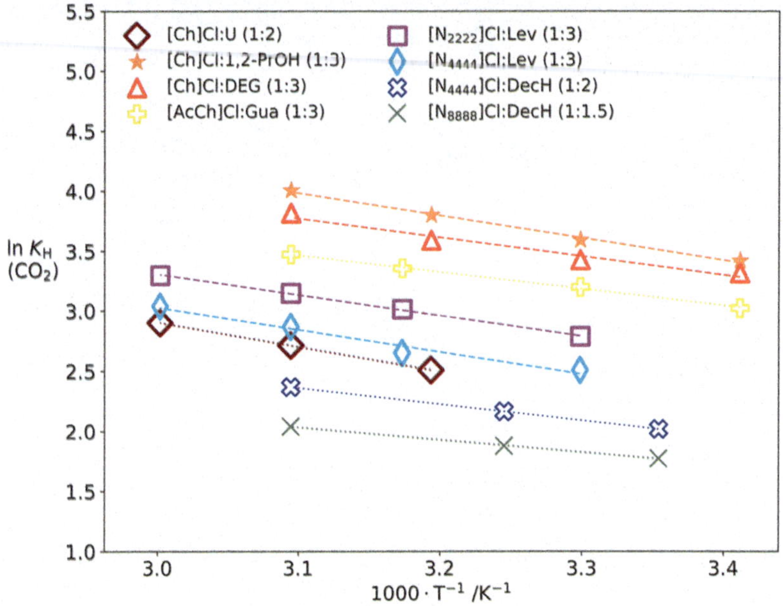

Fig. 4.5 Henry's law constant, K_H, for CO_2 in eight deep eutectic solvents as a function of temperature in the range 298 K to 333 K (◇ [Ch]Cl:U (1:2), ★[Ch]Cl:1,2-PrOH (1:3), △ [Ch]Cl:DEG (1:3), + [AcCh]Cl:Gua(1:3), □ [N_{2222}]Cl:Lev (1:3), ◇[N_{4444}]Cl:Lev(1:3), ✖✕ [N_{4444}]Cl:DecH (1:2), [N_{8888}]Cl:DecH (1:1.5))

4.4 Conclusions

This chapter reviews the published studies of gas absorption in deep eutectic solvents. The bibliographic research covers publications available up to the 3rd of June 2019 and their references. The publications considered for this review were selected from a mix and match of the keywords: gas, eutectic, low melting mixture, carbon dioxide, ethane, ethylene, propane, propene, propylene, acetylene, propyne, methylacetylene, olefin, paraffin, carbon monoxide, oxygen, hydrogen, hydrogen sulfide, ammonia, nitric oxide, nitrogen, sulfur dioxide, N_2, SO_2, CO_2, and solubility.

A huge variety in the behavior of the mixtures was identified from the analysis of 87 different eutectic mixtures with 10 different gases. The gas solubility, expressed as Henry's law coefficient, varies up to four orders of magnitude. This is surely the result of the difference in interactions between the gases and the eutectic mixtures. Amongst the reviewed works, there are without question some reactive mixtures where the gas reacts chemically with one or both of the components of the eutectic mixture; however, it was decided to report the gas absorption identically as in the original papers, even if this means that the concept of Henry's law constant is not applicable. Another reason for the wide extent of the gas absorption values reported might be the presence of non-accounted impurities present in the eutectic mixture. It is known that a large number of components used for deep eutectic solvents are hygroscopic, and that water acts as a third component in the mixtures, changing the physical and sometimes chemical properties. A critical review of the published data is then necessary as care should be taken to have similar amounts of water in molar fractions reported in order to accurately compare the gas solvation capacities of the different eutectic mixtures.

Unfortunately, no evident descriptor could be identified to explain the variety of equilibrium gas absorptions found for the different deep eutectic solvents. By far, the larger solubilities were found for SO_2 and NO in all the eutectic mixtures studied with these gases. The larger absorption for SO_2 is reported for the common [Ch]Cl:U mixture, while for NO the larger absorption was measured for mixtures based in tetraalkylphosphonium salts. H_2S is one order of magnitude less soluble than the two previous gases, the larger absorptions being found also in mixtures based on tetraalkylphosphonium salts.

Carbon dioxide is the most studied gas in eutectic mixtures, and its absorption spans a wide range of concentrations. The most striking values are found for mixtures based in allyltriphenylphosphonium salts with absorptions close to those found for H_2S. Eutectic mixtures containing amines are also reported as excellent absorbers of CO_2, this behavior being probably explained by a chemical reaction between the gas and these components of the eutectic mixture.

References

Abbott AP, Capper G, Davies DL et al (2003) Novel solvent properties of choline chloride/urea mixtures. Chem Commun:70–71

Abbott AP, Boothby D, Capper G et al (2004) Deep eutectic solvents formed between choline chloride and carboxylic acids: versatile alternatives to ionic liquids. J Am Chem Soc 126:9142–9147. https://doi.org/10.1021/ja048266j

Adeyemi I, Abu-Zahra MRM, Alnashef I (2017a) Experimental study of the solubility of CO2 in novel amine based deep eutectic solvents. In: Energy procedia. Elsevier Ltd, pp 1394–1400

Adeyemi I, Abu-Zahra MRM, Alnashef I (2017b) Novel green solvents for CO2 capture. In: Energy procedia. Elsevier Ltd, pp 2552–2560

Akhmetshina AI, Petukhov AN, Mechergui A et al (2018) Evaluation of Methanesulfonate-based deep eutectic solvent for ammonia sorption. J Chem Eng Data 63:1896–1904. https://doi.org/10.1021/acs.jced.7b01004

Ali E, Hadj-Kali MK, Mulyono S et al (2014) Solubility of CO2 in deep eutectic solvents: experiments and modelling using the Peng-Robinson equation of state. Chem Eng Res Des 92:1898–1906. https://doi.org/10.1016/j.cherd.2014.02.004

Ali E, Hadj-Kali MK, Mulyono S, Alnashef I (2016) Analysis of operating conditions for CO2 capturing process using deep eutectic solvents. Int J Greenhouse Gas Control 47:342–350. https://doi.org/10.1016/j.ijggc.2016.02.006

Altamash T, Atilhan M, Aliyan A et al (2017) Rheological, thermodynamic, and gas solubility properties of Phenylacetic acid-based deep eutectic solvents. Chem Eng Technol 40:778–790. https://doi.org/10.1002/ceat.201600475

Altamash T, Nasser MS, Elhamarnah Y et al (2018) Gas solubility and rheological behavior study of betaine and alanine based natural deep eutectic solvents (NADES). J Mol Liq 256:286–295. https://doi.org/10.1016/j.molliq.2018.02.049

Altamash T, Amhamed AI, Aparicio S, Atilhan M (2019) Combined experimental and theoretical study on high pressure methane solubility in natural deep eutectic solvents. Ind Eng Chem Res 58:8097–8111. https://doi.org/10.1021/acs.iecr.9b00702

Bhawna PA, Pandey S (2017) Superbase-added choline chloride-based deep eutectic solvents for CO2 capture and sequestration. ChemistrySelect 2:11422–11430. https://doi.org/10.1002/slct.201702259

Chemat F, Gnanasundaram N, Shariff AM, Murugesan T (2016) Effect of L-arginine on solubility of CO2 in choline chloride + glycerol based deep eutectic solvents. In: Procedia engineering. Elsevier Ltd, pp 236–242

Chen Y, Ai N, Li G et al (2014) Solubilities of carbon dioxide in eutectic mixtures of choline chloride and dihydric alcohols. J Chem Eng Data 59:1247–1253. https://doi.org/10.1021/je400884v

Chen Y, Jiang B, Dou H et al (2018) Highly efficient and reversible capture of low partial pressure SO2 by functional deep eutectic solvents. Energy Fuel 32:10737–10744. https://doi.org/10.1021/acs.energyfuels.8b01794

Cui G, Liu J, Lyu S et al (2019a) Efficient and reversible SO2 absorption by environmentally friendly task-specific deep eutectic solvents of PPZBr + Gly. ACS Sustain Chem Eng 7:14236–14246. https://doi.org/10.1021/acssuschemeng.9b03245

Cui G, Lv M, Yang D (2019b) Efficient CO2 absorption by azolide-based deep eutectic solvents. Chem Commun 55:1426–1429. https://doi.org/10.1039/c8cc10085c

Deng D, Han G, Jiang Y (2015) Investigation of a deep eutectic solvent formed by levulinic acid with quaternary ammonium salt as an efficient SO2 absorbent. New J Chem 39:8158–8164. https://doi.org/10.1039/c5nj01629k

Deng D, Jiang Y, Liu X et al (2016) Investigation of solubilities of carbon dioxide in five levulinic acid-based deep eutectic solvents and their thermodynamic properties. J Chem Thermodyn 103:212–217. https://doi.org/10.1016/j.jct.2016.08.015

Deng D, Liu X, Gao B (2017a) Physicochemical properties and investigation of azole-based deep eutectic solvents as efficient and reversible SO2 absorbents. Ind Eng Chem Res 56:13850–13856

Deng R, Sun Y, Bi H et al (2017b) Deep eutectic solvents as tuning media dissolving Cu+ used in facilitated transport supported liquid membrane for ethylene/ethane separation. Energy Fuel 31:11146–11155. https://doi.org/10.1021/acs.energyfuels.7b01305

Deng D, Duan X, Gao B et al (2019a) Efficient and reversible absorption of NH3 by functional azole-glycerol deep eutectic solvents. New J Chem 43:11636–11642. https://doi.org/10.1039/c9nj01788g

Deng D, Gao B, Zhang C et al (2019b) Investigation of protic NH4SCN-based deep eutectic solvents as highly efficient and reversible NH3 absorbents. Chem Eng J 358:936–943. https://doi.org/10.1016/j.cej.2018.10.077

Dou H, Jiang B, Zhang L et al (2018) Synergy of high permeability, selectivity and good stability properties of silver-decorated deep eutectic solvent based facilitated transport membranes for efficient ethylene/ethane separation. J Membr Sci 567:39–48. https://doi.org/10.1016/j.memsci.2018.09.027

Dou J, Zhao Y, Yin F et al (2019) Mechanistic study of selective absorption of NO in flue gas using EG-TBAB deep eutectic solvents. Environ Sci Technol 53:1031–1038. https://doi.org/10.1021/acs.est.8b05408

Duan E, Sun H, Wang Y, Shi L (2011) Absorption of NO and NO2 in caprolactam tetrabutyl ammonium halide ionic liquids. J Air Waste Manag Assoc 61:1393–1397. https://doi.org/10.1080/10473289.2011.623635

Duan X, Gao B, Zhang C, Deng D (2019) Solubility and thermodynamic properties of NH3 in choline chloride-based deep eutectic solvents. J Chem Thermodyn 133:79–84. https://doi.org/10.1016/j.jct.2019.01.031

Francisco M, van den Bruinhorst A, Zubeir LF et al (2013) A new low transition temperature mixture (LTTM) formed by choline chloride+lactic acid: characterization as solvent for CO2 capture. Fluid Phase Equilib 340:77–84. https://doi.org/10.1016/j.fluid.2012.12.001

Ghaedi H, Ayoub M, Sufian S et al (2017a) Measurement and correlation of physicochemical properties of phosphonium-based deep eutectic solvents at several temperatures (293.15 K–343.15 K) for CO2 capture. J Chem Thermodyn 113:41–51. https://doi.org/10.1016/j.jct.2017.05.020

Ghaedi H, Ayoub M, Sufian S et al (2017b) CO2 capture with the help of Phosphonium-based deep eutectic solvents. J Mol Liq 243:564–571. https://doi.org/10.1016/j.molliq.2017.08.046

Ghaedi H, Ayoub M, Sufian S et al (2017c) Investigation of various process parameters on the solubility of carbon dioxide in phosphonium-based deep eutectic solvents and their aqueous mixtures: experimental and modeling. Int J Greenhouse Gas Control 66:147–158. https://doi.org/10.1016/j.ijggc.2017.09.020

Ghaedi H, Ayoub M, Sufian S et al (2018) Density and refractive index measurements of transition-temperature mixture (deep eutectic analogues) based on potassium carbonate with dual hydrogen bond donors for CO2 capture. J Chem Thermodyn 118:147–158. https://doi.org/10.1016/j.jct.2017.11.008

Haider MB, Jha D, Marriyappan Sivagnanam B, Kumar R (2018) Thermodynamic and kinetic studies of CO2 capture by glycol and amine-based deep eutectic solvents. J Chem Eng Data 63:2671–2680. https://doi.org/10.1021/acs.jced.8b00015

Huang Z, Jiang B, Yang H et al (2017) Investigation of glycerol-derived binary and ternary systems in CO2 capture process. Fuel 210:836–843. https://doi.org/10.1016/j.fuel.2017.08.043

Ji Y, Hou Y, Ren S et al (2016) Phase equilibria of high pressure CO2 and deep eutectic solvents formed by quaternary ammonium salts and phenol. Fluid Phase Equilib 429:14–20. https://doi.org/10.1016/j.fluid.2016.08.020

Jiang B, Lin W, Zhang L et al (2016) 1,3-Dimethylurea Tetrabutylphosphonium bromide ionic liquids for NO efficient and reversible capture. Energy Fuels 30:735–739. https://doi.org/10.1021/acs.energyfuels.5b01826

Jiang B, Dou H, Wang B et al (2017a) Silver-based deep eutectic solvents as separation media: supported liquid membranes for facilitated olefin transport. ACS Sustain Chem Eng 5:6873–6882. https://doi.org/10.1021/acssuschemeng.7b01092

Jiang B, Dou H, Zhang L et al (2017b) Novel supported liquid membranes based on deep eutectic solvents for olefin-paraffin separation via facilitated transport. J Membr Sci 536:123–132. https://doi.org/10.1016/j.memsci.2017.05.004

Jiang B, Tao W, Dou H et al (2017c) A novel supported liquid membrane based on binary metal chloride deep eutectic solvents for ethylene/ethane separation. Ind Eng Chem Res 56:15153–15162. https://doi.org/10.1021/acs.iecr.7b03843

Jiang WJ, Zhong FY, Liu Y, Huang K (2019) Effective and reversible capture of NH3 by ethylamine hydrochloride plus glycerol deep eutectic solvents. ACS Sustain Chem Eng 7:10552–10560. https://doi.org/10.1021/acssuschemeng.9b01102

Karimi M, Jodaei A, Khajvandi A et al (2018) In-situ capture and conversion of atmospheric CO2 into nano-CaCO3 using a novel pathway based on deep eutectic choline chloride-calcium chloride. J Environ Manag 206:516–522. https://doi.org/10.1016/j.jenvman.2017.11.005

Li X, Hou M, Han B et al (2008a) Solubility of CO2 in a choline chloride + urea eutectic mixture. J Chem Eng Data 53:548–550. https://doi.org/10.1021/je700638u

Li X, Hou M, Zhang Z et al (2008b) Absorption of CO2 by ionic liquid/polyethylene glycol mixture and the thermodynamic parameters. Green Chem 10:879–884. https://doi.org/10.1039/b801948g

Li Y, Ali MC, Yang Q et al (2017) Hybrid deep eutectic solvents with flexible hydrogen-bonded supramolecular networks for highly efficient uptake of NH3. ChemSusChem 10:3368–3377. https://doi.org/10.1002/cssc.201701135

Li X, Liu X, Deng D (2018) Solubilities and thermodynamic properties of CO2 in four azole-based deep eutectic solvents. J Chem Eng Data 63:2091–2096. https://doi.org/10.1021/acs.jced.8b00098

Li Z, Wang L, Li C et al (2019) Absorption of carbon dioxide using ethanolamine-based deep eutectic solvents. ACS Sustain Chem Eng 7:10403–10414. https://doi.org/10.1021/acssuschemeng.9b00555

Liu B, Wei F, Zhao J, Wang Y (2013a) Characterization of amide-thiocyanates eutectic ionic liquids and their application in SO2 absorption. RSC Adv 3:2470–2476. https://doi.org/10.1039/c2ra22990k
Liu B, Zhao J, Wei F (2013b) Characterization of caprolactam based eutectic ionic liquids and their application in SO2 absorption. J Mol Liq 180:19–25. https://doi.org/10.1016/j.molliq.2012.12.024
Liu X, Gao B, Jiang Y et al (2017) Solubilities and thermodynamic properties of carbon dioxide in Guaiacol-based deep eutectic solvents. J Chem Eng Data 62:1448–1455. https://doi.org/10.1021/acs.jced.6b01013
Liu X, Gao B, Deng D (2018) SO2 absorption/desorption performance of renewable phenol-based deep eutectic solvents. Sep Sci Technol 53:2150–2158. https://doi.org/10.1080/01496395.2018.1446026
Liu F, Chen W, Mi J et al (2019) Thermodynamic and molecular insights into the absorption of H2S, CO2, and CH4 in choline chloride plus urea mixtures. AICHE J 65. https://doi.org/10.1002/aic.16574
Lu M, Han G, Jiang Y et al (2015) Solubilities of carbon dioxide in the eutectic mixture of levulinic acid (or furfuryl alcohol) and choline chloride. J Chem Thermodyn 88:72–77. https://doi.org/10.1016/j.jct.2015.04.021
Ma C, Sarmad S, Mikkola JP, Ji X (2017) Development of low-cost deep eutectic solvents for CO 2 capture. In: Energy procedia. Elsevier Ltd, pp 3320–3325
Mirza NR, Nicholas NJ, Wu Y et al (2015) Experiments and thermodynamic modeling of the solubility of carbon dioxide in three different deep eutectic solvents (DESs). J Chem Eng Data 60:3246–3252. https://doi.org/10.1021/acs.jced.5b00492
Mirza N, Mumford K, Wu Y et al (2017a) Improved eutectic based solvents for capturing carbon dioxide (CO2). In: Energy procedia. Elsevier Ltd, pp 827–833
Mirza NR, Nicholas NJ, Wu Y et al (2017b) Viscosities and carbon dioxide solubilities of guanidine carbonate and malic acid-based eutectic solvents. J Chem Eng Data 62:348–354. https://doi.org/10.1021/acs.jced.6b00680
Ren H, Lian S, Wang X et al (2018) Exploiting the hydrophilic role of natural deep eutectic solvents for greening CO2 capture. J Clean Prod 193:802–810. https://doi.org/10.1016/j.jclepro.2018.05.051
Sarmad S, Xie Y, Mikkola JP, Ji X (2016) Screening of deep eutectic solvents (DESs) as green CO2 sorbents: from solubility to viscosity. New J Chem 41:290–301. https://doi.org/10.1039/c6nj03140d
Shukla SK, Mikkola JP (2018) Intermolecular interactions upon carbon dioxide capture in deep-eutectic solvents. Phys Chem Chem Phys 20:24591–24601. https://doi.org/10.1039/c8cp03724h
Shukla SK, Mikkola JP (2019) Unusual temperature-promoted carbon dioxide capture in deep-eutectic solvents: the synergistic interactions. Chem Commun 55:3939–3942. https://doi.org/10.1039/c9cc00831d
Su WC, Wong DSH, Li MH (2009) Effect of water on solubility of carbon dioxide in (aminomethanamide + 2-hydroxy-N,N,N-trimethylethanaminium chloride). J Chem Eng Data 54:1951–1955. https://doi.org/10.1021/je900078k
Sun S, Niu Y, Xu Q et al (2015) Efficient SO2 absorptions by four kinds of deep eutectic solvents based on choline chloride. Ind Eng Chem Res 54:8019–8024. https://doi.org/10.1021/acs.iecr.5b01789
Sun Y, Wei G, Tantai X et al (2017a) Highly efficient nitric oxide absorption by environmentally friendly deep eutectic solvents based on 1,3-Dimethylthiourea. Energy Fuel 31:12439–12445. https://doi.org/10.1021/acs.energyfuels.7b02142
Sun Y, Wei G, Tantai X et al (2017b) Highly efficient nitric oxide absorption by environmentally friendly deep eutectic solvents based on 1,3-Dimethylthiourea. Energy Fuel 31:12439–12445. https://doi.org/10.1021/acs.energyfuels.7b02142

Sze LL, Pandey S, Ravula S et al (2014) Ternary deep eutectic solvents tasked for carbon dioxide capture. ACS Sustain Chem Eng 2:2117–2123. https://doi.org/10.1021/sc5001594
Trivedi TJ, Lee JH, Lee HJ et al (2016) Deep eutectic solvents as attractive media for CO2 capture. Green Chem 18:2834–2842. https://doi.org/10.1039/c5gc02319j
Ullah R, Atilhan M, Anaya B et al (2015) A detailed study of cholinium chloride and levulinic acid deep eutectic solvent system for CO2 capture via experimental and molecular simulation approaches. Phys Chem Chem Phys 17:20941–20960. https://doi.org/10.1039/c5cp03364k
Wu H, Kang D, Perng S (2015) Int Conf Appl Energy 2015(00):2015
Wu H, Shen M, Chen X et al (2019) New absorbents for hydrogen sulfide: deep eutectic solvents of tetrabutylammonium bromide/carboxylic acids and choline chloride/carboxylic acids. Sep Purif Technol 224:281–289. https://doi.org/10.1016/J.SEPPUR.2019.04.082
Xie Y, Dong H, Zhang S et al (2013) Thermophysical properties and gas solubilities in choline chloride/urea for CO2 separation. Int Conf Appl Energy:2013–2066
Xie Y, Dong H, Zhang S et al (2014) Effect of water on the density, viscosity, and CO2 solubility in choline chloride/urea. J Chem Eng Data 59:3344–3352. https://doi.org/10.1021/je500320c
Xie Y, Dong H, Zhang S et al (2016) Solubilities of CO2, CH4, H2, CO and N2 in choline chloride/urea. Green Energy Environ 1:195–200. https://doi.org/10.1016/j.gee.2016.09.001
Yang D, Hou M, Ning H et al (2013) Efficient SO2 absorption by renewable choline chloride-glycerol deep eutectic solvents. Green Chem 15:2261–2265. https://doi.org/10.1039/c3gc40815a
Yang D, Han Y, Qi H et al (2017) Efficient absorption of SO2 by EmimCl-EG deep eutectic solvents. ACS Sustain Chem Eng 5:6382–6386. https://doi.org/10.1021/acssuschemeng.7b01554
Yang D, Zhang S, Jiang DE, Dai S (2018) SO2 absorption in EmimCl-TEG deep eutectic solvents. Phys Chem Chem Phys 20:15168–15173. https://doi.org/10.1039/c8cp02250j
Yang D, Zhang S, Jiang DE (2019) Efficient absorption of SO2 by deep eutectic solvents formed by biobased aprotic organic compound Succinonitrile and 1-Ethyl-3-methylimidazolium chloride. ACS Sustain Chem Eng 7:9086–9091. https://doi.org/10.1021/acssuschemeng.9b00851
Zhang K, Hou Y, Wang Y et al (2018) Efficient and reversible absorption of CO2 by functional deep eutectic solvents. Energy Fuel 32:7727–7733. https://doi.org/10.1021/acs.energyfuels.8b01129
Zhao T, Liang J, Zhang Y et al (2018) Unexpectedly efficient SO2 capture and conversion to sulfur in novel imidazole-based deep eutectic solvents. Chem Commun 54:8964–8967. https://doi.org/10.1039/c8cc04349c
Zhong FY, Huang K, Peng HL (2019a) Solubilities of ammonia in choline chloride plus urea at (298.2–353.2) K and (0–300) kPa. J Chem Thermodyn 129:5–11. https://doi.org/10.1016/j.jct.2018.09.020
Zhong FY, Peng HL, Tao DJ et al (2019b) Phenol-based ternary deep eutectic solvents for highly efficient and reversible absorption of NH3. ACS Sustain Chem Eng 7:3258–3266. https://doi.org/10.1021/acssuschemeng.8b05221
Zhong FY, Zhou L, Shen J, et al (2019c) Rational design of azole-based deep eutectic solvents for highly efficient and reversible capture of ammonia ACS Sustainable Chem. Eng. 2019, (7)16: 14170–14179. https://doi.org/10.1021/acssuschemeng.9b02845
Zubeir LF, Lacroix MHM, Kroon MC (2014) Low transition temperature mixtures as innovative and sustainable CO2 capture solvents. J Phys Chem B 118:14429–14441. https://doi.org/10.1021/jp5089004
Zubeir LF, Van Osch DJGP, Rocha MAA et al (2018) Carbon dioxide Solubilities in Decanoic acid-based hydrophobic deep eutectic solvents. J Chem Eng Data 63:913–919. https://doi.org/10.1021/acs.jced.7b00534

Chapter 5
Hydrophobic Deep Eutectic Solvents

Emily L. Byrne, Mark Gilmore, Leila Moura, Małgorzata Swadźba-Kwaśny, and John D. Holbrey

Contents

Abstract Hydrophobic deep eutectic solvents combine the typical, desirable features of deep eutectic solvents; inexpensive, benign, often naturally occurring components with high hydrogen bond-forming tendencies with hydrophobicity; and the ability to form biphases with water. This implies an intrinsically challenging design, where hydrophobicity is obtained in mixtures of components that can form hydrogen bonds and so have hydrophilic tendencies. Moreover, this is usually achieved without resorting to the use of expensive perfluorinated components. This chapter outlines design strategies of hydrophobic deep eutectic solvents, followed by an overview of the key applications as extraction media for organic molecules from fermentation broths and wastewater streams, metal ions from aqueous media, and other uses including microextraction for analysis and CO_2 capture.

E. L. Byrne · M. Gilmore · L. Moura · M. Swadźba-Kwaśny · J. D. Holbrey (✉)
The QUILL Research Centre, School of Chemistry and Chemical Engineering, Queen's University Belfast, Belfast, UK
e-mail: j.holbrey@qub.ac.uk

S. Fourmentin et al. (eds.), *Deep Eutectic Solvents for Medicine, Gas Solubilization and Extraction of Natural Substances*, Environmental Chemistry for a Sustainable World 56, https://doi.org/10.1007/978-3-030-53069-3_5

Keywords Deep eutectic solvents · DES · Hydrophobic · Extractions · Separations

5.1 Introduction

Deep eutectic solvents (DES) evolved from ionic liquids initially driven by the search for lower cost and more readily available alternatives to "standard" ionic liquids. From ionic liquids, deep eutectic solvents inherited ionic components, most commonly organic halide salts such as choline chloride, that were paired with hydrogen bond donors such as urea, alcohols, or acids (Abbott et al. 2003, 2004). This design of mixtures, based on abundant and largely benign components, has proven to be a successful strategy to produce liquids that have been explored across a broad range of applications, mirroring the interest in ionic liquids in the preceding decades. From their initial design, the components of deep eutectic solvents were defined by their hydrogen bond donating and accepting abilities which, by definition, gave them high affinity to water (Abbott et al. 2003, 2004). Discussing the structure of deep eutectic solvents in their liquid state, multiple competing Coulombic and hydrogen bonding interactions are cited, which are synonymous with hydrophilicity (Perkins et al. 2014; Wagle et al. 2015; Ashworth et al. 2016; Hammond et al. 2016; Araujo et al. 2017; Stefanovic et al. 2017; Florindo et al. 2018a; Gilmore et al. 2018b).

Consequently, one area where the applications of deep eutectic solvents have been less prominent than those of ionic liquids is in the generation of hydrophobic deep eutectic solvents with controllable solvent characteristics, to be used as alternatives to volatile organic compounds for separations and biphasic reaction systems. In ionic liquids, hydrophobicity was often introduced through a non-coordinating anion: early on with hexafluorophosphate, $[PF_6]^-$, and later with bis(trifluoromethanesulfonate)imide, $[NTf_2]^-$, and other perfluorinated anions. This had significant impact in driving the exploration of hydrophobic ionic liquids as separation solvents (Huddleston et al. 2001). The design of deep eutectic solvents called for the anion to be inexpensive, environmentally benign, and a good hydrogen bond acceptor, which effectively banned the use of perfluorinated anions, in favor of hydrophilicity-inducing halides.

5.2 Design of Hydrophobic Deep Eutectic Solvents

5.2.1 From Hydrophobic Ionic Liquids to Hydrophobic Deep Eutectic Solvents

The design of hydrophobic deep eutectic solvents, although hindered by the avoidance of perfluorinated anion, was nevertheless inspired by strategies that have been used for producing hydrophobic ionic liquids. With the hindsight of two decades of ionic liquid development, it has been known that the once-popular $[PF_6]^-$ anion hydrolyzed in the presence of water generating HF, and other perfluorinated anions, represented by $[NTf_2]^-$, had prohibitively high cost for most applications – definitely for those where deep eutectic solvents would be replacing conventional solvents (George et al. 2015). In consequence, the strategies used to generate hydrophobic deep eutectic solvents have relied on increasing lipophilicity (hydrophobicity) by introduction of long hydrocarbon chains. The prominent ionic liquid example of this is the tetraalkylphosphonium ionic liquid trihexyl(tetradecyl)phosphonium chloride, $[P_{6\,6\,6\,14}]Cl$. It features a very hydrophilic in nature chlorideanion, but the addition of long alkyl chains to the quaternary cation center enhances hydrophobicity. Unfortunately, this came at the price of increasing viscosity of the produced liquids, which could be detrimental to mass transport in extraction systems (Souza et al. 2019).

Another approach to use non-fluorinated ionic liquids for extractions from aqueous media was based on ionic liquids being effectively salts, therefore prone to *salting-out* effect. Generating biphasic liquid mixtures between hydrophilic ionic liquids and water was first described in 2003 (Gutowski et al. 2003) where the formation of ionic liquid-aqueous biphasic systems induced by the addition of kosmotropic agents such as K_3PO_4 was demonstrated. The more hydrophobic ion pair, typically the organic cation from the ionic liquid and the anion that sits lower in the Hofmeister series (Hyde et al. 2017), is *salted-out*, forming a biphasic mixture in which both phases contain water and a salt. Aqueous biphasic system formation has been demonstrated across a range of ionic liquids and kosmotropic agents (Freire et al. 2012), allowing "soft" partitioning and extraction of biomolecules including proteins (Pei et al. 2009) and pharmaceuticals (McQueen and Lai 2019; Oppermann et al. 2011). It has been reported that this approach offers the advantage of mild operating conditions, necessary to maintain protein structure and functionality without inducing denaturization (Shukla et al. 2018). Further benefit came from tunability of these ionic liquid systems, where the effectiveness would be manipulated by changing the phase-forming components (ions) and concentrations. However, in ionic liquid aqueous biphasic system media, both phases are water-rich and hydrophilic, which poses an intrinsic limitation, restricting their applicability to hydrophilic substrates.

Shortcomings associated with each type of hydrophobic ionic liquid system drove the development of hydrophobic deep eutectic solvents, targeted at benign, inexpensive, and preferably low viscosity media.

5.2.2 Hydrophobicity Through Organic Salts with Hydrophobic Hydrogen Bond Donors

In 2015, Kroon and co-workers (van Osch et al. 2015) described the first family of hydrophobic deep eutectic solvents containing quaternary ammonium salts, $[N_{4444}]Cl$ and $[N_{8888}]Cl$, where $[N_{nnnn}]^+$ is a tetraalkylammonium cation with *n* long

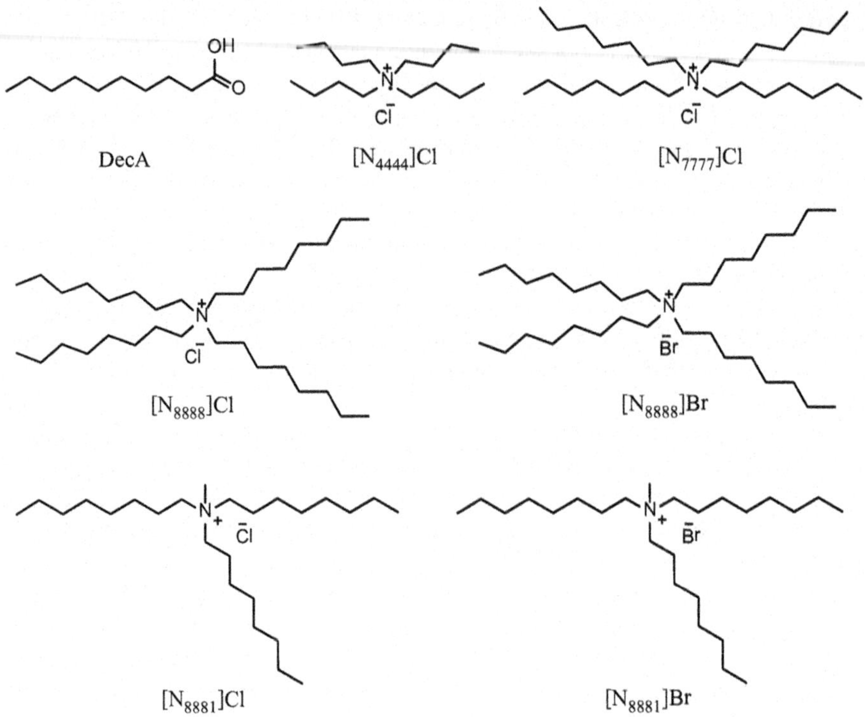

Fig. 5.1 These components were used in the preparation of hydrophobic deep eutectic solvents and were chosen based on the presence of long alkyl chains to induce hydrophobicity. They demonstrate explicit hydrogen bond donating (DecA) and accepting (ammonium salt) properties. (Modified after van Osch et al. 2015)

carbon chains (Fig. 5.1). These were combined with decanoic acid, a hydrophobic hydrogen bond donor. Notably, these combinations of a long-chained carboxylic acid with moderately hydrophobic cations were described as eutectics, although only one composition per system, 1:2 ammonium salt:organic acid, was reported.

The approach to the design of deep eutectic solvents, taken by Kroon and co-workers and following that previously proposed by Abbott and co-workers (Abbott et al. 2003), calls for a halide salt and a hydrogen bond donor. At the same time, incorporating large organic salts to achieve hydrophobicity traces its lineage to hydrophobic quaternary ammonium or phosphonium ionic liquids. Indeed, these deep eutectic solvents have been examined as analogues of these ionic liquids, in the context of extraction of metal species (see Section 3.2, "Extraction of Metals"). Like their ionic liquid predecessors, they suffer from relatively high viscosities; for example, the tetraalkylammonium chloride/decanoic acid systems described by Kroon and co-workers have viscosities in the range 173–783 mPa s at 25 °C. This viscosity range is an order of magnitude larger than that of the more fluid examples of ionic liquids such as 1-ethyl-3-methylimidazolium bis(trifluorosulfonyl)amide which has a viscosity of 34 mPa s at 20 °C (Bonhote et al. 1996). However, they are within a similar viscosity range to many ionic liquids such as 1-butyl-3-methylimidazolium hexafluorophosphate ([C_4mim][PF_6]) and 1-butyl-3-methylimidazolium tetrafluoroborate ([C_4mim][BF_4]) which have viscosities of 371 and 154 mPa s^{-1}, respectively, at 20 °C (Seddon et al. 2002).

5.2.3 Losing the Salt, Abandoning the Hydrogen Bond Acceptor-Hydrogen Bond Donor Strategy

Despite discussions about definitions and nomenclature of what specifically a deep eutectic solvent is (Francisco et al. 2013; Smith et al. 2014; Silva et al. 2018; Martins et al. 2019), it can be easily recognized that liquids can be produced by judicious combination of two components, which leads to melting point depression and formation of a eutectic composition, with a melting point lower than that of either constituent. Such behavior is entirely normal – in fact, it is typical in two-component mixtures – and does not require the presence of an organic salt, or explicit hydrogen bond donor/acceptor pairing. At the same time that Kroon and co-workers were first describing hydrophobic tetraalkylammonium chloride/decanoic acid deep eutectic solvents (van Osch et al. 2015), Marrucho and co-workers (Ribeiro et al. 2015) reported that hydrophobic deep eutectic solvents produced from DL-menthol and naturally occurring carboxylic acids (Fig. 5.2) could be formed and used to extract model biomolecules, e.g., caffeine, tryptophan, isophthalic acid, and vanillin. These liquids do not contain an organic salt component and do not rely on an explicit

Fig. 5.2 Menthol-based hydrophobic deep eutectic solvents whereby menthol is paired with an acid such as pyruvic acid, acetic acid, L-lactic acid, or lauric acid. Menthol was chosen to induce hydrophobicity, and these systems are without explicit hydrogen bond donor/acceptor pairing. (Reprinted with permission from Ribeiro et al. 2015. Copyright (2015) American Chemical Society)

hydrogen bond donor-acceptor pairing strategy in their design, although interactions between the components are presumably hydrogen bonding with both alcohol and carboxylic acid functionalities displaying donor and acceptor ability.

These DL-menthol/carboxylic acid hydrophobic deep eutectic solvents are radically different in composition to the ammonium salt-based deep eutectic solvents from Kroon and co-workers (van Osch et al. 2015) and from typical choline chloride-based deep eutectic solvents (Abbott et al. 2003). They also do not build on traditional ionic liquid components, but crucially demonstrate the formulation of functional hydrophobic liquids containing desirable properties – in this case, derived from bio-sourced menthol and incorporating low cytotoxic biocompatible components.

Subsequently, a very interesting variety of hydrophobic deep eutectic solvents was proposed, comprised solely of long-chain carboxylic acids (Florindo et al. 2018a, b). These liquids take advantage of melting point depression in mixtures of octanoic, nonanoic, decanoic, and dodecanoic acid (C_8-C_{12}) that form room temperature liquids with eutectic points between 9 and 18 °C. Their design was inspired by the behavior of mixtures of longer chain fatty acids (C_{10}-C_{18}), which have slight melt depression and eutectic points between 22 and 56 °C, and have been considered as phase change materials for thermal energy storage (Zhang et al. 2015).

Gilmore et al. (Gilmore et al. 2018b) have based their design on the widely used hydrophobic extracting agent, trioctylphosphine, used for separating both metals and organic molecules from aqueous media. Typically used in kerosene solution, trioctylphosphine oxide was shown to form hydrophobic deep eutectic solvents, acting as a hydrogen bond acceptor, when liquefied by the addition of 1–2 equivalents of a hydrogen-bonding component, such as phenol. Room temperature liquids were formed over a wide compositional range (χ_{TOPO} = 0.2–0.5), and trioctylphosphine oxide/phenol mixtures were significantly more fluid (with viscosities in the range 12–43 mPa s at 25 °C) when compared to the ionic tetraalkylammonium chloride/decanoic acid hydrophobic deep eutectic solvents (viscosities 173–783 mPa s at 25 °C) from Kroon and co-workers (van Osch et al. 2015). Although the liquids were glass-forming, a eutectic point at $\chi_{TOPO} = 0.33$ was extrapolated from the variations in solid/liquid transition temperatures. The trioctylphosphine oxide/phenol deep eutectic solvent was demonstrated as an effective extractant for uranyl ($[UO_2]^{2+}$) species from aqueous acid due to the high concentration of coordinating trioctylphosphine oxide in the liquids, much higher than that attainable in kerosene. Furthermore, the eutectic formation strategy was also suggested as a potential route to extract phenolic and other acidic hydrogen bond donors from water, based on the positive partition coefficient of phenol from water.

Koon and co-workers (van den Bruinhorst et al. 2019) subsequently showed that trioctylphosphine oxide could form room temperature liquids over a wide compositional range with *N,N′*-dihexylthiourea (χ_{TOPO} = 0.3–0.5) and over a short compositional range with dodecanoic acid (χ_{TOPO} = 0.4–0.5), both of these components acting as hydrogen bond donors towards trioctylphosphine oxide. The eutectic point was determined around $\chi_{TOPO} = 0.4$ in both cases.

5.3 Applications

A common theme for all the hydrophobic deep eutectic solvent systems that have been reported to date is that hydrogen bond donor and acceptor capability of the deep eutectic solvent forming components is retained. For example, using either carboxylic acids, alcohols, or phenols as one component and establishing immiscibility with water through the inclusion of long alkyl groups, e.g., tetraalkylammonium/phosphonium salts and trioctylphosphine oxide, fatty acids, and linear alcohols.

The drivers for development and investigation of hydrophobic deep eutectic solvents have been to enable separations and partitioning from aqueous media, and efforts along these lines can be generally divided between extraction of organic compounds and inorganic metal species.

5.3.1 Extraction of Organic Compounds

The "gold standard" for understanding extraction and solute distribution between aqueous and hydrophobic extracting phases is the octanol/water partition coefficient, LogP (Sangster 1989). The more hydrophobic a solute is, the greater its distribution coefficient is for partitioning to octanol from water. A recent work from Dietz and co-workers (Kaul et al. 2019) reports on the measurement of distribution coefficients in a range of representative organic molecules between hydrophobic menthol/decanoic acid deep eutectic solvents and water using ^{14}C-labelled radiotracers and scintillation counting. They compared the results to the corresponding octanol-water partition coefficients (Fig. 5.3) and showed that partitioning of solutes in this hydrophobic deep eutectic solvent/water and in octanol/water systems were strongly correlated. This means that distributions of organic solutes between hydrophobic deep eutectic solvents and water can be predicted and understood.

The authors have highlighted the potential to tune hydrophobic deep eutectic solvents to modify the distribution ratios, suggesting that the perspectives are at least comparable to those previously demonstrated with hydrophobic ionic liquids. In this context, it must be noted that extraction of organic compounds with hydrophobic deep eutectic solvents is still in its very naissance, and even if the initial results, discussed below, did not exceed industrial benchmarks, they must be taken as a "learning set" for further improvements and fine-tuning, rather than a negative outcome discouraging further trials.

In 2016, Kroon and co-workers (van Osch et al. 2016) first described the use of their hydrophobic tetraalkylammonium salt-containing deep eutectic solvents to extract volatile fatty acids (C_2-C_4) from aqueous media. Extraction efficiencies reported were greater than those using trioctylamine as a benchmark extractant as shown in Table 5.1. In the trioctylamine system, extraction would be expected to benefit from partial proton transfer from acids to the amine, resulting in the

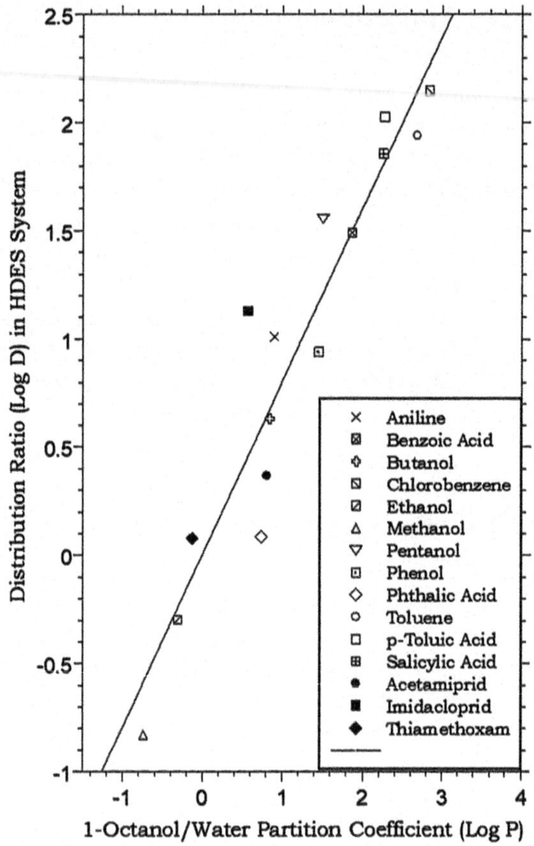

Fig. 5.3 Comparison of octanol/water and hydrophobic deep eutectic solvent/water (deep eutectic solvent = menthol/decanoic acid) partitioning of organic solutes showing a linear correlation of partition ratios (Kaul et al. 2019). Published by the Royal Society of Chemistry

formation of a "poor" protic ionic liquid (Yoshizawa et al. 2003; MacFarlane and Seddon 2007). In the case of hydrophobic deep eutectic solvents, enhancement was attributed to increased hydrogen bonding ability in the deep eutectic solvent. Extraction efficiency (Table 5.1) increased with increasing hydrophobicity of the acids extracted, reaching 38.0, 70.5, and 89.8% for acetic, propionic, and butyric acids, respectively. However, there was no compelling correlation to the hydrophobicity of the ammonium salt components of the deep eutectic solvents.

Following on from the first studies by Kroon and co-workers (van Osch et al. 2016), and the subsequent identification of different hydrophobic deep eutectic solvent motifs (discussed above), the majority of work has been directed at the use of hydrophobic deep eutectic solvents as extractants of biomolecules from aqueous media (Ribeiro et al. 2015; Krízek et al. 2018). The key classes of molecules that have been explored are listed in Table 5.2.

Hydrophobic deep eutectic solvents based on menthol and natural organic acids were proposed as solvents to extraction of caffeine, vanillic acid, tetracycline, and tryptophan, tested as model biomolecules of commercial interest. Unfortunately, partition coefficients were poorer than the corresponding highest performing

Table 5.1 Comparison of volatile fatty acid extraction using novel hydrophobic deep eutectic solvents and trioctylamine benchmark. The extraction efficiency increased with increasing acid component hydrophobicity

	Extraction efficiency / %		
Deep eutectic solvent	Acetic acid	Propionic acid	Butyric acid
DecA:[N_{8881}]Cl (2:1)	38.0	70.5	89.8
DecA:[N_{7777}]Cl (2:1)	32.0	76.5	91.5
DecA:[N_{8888}]Cl (2:1)	25.0	52.7	81.3
DecA:[N_{8881}]Br (2:1)	29.7	63.4	83.1
DecA:[N_{8888}]Br (2:1)	30.6	65.9	87.4
Trioctylamine	18.6	45.9	73.5

Data taken from van Osch et al. (2015)

Table 5.2 The key classes of biomolecule partitioning from aqueous media using hydrophobic deep eutectic solvents

Biomolecule class	References
Short-chain alcohols	Verma and Banerjee (2018) and Verma et al. (2018)
Furfurals	Dietz et al. (2019b)
Caffeine	Ribeiro et al. (2015) and Krízek et al. (2018)
Vanillin	Ribeiro et al. (2015) and Krízek et al. (2018)
Lycopene	Silva et al. (2019)
Pesticides (e.g., neonicotinoids)	Florindo et al. (2017)
Phenolics	Florindo et al. (2018a, b) and Sas et al. (2019)
Volatile fatty acid	van den Bruinhorst et al. (2019)

literature benchmarks (Ribeiro et al. 2015). Tryptophan was the only molecule from this study where the partition coefficient between a 2:1 DL-menthol/lauric acid deep eutectic solvent and water was positive to the extractant from water ($K_i = 14.5$), falling within the range of other benchmark solvent that ranged from $K_i = 4.5$ with 1-ethyl-3-methylimidazolium bis(trifluorosulfonylimide) [C_2mim][NTf_2] (Tomé et al. 2010) to $K_i = 124$ with a 1-allyl-3-methylimidazolium chloride/tripotassium phosphate [Amim]Cl/K_3PO_4 aqueous biphasic extraction system (Neves et al. 2009; Ventura et al. 2009; Ribeiro et al. 2015). Nevertheless, such hydrophobic deep eutectic solvents based on menthol and organic acids have been reported to be superior to organic solvents (e.g., methanol, ethanol, methanol:chloroform mixtures) for the extraction of phytocannabinoids (Krízek et al. 2018), highlighting the opportunities to use new, biodegradable, and pharmaceutically acceptable deep eutectic solvents as alternatives to organic solvents in biomolecule extraction.

Verma and co-workers examined short-chain alcohol extraction from aqueous media with menthol/carboxylic acid deep eutectic solvents, with interest in renewable biofuel production from acetone-butanol-ethanol fermentation where water-rich azeotropes are formed (Verma and Banerjee 2018; Verma et al. 2018). Distribution coefficients for lower alcohols followed the order butanol (8.9) > propanol (3.2) > ethanol (0.0505), consistent with the hydrophobicity of the alcohols. Molecular dynamics simulations demonstrated that menthol has a much greater role as an active extractant than the carboxylic component. A flow system for extraction and isolation of 1-butanol with recycling of the extractant deep eutectic solvent was described (Fig. 5.4).

Other target molecules from biorefinery sources, furfural and hydroxymethylfurfural, were isolated using hydrophobic deep eutectic solvents combining menthol, lidocaine, thymol, or atropine with decanoic or dodecanoic acids (Fig. 5.5). Promisingly, all these systems outperformed toluene as the benchmark solvent for hydroxymethylfurfural extraction (Dietz et al. 2019a, b). The thymol/decanoic acid deep eutectic solvent was identified as the best medium for both furfural and hydroxymethylfurfural extractions, with partition coefficients of K ≈ >10 and 1.75, respectively, compared to values of K ≈ 4.2 and 0.0 with toluene.

Hydrophobic deep eutectic solvents containing long-chain alcohols instead of carboxylic acids have also been investigated as extractants and used to access

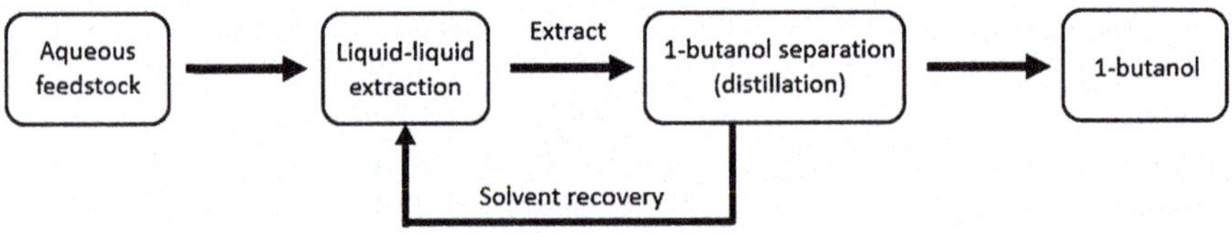

Fig. 5.4 Flow chart for experimental cycle of short-chain alcohol extraction from aqueous media (Ishii et al. 1985; Qureshi and Maddox 2005; Verma et al. 2018). (Reprinted with permission from Verma et al. 2018. Copyright 2018 American Chemical Society)

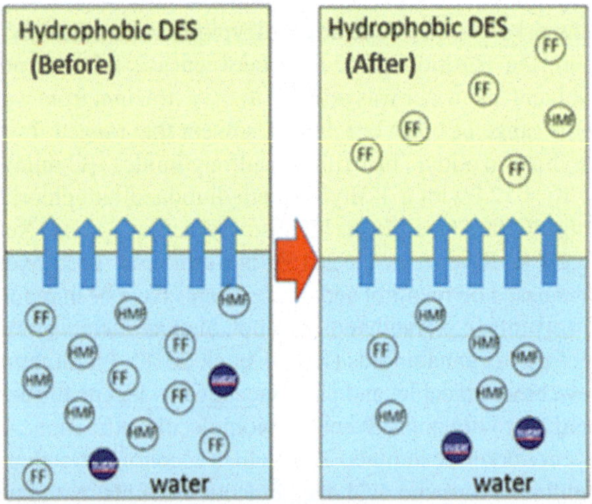

Fig. 5.5 Partitioning of furfural (FF) and hydroxymethylfurfural (HMF) between aqueous and deep eutectic solvent phases. A thymol/decanoic acid hydrophobic deep eutectic solvent had the best extraction performance. (Reprinted with permission from Dietz et al. 2019b. Copyright (2019) American Chemical Society)

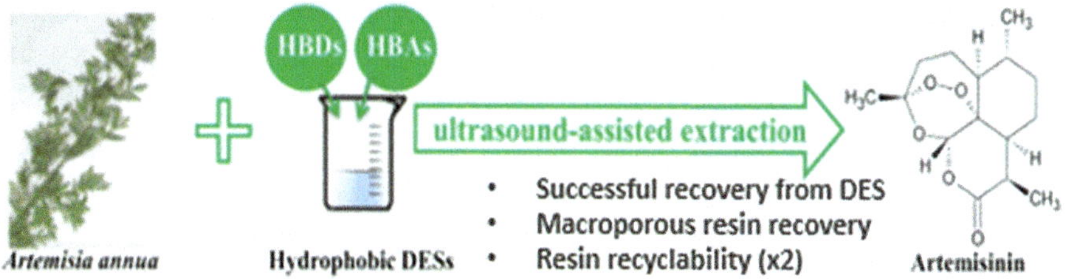

Fig. 5.6 Simplistic representation of extraction procedure to remove artemisinin from *Artemisia annua* leaves. 1:4 methyltrioctylammonium chloride/1-butanol gave the best extraction efficiency. (Reprinted with permissions from Cao et al. 2017b. Copyright (2017) American Chemical Society)

bioactive compounds from biomass (Cao et al. 2017a, b, 2018; de Faria et al. 2017) and to extract pesticides (e.g., neonicotinoids) known to have harmful effect on bee populations (Florindo et al. 2017). Some encouraging results were reported, such as the extraction of an antimalarial agent, artemisinin, from sweet wormwood leaves (*Artemisia annua*). Combining a tetraalkylammonium chloride/butanol deep eutectic solvent (Cao et al. 2017b) with ultrasound treatment (Fig. 5.6), higher yields were achieved compared to conventional solvent extraction such as with petroleum ether (Lapkin et al. 2006; Zhang et al. 2018).

Following on with the theme of contaminants removal, hydrophobic deep eutectic solvents were studied for the extraction of model phenolic pollutants (4-nitrophenol, 2,4-dinitrophenol, and phenol red) from a simulated wastewater stream (Tiecco et al. 2019). Combinations of a range of potential hydrogen bond donors (e.g., thymol, phenylacetic acid, and glycine) and hydrogen bond acceptors (e.g., trimethylglycine and N,N-dimethyl-N,N-didodecylammonium chloride) were considered first for formation of hydrophobic deep eutectic solvents (Fig. 5.7). Biphasic hydrophobic deep eutectic solvent/water mixtures were achieved combining both N,N-dimethyl-N,N-didodecylammonium chloride and trimethylglycine with thymol. 2,4-Dinitrophenol extraction from water to the tetraalkylammonium salt-containing deep eutectic solvent was more effective than that to the zwitterionic trimethylglycine-containing deep eutectic solvents, with extraction efficiencies reaching 98–100% within 1 min of contact time.

The concurrent publication of two papers describing hydrophobic deep eutectic solvents containing trioctylphosphine oxide in 2018 by Gilmore et al. (Gilmore et al. 2018a) and by Kroon and co-workers (van den Bruinhorst et al. 2019) introduced a new component into the matrix for deep eutectic solvent forming materials: the basic extractant, trioctylphosphine oxide. The application of a trioctylphosphine oxide-phenol deep eutectic solvent (Gilmore et al. 2018a) to metal extraction is discussed later on in this chapter. Trioctylphosphine oxide/*N*,*N*-dihexylthiourea deep eutectic solvents (van den Bruinhorst et al. 2019) were examined as media to extract volatile fatty acids. Undissociated acids were extracted with distribution coefficients in the range $K_i = 0–5$, but the extraction efficiencies were lower than those using just trioctylphosphine oxide in a diluent, which is, as a matter of fact, one of the commercial applications of trioctylphosphine oxide. This lower

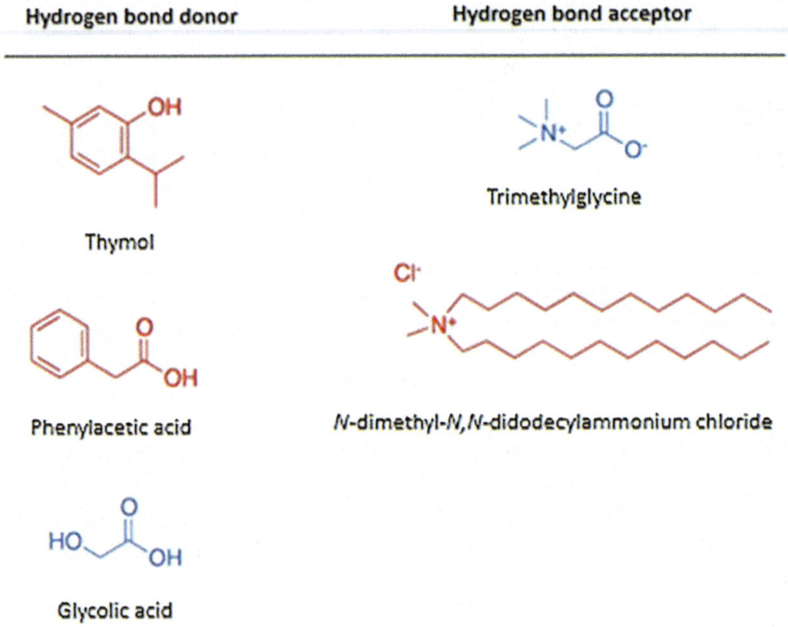

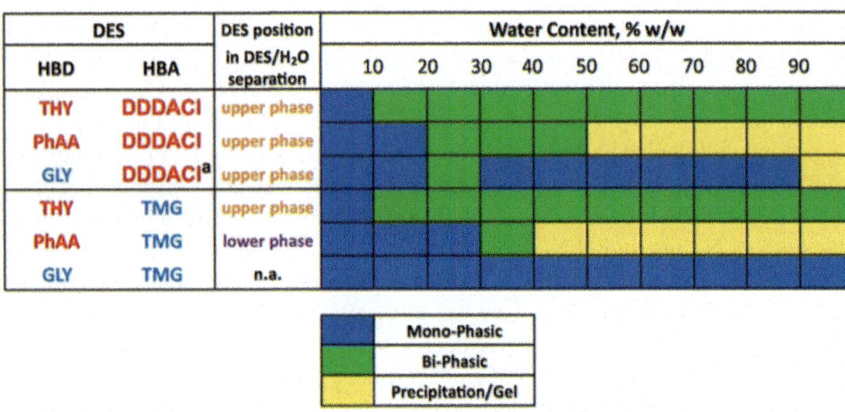

Fig. 5.7 Hydrogen bond acceptor (HBA) and donor (HBD) components (top) screened for hydrophobic deep eutectic solvents (DES) formation for targeting phenolic removal from water and (bottom) matrix of success for hydrophobic deep eutectic solvent/water biphasic mixture formation at $T = 25$ °C. The glycolic acid/trimethylglycine deep eutectic solvent did not form a biphase with water and was not possible to separate. Therefore, its position in deep eutectic solvent/water separation is not applicable (n.a.). Thymol (THY), phenylacetic acid (PhAA), glycolic acid (GLY), trimethylglycine (TMG), and *N*-dimethyl-*N*,*N*-didodecylammonium chloride (DDDACl). (Reprinted with permission from Tiecco et al. 2019. Copyright (2019) Elsevier B.V.)

extraction capability of deep eutectic solvent compared to trioctylphosphine oxide in a hydrocarbon diluent was attributed to reduced accessibility of trioctylphosphine oxide due to competitive hydrogen bonding with the *N,N*-dihexylthiourea component. However, given that urea-carboxylic acid hydrogen bonding is also a key molecular recognition process, it is possible that other competing factors must come into play (Jones et al. 2014).

Finally, deep eutectic solvents based on mixed long-chain carboxylic acid (Florindo et al. 2018a, b) have been examined for removal of bisphenol A, a common water micropollutant, from a water feed. Binary mixtures of dodecanoic acid with C_8-C_{10} fatty acids (Fig. 5.8) in 1:2 or 1:3 molar ratios were able to achieve 76–88% extraction of bisphenol A, with the C_{12}:C_9 (1:3 ratio) deep eutectic solvents displaying the best performance. Ternary mixtures of fatty acids were also tested and higher extraction efficiencies, in the range 78–92%, were reported.

Although the extraction performance of the mixed acid deep eutectic solvents was less effective than comparative K_3PO_4/ionic liquid extractants (Passos et al. 2012), only a minimal leaching of free fatty acids to the aqueous phase was observed. This advantage could potentially reduce the degree of subsequent downstream water treatment required, although in this work the stripping of bisphenol A and recovery of the deep eutectic solvent phase was not reported.

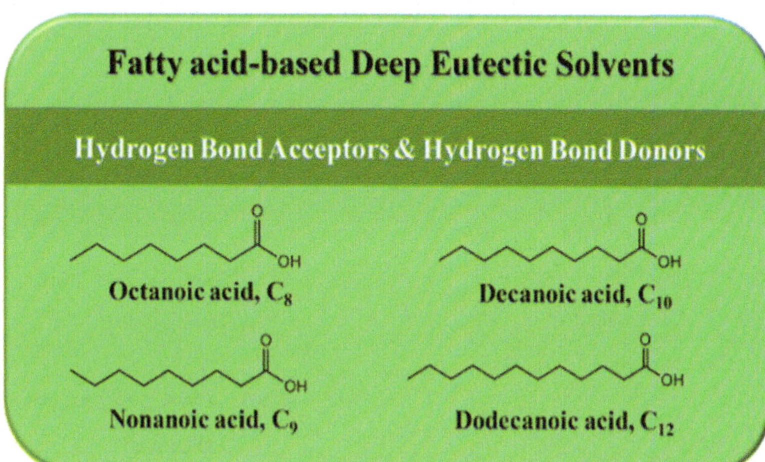

Fig. 5.8 Components of binary and ternary mixtures of fatty acid–based hydrophobic deep eutectic solvents, without explicit hydrogen bond donors/acceptors. Ternary mixtures gave highest extraction efficiencies. Reprinted with permission from Florindo et al. 2018a, b. Copyright (2018) American Chemical Society

5.3.2 Extraction of Metals

Extraction of metal species into nonaqueous media usually requires the presence of either a metal complexing agent or a sacrificial ion exchange component. Both strategies have been used to achieve metal ion partitioning with hydrophobic ionic liquids (Han and Armstrong 2007; Papaiconomou et al. 2008). Although the performance of ionic liquids with small, hydrophobic, perfluorinated anions (hexafluorophosphate, bistriflimide, etc.) is very good, their hydrolytic instability and/or very high cost effectively banishes practical uses in most cases. Lower cost ionic liquids, featuring hydrophobic tetraalkylammonium or tetraalkylphosphonium cations, have been demonstrated to perform well, although practical challenges associated with the high viscosities of these materials still have to be addressed (Binnemans 2007).

Thus, the objectives in examining metal extraction with hydrophobic deep eutectic solvents are to eliminate the need for fluorous anions and to address some of the high viscosity characteristics of the typical tetraalkylammonium and phosphonium ionic liquids.

The first study of metal partitioning with hydrophobic deep eutectic solvents (Tereshatov et al. 2016) described the extraction of anionic indate(III) species from hydrochloric acid and oxalic acid solutions using hydrophobic deep eutectic solvents comprising either tetraalkylammonium chloride or DL-menthol with carboxylic acids such as lauric, decanoic, and oleic acid and ibuprofen as shown in Fig. 5.9.

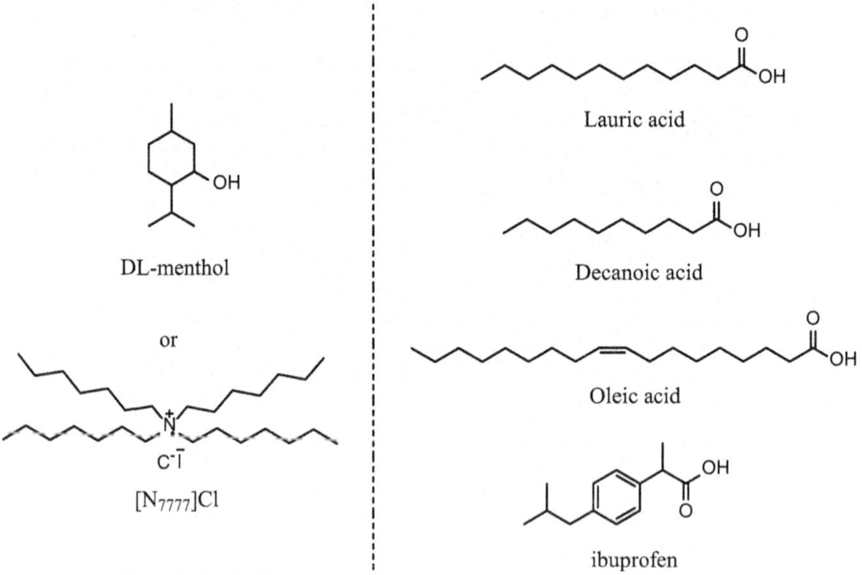

Fig. 5.9 Components used in the preparation of hydrophobic deep eutectic solvents studied for indium extraction from aqueous media. Reprinted with permission from Tereshatov et al. 2016. Copyright (2016) Royal Society of Chemistry

This work follows directly from, and complements, previous exploration of indate(III) anion extraction using tetraalkylphosphonium carboxylate ionic liquids, and one aspect of this study was to explore hydrophobic deep eutectic solvents with lower viscosities than the phosphonium ionic liquids. Although it must be noted that data available tends to relate to "pure" viscosities rather than those of water-saturated deep eutectic solvents, i.e., under operating conditions, deep eutectic solvents containing quaternary ammonium salts showed good extraction through ion exchange of anionic indium species across a wide acid concentration. In contrast, with the menthol-based deep eutectic solvents, poorer indium distribution was observed (Fig. 5.10). Successful back extraction from ammonium-based deep eutectic solvents was achieved by complexation with aqueous diethylenetriaminepentaacetic acid with D_{In} in the range 10^{-1} to 10^{-2}.

Similarly, the extraction of tracer levels of relatively hydrophobic pertechnetate ($[^{99m}TcO_4]^-$) anions from aqueous sources with >99% efficiency has been described using hydrophobic deep eutectic solvents containing 1:2 molar ratios of quaternary ammonium or phosphonium halide salts combined with long-chain fatty acids (Phelps et al. 2018) as shown in Fig. 5.11.

Pertechnetate contamination in wastewater is a major concern at nuclear reactor sites, while aqueous pertechnetate is the primary delivery form of ^{99m}Tc for use of medical positron emission tomography (PET) imaging and diagnostics. Phelps et al.

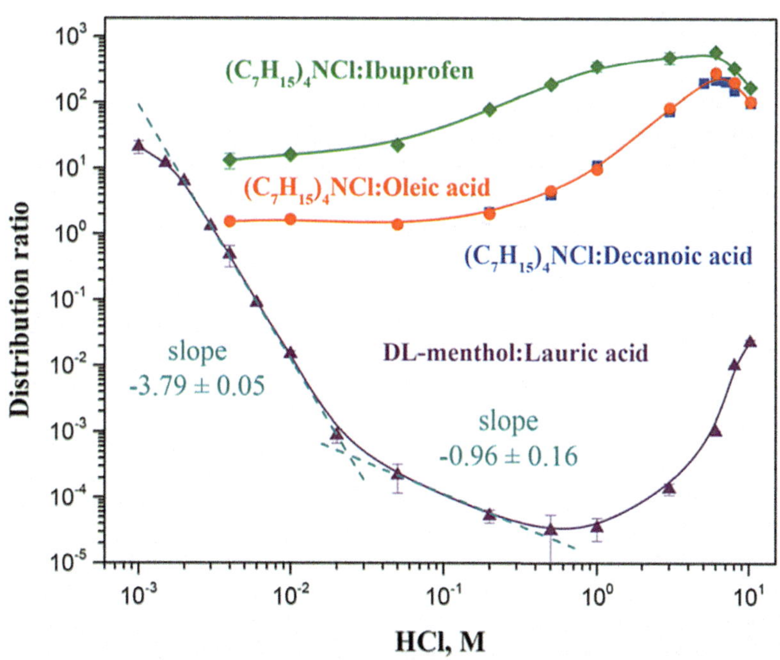

Fig. 5.10 Impact of aqueous hydrochloric acid concentration on the distribution ratio of In into quaternary ammonium and menthol-based hydrophobic deep eutectic solvents. (Reprinted with permission from Tereshatov et al. 2016. Copyright (2016) Royal Society of Chemistry)

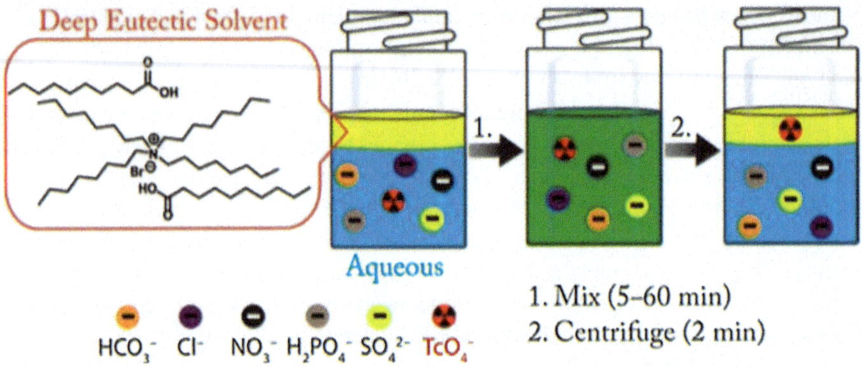

Fig. 5.11 Diagrammatic representation of experimental procedure and outcome from pertechnetate ($[TcO_4]^-$) extraction from electrolyte-rich aqueous media with tetraalkylammonium bromide/long-chain fatty acid deep eutectic solvents highly selective extraction with rejection of other aqueous anions. (Reprinted with permission from Phelps et al. 2018. Copyright (2018) American Chemical Society)

demonstrated distribution coefficients log*D* of 2–3 for $[^{99m}TcO_4]^-$ in the presence of a large excess of competing ions present in typical radioactive wastewater such as bicarbonate (HCO_3^-), chloride (Cl^-), nitrate (NO_3^-), etc. Perrhenate ($[ReO_4]^-$) was also extracted and one could anticipate that these deep eutectic solvents might also be effective for remediation of other hydrophobic anionic pollutants such as perchlorate ($[ClO_4]^-$).

Extraction of transition metal salts from water with lidocaine/decanoic acid mixtures was reported by Kroon and co-workers (van Osch et al. 2016). Lidocaine is an ionizable amine component which when combined with decanoic acid in a 1:2 molar ratio is approximately 25% ionized (Bica et al. 2011; Griffin et al. 2014), and consequently, these extractants bridge between non-ionizable deep eutectic solvents and ionic liquids. Transition metal chlorides (e.g., $CoCl_2$, $FeCl_2$, $MnCl_2$) were extracted non-selectively with high distribution coefficients ($D \approx 0.78$–1.0) in both individual and mixed metal samples through ion exchange mechanisms. In contrast, alkali metal ions such as K^+ had low distribution coefficients ($D \approx 0.13$–0.46) that were attributed to the lack of coordination of alkali metals by long-chain fatty acids in solution (Preston 1985).

Metal ion extraction with DL-menthol and thymol-based deep eutectic solvents containing long-chain carboxylic acids showed good selectivity to Cu(II) and Fe(II), while the extraction of Mg(II), Ca(II), Cr(III), Mn(II), Co(II), and Ni(II) was negligible (Schaeffer et al. 2018). This contrasts with non-selective extraction of all transition metal ions using lidocaine/decanoic acid deep eutectic solvents containing an ionizable amine component (van Osch et al. 2016).

The development of a nonionic hydrophobic deep eutectic solvent from trioctylphosphine oxide/phenol mixtures (Gilmore et al. 2018a) with a significantly lower viscosity than those of tetraalkylammonium or phosphonium salt containing ionic liquids or deep eutectic solvents demonstrates a further way in which

Table 5.3 Extent of uranyl nitrate extraction from aqueous feedstock as a function of nitric acid and uranyl nitrate concentrations. 1 cm^3 trioctylphosphine oxide:phenol (χ_{TOPO} = 0.50) was pre-equilibrated with nitric acid solution and then contacted with 1 cm^3 of a uranyl-containing feed (shaken 10 mins, ambient temperature) and separated by centrifuging. In all cases, extraction to the limits of detection was achieved. (Reprinted with permission from Gilmore et al. 2018a. Copyright (2018) American Chemical Society)

Feed acidity / M	$[UO_2^{2+}]_{init}$ / ppm	$[UO_2^{2+}]_{final}$ / ppm
0.01	250	<0.50
1.00	250	<0.50
3.00	250	<0.50
1.00	2350	<0.50

hydrophobic deep eutectic solvents can be used for the extraction of metal species. Trioctylphosphine oxide has many uses as an extractant, including primary processing of nuclear materials and treatment of radioactive waste to extract uranyl ($[UO_2]^{2+}$) ions (Jianchen and Chongli 2001; Bayliss and Langley 2003), and is usually used as <0.5 M solutions in kerosene or other hydrocarbon diluents. In contrast, the hydrophobic trioctylphosphine oxide/phenol deep eutectic solvent contains *ca.* 2 M trioctylphosphine oxide, and Gilmore et al. were able to demonstrate the use of this deep eutectic solvent as an efficient extractant for uranyl ions from simulated spent nuclear waste with an estimated extraction coefficient of 5×10^3 (Table 5.3).

5.3.3 *Other Applications*

Hydrophobic deep eutectic solvents have further uses and applications. Two of the most significant being the potential to be used in microextraction methods for pre-concentration of compounds and as media for CO_2 capture (Dietz et al. 2017; Zubeir et al. 2018); most examples use hydrophobic deep eutectic solvents formed by at least one ionic component. The use of hydrophobic deep eutectic solvents incorporating decanoic acid as the hydrogen bond donor species has been described for microextraction of pigments from drinks (Zhu et al. 2018) and dyes from colored wastewater (Ahmadi et al. 2019). Microextraction applications for hydrophobic deep eutectic solvents of this kind also extend as far as medical uses where they are used as solvents for the total determination of blood selenium content in combination with a diethyldithiophosphoric acid chelating agent (Akramipour et al. 2019). Other hydrophobic deep eutectic solvents with an ionic component have been used in microextraction systems to determine lead (Naeemullah and Tuzen 2019) and synthetic dyes (Ravandi and Fat'hi 2018; Faraji 2019) in food and/or water samples and nitrite in water and biological samples (Zhang et al. 2019). The isolation of antibiotics from environmental water (Tang et al. 2018) and determination of trace polycyclic aromatic hydrocarbons in simulated environmental water samples (Yousefi et al. 2018) have both been described using a $[N_{8881}]Cl$/1-octanol deep eutectic solvent. In addition, zwitterionic components combined with fluorinated alcohols (e.g., hexafluoroisopropanol) which increase hydrophobicity have been reported for microextraction of pyrethroid insecticides from tea and fruit juices (Deng et al. 2019), while liquid-liquid microextraction for the pre-concentration of pyrethroid (Liu et al. 2019) with hydrophobic quaternary phosphonium salt/straight chain carboxylic acid deep eutectic solvents has also been reported.

The use of choline chloride/phenol mixtures as deep eutectic solvents for pre-concentration and analysis of ionic and organomercury species water and freshwater fish samples after complexation with dithizone (for ionic mercury) or direct extraction (organomercury species) has been described (Thongsaw et al. 2019). Detection limits for Hg^{2+} and $[CH_3Hg]^+$ were 0.073 and 0.091 ng mL^{-1} and enrichment factors were 34.0 and 18.3, respectively. Microextraction for concentration and detection of trace concentrations of nickel in water samples (Rad et al. 2019) has employed a related choline chloride/bromophenol deep eutectic solvent, while other choline chloride/phenolic mixtures have been applied to microextraction of amphetamine and methamphetamine from human blood plasma and pharmaceutical wastewater using a choline chloride/2-phenylethanol deep eutectic solvent (Rajabi et al. 2018). Rajabi et al. also reported the use of hollow fiber microextraction with choline chloride/1-phenylethanol deep eutectic solvent for the concentration and detection of antiarrhythmic agents in biological and environmental samples (Rajabi et al. 2019). However, it is important to note that although these deep eutectic solvents form aqueous biphases, there is no data currently available on the leaching levels of either components of these deep eutectic solvents. While it is known that choline chloride is inherently hygroscopic and that phenol has been found to leach in small amounts into an aqueous phase from a trioctylphosphine oxide-based hydrophobic eutectic (Gilmore et al. 2018a), it is likely that extensive leaching of one or both components may occur.

Hydrophobic deep eutectic solvents comprised of nonionic components have also been applied to a number of microextractions. A DL-menthol/phenyl salicylate mixture has been used to concentrate nitroaromatic explosive compounds in the deep eutectic solvent phase, removing them from aqueous solution (Nedaei et al. 2018). Hydrophobic magnetic ferrofluids, prepared by dispersing magnetic nanoparticles in DL-menthol/carboxylic acid hydrophobic deep eutectic solvents, have been reported as extractants for explosive residues from soil and water samples and for mefenamic acid, an anti-inflammatory analgesic, from urine samples (Zarei et al. 2018; Dil et al. 2019). Finally, these nonionic component-based hydrophobic deep eutectic solvents have been utilized in combination with hydrophilic deep eutectic solvents in biphasic deep eutectic solvent systems consisting of hydrophilic and hydrophobic deep eutectic solvents, i.e., (−)-menthol/L(+)-lactic acid and choline chloride/urea, respectively, for enantiomeric separation in chiral extraction of threonine, achieving a 31.6% enantiomeric excess in a single stage separation (Wang et al. 2019).

5.4 Outlook

In conclusion, some encouraging preliminary results on the design and application of hydrophobic deep eutectic solvents have been reported in this new and rapidly developing field. It has a significant potential and is expected to flourish in the future, developing innovative hydrophobic deep eutectic solvents, with better tailored functionalities and improved performances. It has been demonstrated that functional tailored-to-application liquids can be formulated by mixing two or more components to obtain a room temperature liquid, expanding beyond definitions for either ionic liquids or deep eutectic solvents. Going forward, it is crucial to maintain continuity with the decades of knowledge accumulated in areas such as liquid-liquid extraction of biomolecules from aqueous feeds. For example, to understand how specific hydrophobic deep eutectic solvents perform in relation to octanol-water partitioning in order to build on the past knowledge of efficient extractants, combined with ionic liquids/deep eutectic solvents knowledge, to formulate functional liquids.

References

Abbott AP, Capper G, Davies DL, Rasheed RK, Tambyrajah V (2003) Novel solvent properties of choline chloride/urea mixtures. Chem Commun:70–71. https://doi.org/10.1039/b210714g

Abbott AP, Boothby D, Capper G, Davies DL, Rasheed R (2004) Deep eutectic solvents formed between choline chloride and carboxylic acids: versatile alternatives to ionic liquids. J Am Chem Soc 126(29):9142–9147. https://doi.org/10.1021/ja048266j

Ahmadi R, Kazemi G, Ramezani AM, Safavi A (2019) Shaker-assisted liquid-liquid microextraction of methylene blue using deep eutectic solvent followed by back-extraction and spectrophotometric determination. Microchem J 145:501–507. https://doi.org/10.1016/j.microc.2018.11.005

Akramipour R, Golpayegani MR, Ghasemi M, Noori N, Fattahi N (2019) Development of an efficient sample preparation method for the speciation of Se (IV)/ Se (VI) and total inorganic selenium in blood of children with acute leukemia. New J Chem 43:6951–6958. https://doi.org/10.1039/c9nj00979e

Araujo CF, Coutinho JAP, Nolasco MM, Parker SF, Ribeiro-Claro PJA, Rudic S, Soares BIG, Vaz PD (2017) Inelastic neutron scattering study of reline: shedding light on the hydrogen bonding network of deep eutectic solvents. Phys Chem Chem Phys 19:17998–18009. https://doi.org/10.1039/c7cp01286a

Ashworth CR, Matthews RP, Welton T, Hunt PA (2016) Doubly ionic hydrogen bond interactions within the choline chloride – urea deep eutectic solvent. Phys Chem Chem Phys 18:18145–18160. https://doi.org/10.1039/C6CP02815B

Bayliss C, Langley K (2003) Nuclear decommissioning, waste management, and environmental site remediation. Elsevier, Burlington

Bica K, Shamshina J, Hough WL, MacFarlane DR, Rogers RD (2011) Liquid forms of pharmaceutical co-crystals: exploring the boundaries of salt formation. Chem Commun 47:2267–2269. https://doi.org/10.1039/c0cc04485g

Binnemans K (2007) Lanthanides and actinides in ionic liquids. Chem Rev 107(6):2592–2614. https://doi.org/10.1021/cr050979c

Bonhote P, Dias A-P, Papageorgiou N, Kalyanasundaram K, Gratzel M (1996) Hydrophobic, highly conductive ambient-temperature molten salts. Inorg Chem 35(5):1168–1178. https://doi.org/10.1021/ic951325x

Cao J, Yang M, Cao F, Wang J, Su E (2017a) Tailor-made hydrophobic deep eutectic solvents for cleaner extraction of polyprenyl acetates from Ginkgo biloba leaves. J Clean Prod 152:399–405. https://doi.org/10.1016/j.jclepro.2017.03.140

Cao J, Yang M, Cao F, Wang J, Su E (2017b) Well-designed hydrophobic deep eutectic solvents as green and efficient Media for the Extraction of Artemisinin from Artemisia annua leaves. ACS Sustain Chem Eng 5(4):3270–3278. https://doi.org/10.1021/acssuschemeng.6b03092

Cao J, Chen L, Li M, Cao F, Zhao L, Su E (2018) Two-phase systems developed with hydrophilic and hydrophobic deep eutectic solvents for simultaneously extracting various bioactive compounds with different polarities. Green Chem 20:1879–1886. https://doi.org/10.1039/C7GC03820H

de Faria ELP, do Carmo RS, Cláudio AFM, Freire CSR, Freire MG, Silvestre AJD (2017) Deep eutectic solvents as efficient Media for the Extraction and Recovery of Cynaropicrin from Cynara cardunculus L. leaves. Int J Mol Sci 18(11):2276–2285. https://doi.org/10.3390/ijms18112276

Deng W, Yu L, Li X, Chen J, Wang X, Deng Z, Xiao Y (2019) Hexafluoroisopropanol-based hydrophobic deep eutectic solvents for dispersive liquid-liquid microextraction of pyrethroids in tea beverages and fruit juices. Food Chem 274:891–899. https://doi.org/10.1016/j.foodchem.2018.09.048

Dietz CHJT, van Osch DJGP, Kroon MC, Sadowski G, Annaland MVS, Gallucci F, Zubeir LF, Held C (2017) PC-SAFT modeling of CO2 solubilities in hydrophobic deep eutectic solvents. Fluid Phase Equilib 448:94–98. https://doi.org/10.1016/j.fluid.2017.03.028

Dietz CHJT, Erve A, Kroon MC, van Sint Annaland M, Gallucci F, Held C (2019a) Thermodynamic properties of hydrophobic deep eutectic solvents and solubility of water and HMF in them: measurements and PC-SAFT modeling. Fluid Phase Equilib 489:75–82. https://doi.org/10.1016/j.fluid.2019.02.010

Dietz CHJT, Gallucci F, Annaland M v S, Held C, Kroon MC (2019b) 110th anniversary: distribution coefficients of furfural and 5-Hydroxymethylfurfural in hydrophobic deep eutectic solvent + water systems: experiments and perturbed-chain statistical associating fluid theory predictions. Ind Eng Chem Res 58(10):4240–4247. https://doi.org/10.1021/acs.iecr.8b06234

Dil EA, Ghaedi M, Asfaram A (2019) Application of hydrophobic deep eutectic solvent as the carrier for ferrofluid: a novel strategy for pre-concentration and determination of mefenamic acid in human urine samples by high performance liquid chromatography under experimental design optimization. Talanta 202:526–530. https://doi.org/10.1016/j.talanta.2019.05.027

Faraji M (2019) Determination of some red dyes in food samples using a hydrophobic deep eutectic solvent-based vortex assisted dispersive liquid-liquid microextraction coupled with high performance liquid chromatography. J Chromatogr A 1591:15–23. https://doi.org/10.1016/j.chroma.2019.01.022

Florindo C, Branco LC, Marrucho IM (2017) Development of hydrophobic deep eutectic solvents for extraction of pesticides from aqueous environments. Fluid Phase Equilib 448:135–142. https://doi.org/10.1016/j.fluid.2017.04.002

Florindo C, McIntosh AJS, Welton T, Branco LC, Marrucho IM (2018a) A closer look into deep eutectic solvents: exploring intermolecular interactions using solvatochromic probes. Phys Chem Chem Phys 20:206–213. https://doi.org/10.1039/C7CP06471C

Florindo C, Romero L, Rintoul I, Branco LC, Marrucho IM (2018b) From phase change materials to green solvents: hydrophobic low viscous fatty acid – based deep eutectic solvents. ACS Sustain Chem Eng 6(3):3888–3895. https://doi.org/10.1021/acssuschemeng.7b04235

Francisco M, van den Bruinhorst A, Kroon MC (2013) Low-transition-temperature mixtures (LTTMs): a new generation of designer solvents. Angew Chem Int Ed 52:3074–3085. https://doi.org/10.1002/anie.201207548

Freire MG, Cláudio AFM, Araújo JMM, Coutinho JAP, Marrucho IM, Lopes JNC, Rebelo LPN (2012) Aqueous biphasic systems: a boost brought about by using ionic liquids. Chem Soc Rev 41:4966–4995. https://doi.org/10.1039/c2cs35151j

George A, Brandt A, Tran K, Zahari SMSNS, Klein-Marcuschamer D, Sun N, Sathitsuksanoh N, Shi J, Stavila V, Parthasarathi R, Singh S, Holmes BM, Welton T, Simmons BA, Hallett JP (2015) Design of low-cost ionic liquids for lignocellulosic biomass pretreatment. Green Chem 17:1728–1734. https://doi.org/10.1039/c4gc01208a

Gilmore M, Mccourt ÉN, Connolly F, Nockemann P, Swadźba-kwaśny M, Holbrey JD (2018a) Hydrophobic deep eutectic solvents incorporating Trioctylphosphine oxide: advanced liquid Extractants. ACS Sustain Chem Eng 6:17323–17332. https://doi.org/10.1021/acssuschemeng.8b04843

Gilmore M, Moura LM, Turner AH, Swadźba-kwaśny M, Callear SK, Mc Cune JA, Scherman OA, Holbrey JD (2018b) A comparison of choline:urea and choline:oxalic acid deep eutectic solvents at 338 K. J Chem Phys 148:193823–193833. https://doi.org/10.1063/1.5010246

Griffin PJ, Cosby T, Holt AP, Benson RS, Sangoro JR (2014) Charge transport and structural dynamics in carboxylic-acid-based deep eutectic mixtures. J Phys Chem B 118(31):9378–9385. https://doi.org/10.1021/jp503105g

Gutowski KE, Broker GA, Willauer HD, Huddleston JG, Swatloski RP, Holbrey JD, Rogers RD (2003) Controlling the aqueous miscibility of ionic liquids: aqueous biphasic Systems of Water-Miscible Ionic Liquids and Water-Structuring Salts for recycle, metathesis, and separations. J Am Chem Soc 125(22):6632–6633. https://doi.org/10.1021/ja0351802

Hammond OS, Bowron DT, Edler KJ (2016) Liquid structure of the choline chloride-urea deep eutectic solvent (reline) from neutron diffraction and atomistic modelling. *Green Chem* 18:2736–2744. https://doi.org/10.1039/c5gc02914g

Han X, Armstrong DW (2007) Ionic liquids in separations. Acc Chem Res 40(11):1079–1086. https://doi.org/10.1021/ar700044y

Huddleston JG, Visser AE, Reichert WM, Willauer HD, Broker GA, Rogers RD (2001) Characterization and comparison of hydrophilic and hydrophobic room temperature ionic liquids incorporating the imidazolium cation. Green Chem 3:156–164. https://doi.org/10.1039/b103275p

Hyde AM, Zultanski SL, Waldman JH, Zhong Y-L, Shevlin M, Peng F (2017) General principles and strategies for salting-out informed by the Hofmeister series. Org Process Res Dev 21(9):1355–1370. https://doi.org/10.1021/acs.oprd.7b00197

Ishii S, Taya M, Kobayashi T (1985) Production of butanol by Clostridium Acetobutylicum in Extractive fermentation system. J Chem Eng Jpn 18(2):125–130

Jianchen W, Chongli S (2001) Hot test of trialkyl phosphine oxide (TRPO) for removing actinides from highly saline high-level liquid waste (HLLW). Solv Extn Ion Exch 19(2):231–242. https://doi.org/10.1081/SEI-100102693

Jones AOF, Leech CK, McIntyre GJ, Wilson CC, Thomas LH (2014) Engineering short, strong hydrogen bonds in urea di-carboxylic acid complexes. CrystEngComm 16:8177–8184. https://doi.org/10.1039/c4ce00587b

Kaul MJ, Qadah D, Mandella V, Dietz ML (2019) Systematic evaluation of hydrophobic deep-melting eutectics as alternative solvents for the extraction of organic solutes from aqueous solution. RSC Adv 9:15798–15804. https://doi.org/10.1039/c9ra01596e

Krízek T, Bursová M, Horsley R, Kuchar M, Tuma P, Cabala R, Hlozek T (2018) Menthol-based hydrophobic deep eutectic solvents: towards greener and efficient extraction of phytocannabinoids. J Clean Prod 193:391–396. https://doi.org/10.1016/j.jclepro.2018.05.080

Lapkin AA, Plucinski PK, Cutler M (2006) Comparative assessment of Technologies for Extraction of Artemisinin. J Nat Prod 69(11):1653–1664. https://doi.org/10.1021/np060375j

Liu X, Liu C, Qian H, Qu Y, Zhang S, Lu R, Gao H, Zhou W (2019) Ultrasound-assisted dispersive liquid-liquid microextraction based on a hydrophobic deep eutectic solvent for the preconcentration of pyrethroid insecticides prior to determination by high-performance liquid chromatography. Microchem J 146:614–621. https://doi.org/10.1016/j.microc.2019.01.048

MacFarlane DR, Seddon KR (2007) Ionic liquids — Progress on the fundamental issues. Aust J Chem 60:3–5. https://doi.org/10.1071/CH06478

Martins MAR, Pinho SP, Coutinho JAP (2019) Insights into the nature of eutectic and deep eutectic mixtures. J Solut Chem 48(7):962–982. https://doi.org/10.1007/s10953-018-0793-1

McQueen L, Lai D (2019) Ionic liquid aqueous two-phase systems from a pharmaceutical perspective. Front Chem 7:1–9. https://doi.org/10.3389/fchem.2019.00135

Naeemullah, Tuzen M (2019) 'A new robust, deep eutectic-based floating organic droplets microextraction method for determination of lead in a portable syringe system directly couple with FAAS', *Talanta*. Elsevier B.V., 196(October 2018), pp 71–77. doi: https://doi.org/10.1016/j.talanta.2018.12.027

Nedaei M, Zarei AR, Ghorbanian SA (2018) Development of a new emulsification microextraction method based on solidification of settled organic drop: application of a novel ultra-hydrophobic tailor-made deep eutectic solvent. New J Chem 42:12520–12529. https://doi.org/10.1039/c8nj02219d

Neves CMSS, Ventura SPM, Freire MG, Marrucho IM, Coutinho JAP (2009) Evaluation of Cation influence on the formation and extraction capability of ionic-liquid-based aqueous biphasic systems. J Phys Chem B 113(15):5194–5199

Oppermann S, Stein F, Kragl U (2011) Ionic liquids for two-phase systems and their application for purification, extraction and biocatalysis. Appl Microbiol Biotechnol 89:493–499. https://doi.org/10.1007/s00253-010-2933-4

Papaiconomou N, Lee J, Salminen J, von Stosch M, Prausnitz JM (2008) Selective extraction of copper, mercury, silver, and palladium ions from water using hydrophobic ionic liquids. Ind Eng Chem Res 47(15):5080–5086

Passos H, Sousa ACA, Pastorinho MR, Nogueira AJA, Rebelo LPN, Coutinho JAP, Freire MG (2012) Analytical methods ionic-liquid-based aqueous biphasic systems for improved detection of bisphenol a in human fluids. Anal Methods 4:2664–2667. https://doi.org/10.1039/c2ay25536g

Pei Y, Wang J, Wu K, Xuan X, Lu X (2009) Ionic liquid-based aqueous two-phase extraction of selected proteins. Sep Purif Technol 64(3):288–295. https://doi.org/10.1016/j.seppur.2008.10.010

Perkins SL, Painter P, Colina CM (2014) Experimental and computational studies of choline chloride-based deep eutectic solvents. J Chem Eng Data 59(11):3652–3662. https://doi.org/10.1021/je500520h

Phelps TE, Bhawawet N, Jurisson SS, Baker GA (2018) Efficient and selective extraction of 99mTcO4– from aqueous media using hydrophobic deep eutectic solvents. ACS Sustain Chem Eng 6(11):13656–13661. https://doi.org/10.1021/acssuschemeng.8b03950

Preston JS (1985) Solvent extraction of metals by carboxylic acids. Hydrometallurgy 14(2):171–188. https://doi.org/10.1016/0304-386X(85)90032-5

Qureshi N, Maddox IS (2005) Reduction in butanol inhibition by Perstraction: utilization of concentrated lactose/whey permeate by Clostridium acetobutylicum to enhance butanol fermentation economics. Food Bioprod Process 83(1):43–52. https://doi.org/10.1205/fbp.04163

Rad AS, Rahnama R, Zakeri M, Jamali MR (2019) Dispersive liquid – liquid microextraction based on green type solvents — "deep eutectic solvents" — for highly selective separation and efficient preconcentration of nickel in water samples. J Iran Chem Soc 16(8):1715–1722. https://doi.org/10.1007/s13738-019-01643-0

Rajabi M, Ghassab N, Hemmati M, Asghari A (2018) Emulsification microextraction of amphetamine and methamphetamine in complex matrices using an up-to-date generation of eco-friendly and relatively hydrophobic deep eutectic solvent. J Chromatogr A 1576:1–9. https://doi.org/10.1016/j.chroma.2018.07.040

Rajabi M, Ghassab N, Hemmati M, Asghari A (2019) Highly effective and safe intermediate based on deep eutectic medium for carrier less-three phase hollow fiber microextraction of antiarrhythmic agents in complex matrices. J Chromatogr B 1104:196–204. https://doi.org/10.1016/j.jchromb.2018.11.008

Ravandi MG, Fat'hi MR (2018) Green effervescence assisted dispersive liquid–liquid microextraction based on a hydrophobic deep eutectic solvent for determination of sunset yellow and brilliant blue FCF in food samples. New J Chem 42:14901–14908. https://doi.org/10.1039/c8nj00782a

Ribeiro BD, Florindo C, Iff LC, Coelho MAZ, Marrucho IM (2015) Menthol-based eutectic mixtures: hydrophobic low viscosity solvents. ACS Sustain Chem Eng 3(10):2469–2477. https://doi.org/10.1021/acssuschemeng.5b00532

Sangster J (1989) Octanol-water partition coefficients of simple organic compounds. J Phys Chem Ref Data 18:1111–1229. https://doi.org/10.1063/1.555833

Sas OG, Castro M, Domínguez Á, González B (2019) Removing phenolic pollutants using deep eutectic solvents. Sep Purif Technol 227:115703–115708. https://doi.org/10.1016/j.seppur.2019.115703

Schaeffer N, Martins MAR, Neves CMSS, Pinho SP, Coutinho JAP (2018) Sustainable hydrophobic terpene-based eutectic solvents for the extraction and separation of metals. Chem Commun 54:8104–8107. https://doi.org/10.1039/c8cc04152k

Seddon KR, Stark A, Torres M (2002) Viscosity and density of 1-Alkyl-3-methylimidazolium ionic liquids. In: Abraham MA, Moens L (eds) Clean solvents. ACS Symposium Series; American Chemical Society, Washington, DC, pp 34–49

Shukla SK, Pandey S, Pandey S (2018) Applications of ionic liquids in biphasic separation: aqueous biphasic systems and liquid-liquid equilibria. J Chromatogr A 1559:44–61. https://doi.org/10.1016/j.chroma.2017.10.019

Silva LP, Fernandez L, Conceicao JHF, Martins MAR, Sosa A, Ortega J, Pinho SP, Coutinho JAP (2018) Design and characterization of sugar-based deep eutectic solvents using conductor-like screening model for real solvents. ACS Sustain Chem Eng 6(8):10724–10734. https://doi.org/10.1021/acssuschemeng.8b02042

Silva YPA, Ferreira TAPC, Jiao G, Brooks MS (2019) Sustainable approach for lycopene extraction from tomato processing by-product using hydrophobic eutectic solvents. J Food Sci Technol 56(3):1649–1654. https://doi.org/10.1007/s13197-019-03618-8

Smith EL, Abbott AP, Ryder KS (2014) Deep eutectic solvents (DESs) and their applications. Chem Rev 114(21):11060–11082. https://doi.org/10.1021/cr300162p

Souza AGO, Aliprandini P, Espinosa DCR, Tenório JAS (2019) Scandium extraction from nickel processing waste using Cyanex 923 in sulfuric medium. JOM 71(6):2003–2009. https://doi.org/10.1007/s11837-019-03427-6

Stefanovic R, Ludwig M, Webber GB, Atkin R, Page AJ (2017) Nanostructure, hydrogen bonding and rheology in choline chloride deep eutectic solvents as a function of the hydrogen bond donor. Phys Chem Chem Phys 19:3297–3306. https://doi.org/10.1039/C6CP07932F

Tang W, Dai Y, Row KH (2018) Evaluation of fatty acid/alcohol-based hydrophobic deep eutectic solvents as media for extracting antibiotics from environmental water. Anal Bioanal Chem 410(28):7325–7336. https://doi.org/10.1007/s00216-018-1346-6

Tereshatov EE, Boloeva MY, Folden III CM (2016) First evidence of metal transfer into hydrophobic deep eutectic and low-transition-temperature mixtures: indium extraction from hydrochloric and oxalic acids. Green Chem 18:4616–4622. https://doi.org/10.1039/c5gc03080c

Thongsaw A, Udnan Y, Ross GM, Chaiyasith WC (2019) Speciation of mercury in water and biological samples by eco-friendly ultrasound-assisted deep eutectic solvent based on liquid phase microextraction with electrothermal atomic absorption spectrometry. Talanta 197:310–318. https://doi.org/10.1016/j.talanta.2019.01.018

Tiecco M, Cappellini F, Nicoletti F, Del Giacco T, Germani R, Di Profio P (2019) Role of the hydrogen bond donor component for a proper development of novel hydrophobic deep eutectic solvents. J Mol Liq 281:423–430. https://doi.org/10.1016/j.molliq.2019.02.107

Tomé LIN, Catambas VR, Teles ARR, Freire MG, Marrucho IM, Coutinho JAP (2010) Tryptophan extraction using hydrophobic ionic liquids. Sep Purif Technol 72(2):167–173. https://doi.org/10.1016/j.seppur.2010.02.002

van den Bruinhorst A, Raes S, Maesara SA, Kroon MC, Esteves ACC, Meuldijk J (2019) Hydrophobic eutectic mixtures as volatile fatty acid extractants. Sep Purif Technol 216:147–157. https://doi.org/10.1016/j.seppur.2018.12.087

van Osch DJGP, Zubeir LF, van den Bruinhorst A, Rocha MAA, Kroon MC (2015) Hydrophobic deep eutectic solvents as water-immiscible extractants. Green Chem 17(9):4518–4521. https://doi.org/10.1039/C5GC01451D

van Osch DJGP, Parmentier D, Dietz CHJT, van den Bruinhorst A, Tuinier R, Kroon MC (2016) Removal of alkali and transition metal ions from water with hydrophobic deep eutectic solvents. Chem Commun 52:11987–11990. https://doi.org/10.1039/C6CC06105B

Ventura SPM, Neves CMSS, Freire MG, Marrucho IM, Oliveira J, Coutinho JAP (2009) Evaluation of anion influence on the formation and extraction capacity of ionic-liquid-based aqueous biphasic systems. J Phys Chem B 113(27):9304–9310. https://doi.org/10.1021/jp903286d

Verma R, Banerjee T (2018) Liquid–liquid extraction of lower alcohols using menthol-based hydrophobic deep eutectic solvent: experiments and COSMO-SAC predictions. Ind Eng Chem Res 57(9):3371–3381. https://doi.org/10.1021/acs.iecr.7b05270

Verma R, Mohan M, Goud VV, Banerjee T (2018) Operational strategies and comprehensive evaluation of menthol based deep eutectic solvent for the extraction of lower alcohols from aqueous media. ACS Sustain Chem 6(12):16920–16932. https://doi.org/10.1021/acssuschemeng.8b04255

Wagle DV, Baker GA, Mamontov E (2015) Differential microscopic mobility of components within a deep eutectic solvent. J Phys Chem Lett 6(15):2924–2928. https://doi.org/10.1021/acs.jpclett.5b01192

Wang R, Sun D, Wang C, Liu L, Li F, Tan Z (2019) Biphasic recognition chiral extraction of threonine enantiomers in a two-phase system formed by hydrophobic and hydrophilic deep-eutectic solvents. Sep Purif Technol 215:102–107. https://doi.org/10.1016/j.seppur.2019.01.022

Yoshizawa M, Xu W, Angell CA (2003) Ionic liquids by proton transfer: vapor pressure, conductivity, and the relevance of ΔpKa from aqueous solutions. J Am Chem Soc 125(50):15411–15419. https://doi.org/10.1021/ja035783d

Yousefi SM, Shemirani F, Ghorbanian SA (2018) Hydrophobic deep eutectic solvents in developing microextraction methods based on solidification of floating drop: application to the trace HPLC/FLD determination of PAHs. Chromatographia 81:1201–1211. https://doi.org/10.1007/s10337-018-3548-7

Zarei AR, Nedaei M, Ghorbanian SA (2018) Ferrofluid of magnetic clay and menthol based deep eutectic solvent: application in directly suspended droplet microextraction for enrichment of some emerging contaminant explosives in water and soil samples. J Chromatogr A 1553:32–42. https://doi.org/10.1016/j.chroma.2018.04.023

Zhang Z, Yuan Y, Zhang N, Cao X (2015) Thermophysical properties of some fatty acids/surfactants as phase change slurries for thermal energy storage. J Chem Eng Data 60(8):2495–2501. https://doi.org/10.1021/acs.jced.5b00371

Zhang Y, Prawang P, Li C, Meng X, Zhao Y, Wang H, Zhang S (2018) Ultrasonic assisted extraction of artemisinin from Artemisia Annua L. using monoether-based solvents. Green Chem 20:713–723. https://doi.org/10.1039/c7gc03191b

Zhang K, Li S, Liu C, Wang Q, Wang Y, Fan J (2019) A hydrophobic deep eutectic solvent based vortex-assisted dispersive liquid-liquid microextraction combined with HPLC for the determination of nitrite in water and biological samples. J Sep Sci 42(2):574–581. https://doi.org/10.1002/jssc.201800921

Zhu S, Zhou J, Jia H, Zhang H (2018) Liquid-liquid microextraction of synthetic pigments in beverages using a hydrophobic deep eutectic solvent. Food Chem 243:351–356. https://doi.org/10.1016/j.foodchem.2017.09.141

Zubeir LF, van Osch DJGP, Rocha MAA, Banat F, Kroon MC (2018) Carbon dioxide Solubilities in Decanoic acid-based hydrophobic deep eutectic solvents. J Chem Eng Data 63(4):913–919. https://doi.org/10.1021/acs.jced.7b00534

Chapter 6
Methods for Extraction of Bioactive Compounds from Plant and Animal Matter Using Deep Eutectic Solvents

Lamia Nakhle, Miriana Kfoury, Isabelle Mallard, David Landy, and Hélène Greige-Gerges

Contents

L. Nakhle
Bioactive Molecules Research Laboratory, Faculty of Sciences, Lebanese University, Jdaidet el-Matn, Lebanon

Unité de Chimie Environnementale et Interactions sur le Vivant (UCEIV, UR 4492), SFR Condorcet FR CNRS 3417, Université du Littoral Côte d'Opale (ULCO), Dunkerque, France
e-mail: lamia.nakhle@univ-littoral.fr

M. Kfoury · H. Greige-Gerges (✉)
Bioactive Molecules Research Laboratory, Faculty of Sciences, Lebanese University, Jdaidet el-Matn, Lebanon
e-mail: hgreige@ul.edu.lb

I. Mallard · D. Landy
Unité de Chimie Environnementale et Interactions sur le Vivant (UCEIV, UR 4492), SFR Condorcet FR CNRS 3417, Université du Littoral Côte d'Opale (ULCO), Dunkerque, France
e-mail: isabelle.mallard@univ-littoral.fr; david.landy@univ-littoral.fr

S. Fourmentin et al. (eds.), *Deep Eutectic Solvents for Medicine, Gas Solubilization and Extraction of Natural Substances*, Environmental Chemistry for a Sustainable World 56, https://doi.org/10.1007/978-3-030-53069-3_6

Abstract Nowadays, scientists are going through developing more eco-friendly analysis methods of simple and complex samples that follow the principles of "green chemistry." One of the followed strategies is by replacing the toxic conventional organic solvents used in the extraction by a new generation of solvents called "deep eutectic solvent" or "natural deep eutectic solvent." These solvents are formed between two or more cheap nontoxic components via hydrogen bonds. This review presents the various extraction methods that use deep eutectic solvents as extraction solvents, mainly the liquid-liquid-phase microextraction methods, solid-phase microextraction methods, and the newly combined techniques. In addition, the advantages and drawbacks of using these green solvents in comparison to the conventional organic solvents used in conventional extraction methods are discussed. It was observed that, with all reported extractions, deep eutectic solvents showed better extraction efficiency and higher recovery values for the studied natural target analytes compared to water and lots of conventional organic solvents. For the protein extraction, these solvents showed around 93–99% extraction efficiency. In addition, new types of deep eutectic solvents, like the ternary deep eutectic solvent molecularly imprinted polymer, were synthesized, improving the solvent characteristics; therefore, lower volume of the solvent was used with shorter extraction time.

Keywords Bioactive compounds · Combined extraction techniques · Deep eutectic solvents · Green chemistry · Imprinted polymer · Liquid-liquid-phase microextraction · Optimization · Physicochemical properties · Proteins · Solid-phase microextraction

6.1 Introduction

Nowadays, pharmaceutical industries intend to replace the synthesized chemo-based or semi-chemo-based active molecules used in the drug formulations with bio-based active molecules present in environmental and biological matrices in order to reduce undesirable side effects (de Cássia da Silveira e Sá et al. 2014).

Regardless of their pharmaceutical applications, natural bioactive molecules are also used in the production of agrochemicals, nutraceuticals, cosmetics, etc. They are mainly found in plant materials as secondary metabolites (Azmir et al. 2013).

On the other hand, proteins are important biomacromolecules because they are considered as multifunctional biomaterials. Therefore, extraction and purification of proteins became very interesting for biotechnology industry and for proteins applications in fields of research and pharmaceuticals (Huang et al. 2015; Xu et al. 2016a, b).

At the same time, direct determination of the analytes in drug and food formulations, or in blood and urine samples, is difficult because of the complexity of the sample and their presence in very small quantity (Yilmaz and Soylak 2018). Therefore, developing an extraction method, for the valuable compounds, from their original matrix into an adequate solvent is the first critical step in the analytical method because it prepares this sample for a further sensitive and selective quantitative analysis (Rezaee et al. 2006). Hence, it is necessary that the developed extraction method follows the 12 principles of the green chemistry (Anastas and Eghbali 2010). In addition, the extraction method should be time-saving, cheap, and reveal a high yield.

Several conventional separation techniques are available such as liquid-liquid extraction, solid-phase extraction, coprecipitation, as well as many exhaustive extraction methods (maceration, steam or hydro-distillation, pressing, decoction, infusion, percolation, and Soxhlet extraction) (Chemat et al. 2012; Yilmaz and Soylak 2018). Despite being expensive, labor-intensive, and time-consuming, these techniques were always associated with a certain degree of toxicity due to the large volume of toxic organic solvents used (Alfonsi et al. 2008; Zhuang et al. 2017; Mohebbi et al. 2018).

Since 1998, ionic liquids have gained much attention to be used as alternative to organic solvents due to their low vapor pressure and variable viscosities (Khataei et al. 2018). However, compared to organic solvents, most ionic liquids are more costly and difficult to prepare. Also they could have a toxic impact on the environment (Gu et al. 2014).

New solvents, so-called deep eutectic solvents, were developed by Abbott since 2003 (Abbott et al. 2003). These solvents are composed of two or more cheap nontoxic components, one of them with the capacity to be a hydrogen bond acceptor, while the other possesses the properties of a hydrogen bond donor (Florindo et al. 2014; Cvjetko et al. 2016; Moura et al. 2017). Due to the formation of intramolecular hydrogen bonds and Van der Waals interactions, these solvents have much lower melting point than that of its individual components (Abbott et al. 2004; Moura et al. 2017).

Natural deep eutectic solvents are prepared from natural metabolites produced by cell metabolism; therefore, they have been synthesized using simple molecules present in living cells such as urea, alcohols, organic acids, amines, amino acids, sugars, choline, and even water (Vanda et al. 2018). Normally, the hydrogen bond acceptor is an ammonium chloride or an amino acid, while the hydrogen bond donor is an organic acid or a carbohydrate (Choi et al. 2011; Dai et al. 2014).

Deep eutectic solvents and natural deep eutectic solvents are known for their various properties: low vapor pressure, nonflammability, high thermal stability, and low thermal conductivity (Nam et al. 2015; Radošević et al. 2016; Ruesgas-Ramón et al. 2017; Moura et al. 2017; Khataei et al. 2018). Deep eutectic solvents are generally hydrophilic; the first hydrophobic one, synthesized from decanoic acid and quaternary ammonium salts, was used for the extraction of volatile fatty acids from aqueous solutions (van Osch et al. 2015). These solvents are considered "green" and excellent alternatives to conventional and nonconventional organic solvents, being effective to extract hydrophilic and hydrophobic compounds (Tang et al. 2014).

New techniques miniaturizing solid-phase extraction or liquid-liquid extraction have arisen such as solid-phase microextraction and liquid-phase microextraction, respectively (Hawthorne et al. 1992; Pedersen-Bjergaard and Rasmussen 1999). These techniques are characterized by using low amounts of sample matrices and small volumes of organic solvents. They are recently recommended because they offer many advantages such as high degree of concentration and minimized extraction time and energy consumption (Aydin et al. 2018). The application of these solvents in these techniques will be discussed later on.

This chapter presents an emphasis on the extraction techniques that used deep eutectic solvents as extraction solvents. Also, the effects of their properties and of the method parameters on the extraction efficiency are discussed. This review examines additionally the advantages and drawbacks of each extraction method. Moreover, the recent combinations of different extraction techniques using deep eutectic solvents in extraction processes are also reviewed.

6.2 Microextraction Techniques: Description and Application with Deep Eutectic Solvents

Liquid-liquid extraction and solid-phase extraction techniques were first introduced in the early 1970s. Liquid-liquid extraction involves adding a solvent to the sample that is immiscible, followed by a selective partitioning of analytes between the two phases. Solid-phase extraction technique consists on passing aqueous samples through a solid sorbent where the analytes will be retained (Okenicová et al. 2016). Therefore, the selection of an appropriate sorbent is very important. This technique is considered better than liquid-liquid extraction because it reduces and even eliminates the use of toxic and inflammable solvents, thus becoming more environmentally friendly (Picó et al. 2007; Ince et al. 2010).

Since then, extraction techniques trends in analytical chemistry have been focusing toward less organic solvent consumption, faster extraction time, automation, and improved quantification, which includes higher recoveries, better reproducibility, and lower method detection limits. That led to the invention of small-scale miniaturizing versions of liquid-liquid extraction and solid-phase extraction techniques: the liquid-phase microextraction and solid-phase microextraction techniques, respectively (Raynie 2004). In this part of the chapter, different types of liquid-phase microextraction and solid-phase microextraction techniques will be described, and their applications using deep eutectic solvents as extraction solvents will be presented.

6.2.1 *Liquid-Phase Microextraction*

In the liquid-phase microextraction techniques, the sample solution containing the target analyte is designated as the donor phase and the extraction solvent as the acceptor phase. These techniques consist on adding few microliters of the solvent into the aqueous sample followed by the collection of the extraction solvent containing the target analytes (Yilmaz and Soylak 2018). Different liquid-phase microextraction techniques exist such as single-drop microextraction, hollow-fiber liquid-phase microextraction, and dispersive liquid-liquid microextraction.

Single-Drop Microextraction

This method consists of a solvent drop suspended at the end of the needle of a microsyringe. Single-drop microextraction can be distinguished in two different techniques: headspace single-drop microextraction or direct immersion single-drop microextraction. Figure 6.1 presents schematically these two techniques.

Headspace single-drop microextraction is applied for the extraction of volatile or semi-volatile compounds from a complex matrix because the suspended solvent drop is exposed only to the headspace of the sample. Table 6.1 presents examples of extractions processes via the headspace single-drop microextraction technique using deep eutectic solvents as extraction solvents.

Tang et al. (2014) applied headspace single-drop microextraction method to extract terpenoids using the optimal deep eutectic solvent choline chloride/ethylene glycol (1:4 molar ratio) because it showed the best extraction efficiency. This method showed a pronounced advantage compared to other conventional methods (ultrasonication and heat reflux extraction) using methanol as extraction solvent (Tang et al. 2014). Also, Yousefi et al. (2018) validated the efficacy of this method in the extraction of aromatic hydrocarbons using a new type of hydrophobic deep eutectic solvent-magnetic bucky gel. The solvent drop was formed by mixing the optimal deep eutectic solvent (choline chloride/chlorophenol 1:2 molar ratio) with magnetic multiwalled carbon nanotubes. This mixture made the solvent drop more stable. Therefore, the heating temperature and the stirring rates were increased, which allowed to decrease the extraction time (Yousefi et al. 2018).

Direct immersion single-drop microextraction is applied for the extraction of nonvolatile compounds and polar analytes. Herein, the acceptor solvent should be immiscible with the donor phase as the solvent drop is immersed in it. Gu et al. (2014) used this method for the extraction of phenolic compounds (phenol, *p*-cresol, and β-naphthol) from crude oils with 10 μL of choline chloride/ethylene glycol (1:3

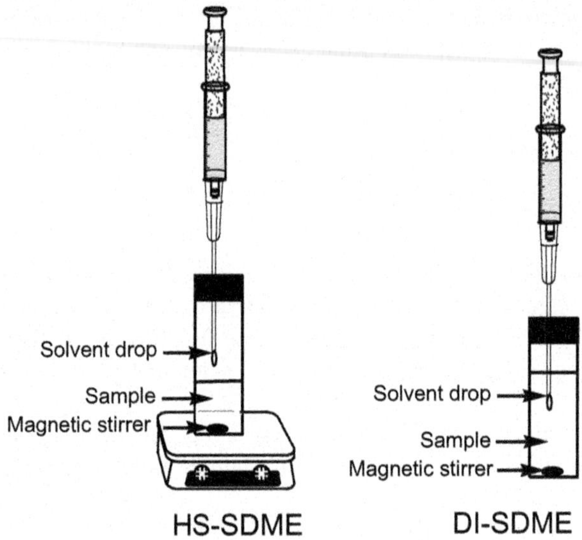

Fig. 6.1 Headspace single-drop microextraction technique (HS-SDME), where the solvent drop is exposed to the headspace of the sample and is not in direct contact with the matrix. The solvent drop adsorbs the target compounds volatilized from the sample matrix heated and stirred for a specific time. Direct immersion single-drop microextraction technique (DI-SDME), where the solvent drop is immersed directly in the sample phase. After extraction, the solvent drop is retracted back into the microsyringe and analyzed

Table 6.1 Examples of the application of the headspace single-drop microextraction technique with deep eutectic solvents as extraction solvents

Extraction medium	Target compounds	DES: HBA/HBD (molar ratio)	References
Chamaecyparis obtusa leaves	Terpenoids: linalool, terpineol, and terpinyl acetate	Choline chloride/ethylene glycol (1:2, 1:3, 1:4, 1:5)	Tang et al. (2014)
Water samples close to gas station spots Morning urine samples from healthy gas station workers	Volatile aromatic hydrocarbons: benzene, toluene, ethylbenzene, and xylene isomers	Hydrophobic magnetic bucky gel formed by mixing each DES: choline chloride/phenol (1:2); choline chloride/chlorophenol (1:2), and choline chloride/resorcinol (1:2) with magnetic multiwalled carbon nanotubes	Yousefi et al. (2018)

DES deep eutectic solvent, *HBA* hydrogen bond acceptor, *HBD* hydrogen bond donor

molar ratio) deep eutectic solvent. Ultrasonication was also applied for obtaining a better yield because polar compounds are being extracted from nonpolar solutions. According to the authors, this solvent exhibited a better extraction than water or than the solutions of the individual components of the deep eutectic solvent (Gu et al. 2014).

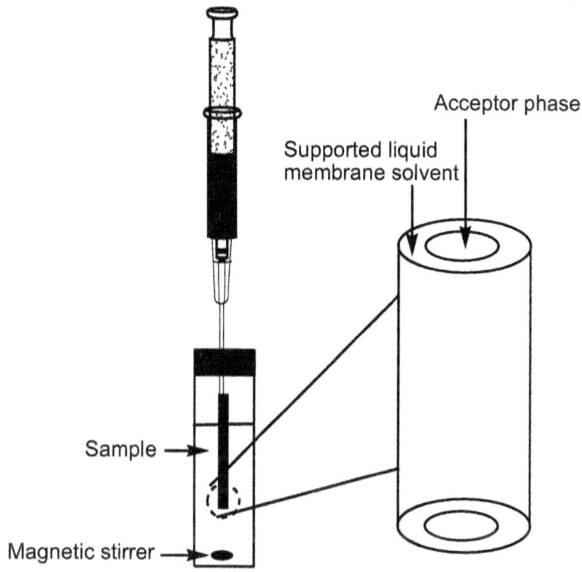

Fig. 6.2 Hollow-fiber liquid-phase microextraction technique. Three phases are present: the first one is the biological sample or the donor phase containing the target analytes; this phase is stirred in a sample vial. The acceptor phase is the second phase located in the lumen of the membrane. The third phase separates the other two phases: a supported liquid membrane solvent (immiscible with the two other phases) designated as the extraction phase; it is located inside the pores of the fiber membrane

Hollow-Fiber Liquid-Phase Microextraction

Hollow-fiber liquid-phase microextraction constitutes another selective method, which can be used for the extraction of components from complicated biological fluids. Interestingly, this technique prevents the diffusion of large molecules from the donor to the acceptor phase (Khataei et al. 2018). This is due to the use of a porous polypropylene hollow-fiber membrane. The hollow-fiber liquid-phase microextraction consists of three phases presented in Fig. 6.2.

Hollow-fiber liquid-phase microextraction was used by Khataei et al. (2018) for the extraction of steroidal hormones (dydrogesterone and cyproterone acetate) from urine and plasma samples using methyltriphenylphosphonium iodide ($Me(Ph)_3PI$)/ ethylene glycol (1:4 molar ratio) plus 20% v/w methanol. n-Dodecane was used as the supported liquid membrane hydrophobic solvent. Compared to other extraction techniques defined later on such as hybrid solid-phase extraction, solid-phase extraction, and hollow-fiber liquid-liquid-liquid microextraction, this technique has the simplest detection instrument, good limit of detection, and wider linear range in both plasma and urine matrices. In addition, clear chromatograms were obtained via this method (Khataei et al. 2018).

Dispersive Liquid-Liquid Microextraction Methods

Dispersive liquid-liquid microextractions are extraction techniques where small droplets of extraction solvent are formed and dispersed throughout the aqueous phase. In this way, the contact surface between both immiscible phases is increased.

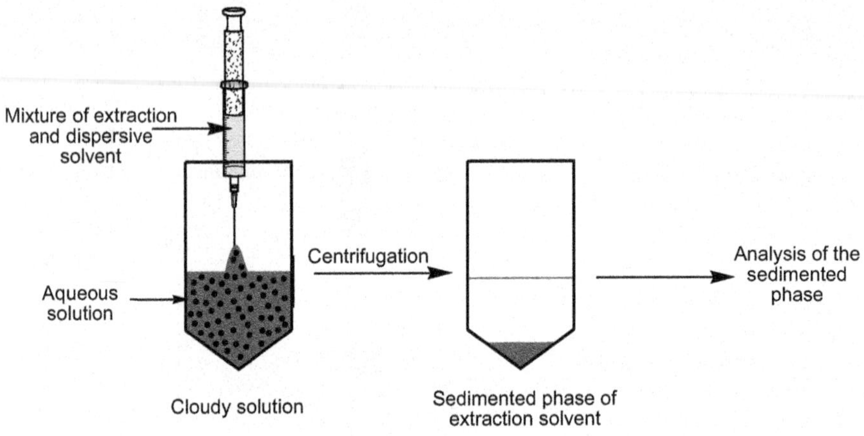

Fig. 6.3 Binary solvents-dispersive liquid-liquid microextraction method. The dispersion of the solvent droplets is done by the help of a dispersive solvent that is miscible in both aqueous and extraction phases. The extraction solvent phase is analyzed after centrifugation. (Figure modified from Jain et al. 2015)

After centrifugation, the deep eutectic solvent can be found in either the upper or the bottom phase of the tube depending on its density. Being simple, fast, and cheap, this method has been highly used (Ribeiro et al. 2015).

Figure 6.3 shows a schematic representation of binary solvents-dispersive liquid-liquid microextraction method where a dispersive solvent is used in order to disperse the solvent drops. When extracting benzoylureas, this method was tested using solidified deep eutectic solvent as extraction solvent (Zeng et al. 2017). Exceptionally, there was no need for a dispersive solvent, and this method presented the advantage of lower minimum detection value and higher enrichment factor than other techniques (dispersive liquid-liquid microextraction using ionic liquids or solidified ionic liquids, ultrasound-assisted hybrid ionic liquid-dispersive liquid-liquid microextraction, and solid-phase extraction using acetonitrile as extraction solvent) (Zeng et al. 2017).

In the gas-associated dispersive liquid-phase microextraction, the dispersing solvent is replaced by a gas. With air as gas, the method is called air-assisted dispersive liquid-phase microextraction or air-assisted emulsification liquid-liquid microextraction (Lamei et al. 2017; Ge et al. 2018). Figure 6.4 explains this technique schematically.

Table 6.2 presents examples of the application of dispersive liquid-liquid microextraction techniques with deep eutectic solvents as extraction solvents.

The linear ranges of this method, when extracting nine pesticides using deep eutectic solvent, were equivalent or wider than other extraction methods (solid-phase microextraction, vortex-assisted low-density solvent liquid-liquid microextraction, and salt-induced demulsification and sequential dispersive liquid-liquid microextraction). Also, higher extraction efficiencies were also observed (Farajzadeh et al. 2017).

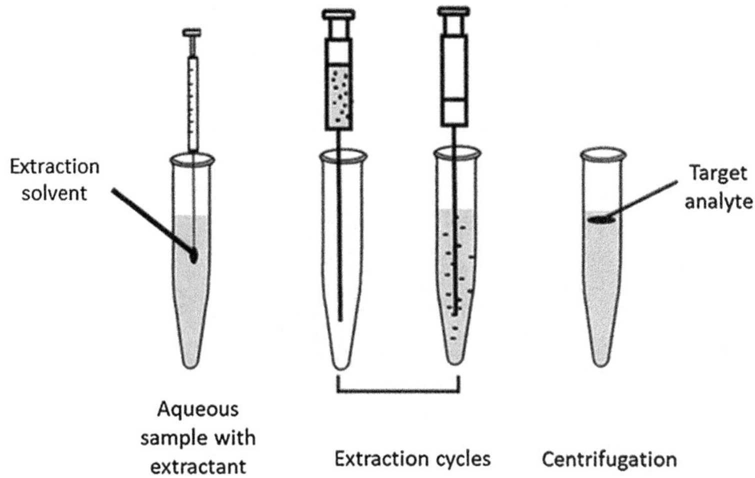

Fig. 6.4 Air-assisted dispersive liquid-phase microextraction method. This technique consists on multiple sucking and injecting processes via a syringe. At this stage, the target analytes are extracted into the fine droplets of the extraction solvent. (Figure modified from Yang et al. 2015)

Also, Moghadam et al. (2018) developed a similar microextraction procedure, air-agitated emulsification microextraction, based on a low-density deep eutectic solvent for the extraction of antidepressant drugs from human plasma samples and pharmaceutical wastewater samples by three types of deep eutectic solvents. Compared to other conventional techniques, this method proved its ability for accurate analysis of trace levels of drugs close to the therapeutic/toxic ranges (Moghadam et al. 2018).

For the extraction of rhodamine B, recoveries experiments showed a minimal efficacy of this method using the optimal deep eutectic solvent (97% extraction efficacy) compared to solid-phase extraction, magnetic solid-phase extraction methods, and magnetic stirring-assisted dispersive liquid-liquid microextraction using 1-octanol as extraction solvent (Yilmaz and Soylak 2018).

Dispersive liquid-liquid microextraction based on the solidified deep eutectic solvent method developed by Habibollahi et al. (2018) consists on the rapid addition of deep eutectic solvent (50 μL): 1-octyl-3-methylimidazolium chloride/1-undecanol (1:2 molar ratio) to the sample solution; this forms an emulsion. Next, NaCl was added to break the emulsion. After vortexing and centrifuging, the fine droplets of deep eutectic solvent were extracted from the upper phase and subjected finally to a graphite furnace atomic absorption spectrophotometry analysis. Compared to other techniques (wet digestion extraction, solid-phase extraction, cloud point extraction, persistent sample circulation microextraction, continuous sample drop flow-based microextraction, and dispersive liquid-liquid microextraction using solidified 1-undecanol organic solvent drop), this method, when extracting metals, offers a higher advantage because no disperser solvents are necessary, which prevents a decrease in the partition coefficients of the metals into the

Table 6.2 Examples of the application of dispersive liquid-liquid microextraction techniques with deep eutectic solvents as extraction solvents

Extraction medium	Target compounds	DES:HBA/HBD (molar ratio)	Remarks	References
Environmental waters	Benzoylureas: diflubenzuron, triflumuron, hexaflumuron, flufenoxuron, and lufenuron	1-octyl-2,3-dimethylimidazolium chloride/1-dodecanol (1:2), 1-decyl-3-methylimidazolium chloride/1-dodecanol (1:2), 1-decyl-2,3-dimethylimidazolium chloride/1-dodecanol (1:2), and 1-dodecyl-3-methylimidazolium chloride/1-dodecanol (1:2)	The optimal DES was 1-decyl-3-methylimidazolium chloride/1-dodecanol (1:2).	Zeng et al. (2017)
Vegetable oils: olive, soy, peanuts, corn, and sunflower oils	Oxyprenylated phenylpropanoids: ferulic acid, umbelliferone, boropinic acid, 7-isopentenyloxycoumarin, 4′ geranyloxyferulic acid, and auraptene	Betaine/glycolic acid (1:2), betaine/phenylacetic acid (1:2), betaine/2-furoic acid (1:2), and betaine/S-(+)-mandelic acid (1:1)	Betaine/phenylacetic acid (1:2) was the selected DES. Isopropanol was used as dispersive solvent.	Ferrone et al. (2018)
Urine and blood samples	Methadone	Choline chloride/5,6,7,8-tetrahydro-5,5,8,8-tetramethylnaphtalen-2-ol (1:1, 1:2, and 1:3)	Optimal molar ratio of DES was 1:2. Tetrahydrofuran: emulsifier agent	Lamei et al. (2017)
Fruit and vegetable juices	Pesticides residues: penconazole, tebuconazole, diazinon, haloxyfop-R–methyl, hexaconazole, diniconazole, clodinafop-propargyl, bromopropylate, and fenazaquin	Choline chloride/4-chlorophenol (1:2)	Air was used as the dispersive agent.	Farajzadeh et al. (2017)
Environmental water samples	6-benzophenones	DL-menthol/octanoic acid (2:1, 1:1, and 1:2), DL-menthol/decanoic acid (2:1, 1:1, and 1:2), DL-menthol/dodecanoic acid (2:1, 1:1, and 1:2), tetrabutylammonium bromide/octanoic acid (1:1), tetrabutylammonium bromide/decanoic acid (1:1), tetrabutylammonium bromide/dodecanoic acid (1:1), tetrabutylammonium chloride/octanoic acid (1:1), tetrabutylammonium chloride/decanoic acid (1:1), and tetrabutylammonium chloride/dodecanoic acid (1:1)	DL-menthol/decanoic acid (1:1) was the best extraction solvent. Air was used to disperse DES in the solution.	Ge et al. (2018)

Human plasma and pharmaceutical wastewater samples	Antidepressant drugs: escitalopram, desipramine, and imipramine	Choline chloride/phenol (1:4), choline chloride/ethylene glycol (1:2), and choline chloride/urea (1:2)	Choline chloride/phenol (1:4) was the selected DES. Tetrahydrofuran was used to disperse DES in the solution.	Moghadam et al. (2018)
Cosmetic products (lipsticks, cologne, and acetone cleaning solution of nail polish samples), sea, and river water samples	Rhodamine B	Choline chloride/phenol (1:2), methyltrioctylammonium chloride/decanoic acid (1:2), and tetrabutylammonium chloride/decanoic acid (1:2, 1:3, and 1:4)	Tetrabutylammonium chloride/decanoic acid (1:2) was the optimal DES. Tetrahydrofuran was used to disperse DES in the solution.	Yilmaz and Soylak (2018)
Soil and vegetables: *Raphanus sativus*, *Spinacia oleracea*, *Coriandrum sativum*, and *Daucus carota*	Heavy metals: Pb, cd, and hg	1-decyl-2,3-dimethylimidazoliumchloride/1-undecanol (1:1, 1:2, 1:3, 2:5, 3:7), 1-octyl-2,3-dimethylimidazolium chloride/1-undecanol (1:1, 1:2, 1:3, 2:5, 3:7), 1-octyl-3-methylimidazolium chloride/1-undecanol (1:1, 1:2, 1:3, 2:5, 3:7), 1-dodecyl-3-methylimidazolium chloride/1-undecanol (1:1, 1:2, 1:3, 2:5, 3:7), and 1-decyl-3-methylimidazolium chloride/1-undecanol (1:1, 1:2, 1:3, 2:5, 3:7)	1-octyl-3-methylimidazolium chloride/1-undecanol (1:2) was the best DES. Diethyldithiophosphoric acid was used as a chelating agent.	Habibollah-i et al. (2018)

DES deep eutectic solvent, *HBA* hydrogen bond acceptor, *HBD* hydrogen bond donor

extraction solvent, hence reducing the environmental pollution. In addition, the used deep eutectic solvent showed high stability, low density, and a melting point near room temperature (Habibollahi et al. 2018).

Deep Eutectic Solvent-Based Subcritical Water Extraction

When increasing the temperature of water above its boiling point (between 100 and 374 °C) but keeping the pressure tuned for liquid state, subcritical water is obtained. At these conditions, water acts like an organic solvent that can dissolve a wide range of analytes of low polarities, hence presenting the advantage of using only water as the extraction solvent. New approaches on using deep eutectic solvent in a certain percentage with subcritical water are being reported. Nevertheless, one should underline the high amount of water added to the deep eutectic solvent (generally superior to 50% volume), thus questioning the remaining structure of the deep eutectic solvent. For example, Machmudah et al. (2018) tested the effect of the addition of 10–30% volume of deep eutectic solvent to pure water on the extraction of xanthone (Machmudah et al. 2018). Table 6.3 presents some examples of the application of subcritical water extraction technique with deep eutectic solvents as extraction solvents.

Saravana et al. (2018a, b) proved that the yield of polysaccharides obtained using the optimal deep eutectic solvent (choline chloride/glycerol (1:2 molar ratio) + 70% water) was at least twice of that obtained from a solution of HCl/water mixture, usually used to extract polysaccharides (Saravana et al. 2018a, b). Through adding 10–30% citric acid/alanine (1:1 molar ratio) deep eutectic solvent to water media, Machmudah et al. (2018) proved the efficacy of the resulting subcritical water extraction method to extract xanthones. Accordingly, scanning electron microscope images showed the disruption of the surface of the pericarps of mangosteen after treatment by this method at high temperature. The formation of pores was obviously observed via the pronounced cleavage of intermolecular and intramolecular bonds in and/or between lignin, cellulose, and hemicellulose by deep eutectic solvent (Machmudah et al. 2018).

Deep Eutectic Solvent-Based Aqueous Two-Phase System

Such system is formed when two polymers, one polymer, and one salt or two salts are mixed in the presence of water, forming two distinct aqueous phases. K_2HPO_4 is the frequently used salt because it demonstrated a better phase-forming ability than other salts (for example, Na_2HPO_4 and KH_2PO_4) (Li et al. 2016). This technique is widely used to extract proteins. It prevents proteins denaturation and preserves their biological activity. Also, it is used for the extraction of genetic materials, drugs, cells, and organelles (Zeng et al. 2014; Li et al. 2016). The detailed process of the aqueous two-phase system is illustrated in Fig. 6.5.

Bovine serum albumin was chosen as a model protein. All tested deep eutectic solvent showed around 93–99% extraction efficiency of this protein (Li et al. 2016). In addition, the presence of relatively high salt concentration resulted in a competition between the salt ions and bovine serum albumin for water molecules. Due to the decrease of the amount of water required for the solubilization of the protein, the solubility of bovine serum albumin in the salt phase was tremendously decreased; this results in the increase of its concentration in the deep eutectic solvent phase. UV-Vis, FT-IR, and circular dichroïsm experiments showed no interaction and no interference between all deep eutectic solvents and bovine serum albumin and that the conformation of bovine serum albumin was maintained during extraction. The extraction process is mainly due to the aggregation of protein molecules that are expected to be surrounded by deep eutectic solvent. Dynamic light scattering experiments showed that before extraction, the particle sizes in deep eutectic solvent solution were around 615 nm and those of bovine serum albumin aqueous protein solution were around 4 nm, 18 nm, and 255 nm. After extraction, the size of the observed particles became about 5560 nm. Apparently, the size of this newly formed aggregate was greater than that of deep eutectic solvent particles and protein particles. Therefore, deep eutectic solvent-protein aggregate could be formed (Xu et al. 2015). Examples of the application of aqueous two-phase system technique with deep eutectic solvents as extraction solvents are presented in Table 6.4.

Li et al. (2016) used six betaine-based deep eutectic solvents differing by their hydrophilic property, viscosity, and density for the extraction of proteins. The extraction efficiency using the optimal deep eutectic solvent was higher than 99% as the bovine serum albumin band (66 kD), obtained by SDS-PAGE analysis, was present in the deep eutectic solvent-rich top phase but was not detectable in the bottom phase. This method was validated for accuracy, repeatability, and environment stability experiments investigated under the following optimized conditions

Table 6.3 Examples of the application of subcritical water extraction technique with deep eutectic solvents as extraction solvents

Extraction medium	Target compounds	DES: HBA/HBD (molar ratio)	References
Brown seaweed from *Saccharina japonica*	Polysaccharides: alginate and fucoidan	Choline chloride/1,2-propanediol (1:2), choline chloride/glycerol (1:2), choline chloride/ethylene glycol (1:2), choline chloride/1,3-butanediol (1:2), choline chloride/1,4-butanediol (1:2), choline chloride/urea (1:2), and choline chloride/propanedioic acid (1:2)	Saravana et al. (2018a, b)
Pericarps of mangosteen (*Garcinia mangostana Linn)*	Xanthones	Citric acid/alanine (1:1)	Machmudah et al. (2018)

DES deep eutectic solvent, *HBA* hydrogen bond acceptor, *HBD* hydrogen bond donor

(betaine/urea deep eutectic solvent (1.4 g)/K_2HPO_4 (0.75 g mL^{-1}, 2.0 mL)/bovine serum albumin (15 mg) and the separation time: 12 min).

Compared to polypropylene glycol-based aqueous biphasic system, the extraction efficiency of three hydrophobic dyes was enhanced with deep eutectic solvent. It was found that the more hydrophobic the pigment is, the more it is extracted in the deep eutectic solvent-rich phase (Zhang et al. 2018).

Deep Eutectic Solvent-Based Flow Method Liquid-Liquid Extraction: Stepwise Injection Analysis System

It is an automated extraction technique developed by Nugbienyo et al. (2017) (Fig. 6.6). The authors reported the extraction of antiarrhythmic agent procainamide from human saliva using choline chloride/glycerol at a 1:2 molar ratio. Acetonitrile, which will extract the other components present in the human saliva, is added because the deep eutectic solvent is water-miscible but insoluble in acetonitrile; therefore, separation of the two phases occurs at room temperature.

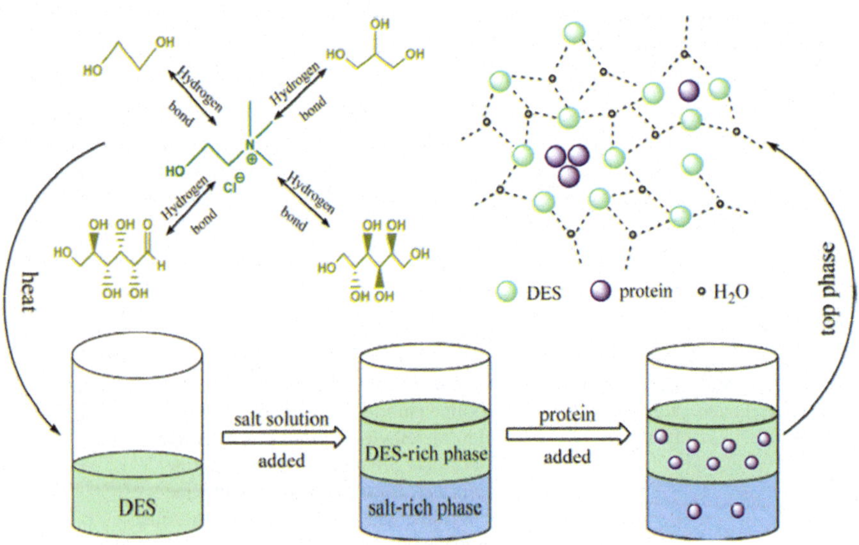

Fig. 6.5 Deep eutectic solvent (DES)-based aqueous two-phase system extraction method. Two aqueous phases are present in this technique: the DES-rich aqueous phase and the salt-rich aqueous phase. When these phase components are mixed above a certain salt critical concentration, they are separated into two unambiguous aqueous phases. It is then where the proteins present in the sample are partitioned between these two phases with a higher affinity for the DES-rich aqueous phase. (Figure reprinted with permission from Xu et al. 2015)

6.2.2 Ultrasound Microextraction

Ultrasound microextraction technique is performed under ultrasonic energy that increases the contact between the sample and the extraction solvent (Fig. 6.7).

Ultrasounds radiation can be applied via water baths or ultrasonic probes. Mainly, cavitation phenomenon is observed with the generation of bubbles, leading to increase in pressure and temperature. This allows the disruption of the cell walls, facilitating solvent penetration into the material and allowing the release of the target compounds (Chanioti and Tzia 2018). The extraction efficiency depends on different factors like ultrasonic conditions (temperature, amplitude, time) and sample features (matrix, amount, particle size) (Yilmaz and Soylak 2015; Huang et al. 2017; Bosiljkov et al. 2017; Zhou et al. 2018). Advantageously, no dispersive solvent is needed. This technique involves lower volume of deep eutectic solvent (<500 μL) and shorter time of extraction (<15 min) than liquid-phase microextraction techniques (Bakirtzi et al. 2016; Khezeli et al. 2016; Mouratoglou et al. 2016). Examples of the application of ultrasound microextraction technique with deep eutectic solvents as extraction solvents are summarized in Table 6.5.

Similar or larger amounts of flavonoids were extracted using this method with deep eutectic solvent with relatively low cost, low vapor pressure, and low toxicity compared to other extraction methods (heating and stirring) using conventional organic solvents (Bi et al. 2013). Also, higher extraction yield of polysaccharides was obtained using this method in the optimized conditions in comparison to hot

Table 6.4 Examples of the application of aqueous two-phase system technique with deep eutectic solvents as extraction solvents

Extraction medium	Target compounds	DES:HBA/HBD (molar ratio)	Selected DES	References
Proteins mixture	Bovine serum albumin	Choline chloride/urea (1:2), tetramethylammonium chloride/urea (1:2), tetrapropylammonium bromide/urea (1:2), and choline chloride/methylurea (1:2)	Choline chloride/urea (1:2)	Zeng et al. (2014)
Proteins mixture	Bovine serum albumin	Choline chloride/ethylene glycol (1:2), choline chloride/glycerol (1:1), choline chloride/D-glucose (2:1), and choline chloride/D-sorbitol (1:1)	Choline chloride/glycerol (1:1)	Xu et al. (2015)
Proteins mixture and calf blood	Bovine serum albumin, tyrosine, and ovalbumin	Betaine/urea/water (1:2:1), betaine/methylurea/water (1:3:1), betaine/D-(+)-glucose/water (1:1:2), betaine/D-sorbitol/water (1:2:1), betaine/ethylene glycol (1:2), and betaine/glycerol (1:1)	Betaine/urea/water (1:2:1)	Li et al. (2016)
Three dyes	Amaranth, sunset yellow FCF, and Sudan III	Tetrabutylammonium bromide/polypropylene glycol 400 (1:2)		Zhang et al. (2018)

DES deep eutectic solvent, *HBA* hydrogen bond acceptor, *HBD* hydrogen bond donor

water extraction and water-based ultrasound extraction (Zhang and Wang 2017). Bajkacz and Adamek (2017) used natural deep eutectic solvent-based ultrasound microextraction procedure that provided high accuracy, sensitivity, and high extraction efficiency for the simultaneous analysis of four isoflavones. This method was faster in the addition of higher recoveries with lower relative standard deviation in comparison with the Soxhlet extraction and microwave extraction methods (Bajkacz and Adamek 2017). In addition, the extraction yield of wine lees anthocyanins was improved in comparison to alternative methods (stirring, heating, and heating+stirring) and to the extraction with conventional solvents (water, methanol, ethanol, 70% methanol, 70% ethanol) (Jeong et al. 2017).

6.2.3 *Microwave Extraction*

This technique uses the non-ionized electromagnetic irradiation in a frequency range of 0.3–300 GHz in order to heat both solvent and samples by movements of ions and rotation of molecular and atomic dipoles, leading to an increase in the extraction kinetics (Fig. 6.8). In addition to the method parameters, the extraction efficiency depends on the pressure, the radiation power and the sample composition. This method is known for many advantages: homogeneous heating, high speed, and high heat efficiency, which may be responsible for a high extraction efficiency and short extraction time (Cui et al. 2015; Peng et al. 2016; Wang et al. 2018). Examples

Fig. 6.6 The stepwise injection analysis manifold for the determination of procainamide in saliva. This method involves several steps: the first one includes the addition of a specific deep eutectic solvent (DES) to a sample (saliva, for instance), followed by formation of a homogeneous solution. The second step is the separation of the DES phase by the addition of acetonitrile at room temperature. In this step, the analyte is extracted into the DES phase, using air-bubbling to promote the extraction process and phase separation. All these steps are controlled by the movement of the syringe that introduces the different solutions. Finally, the bottom DES phase was moved into the flow cell for analysis. (Figure reprinted with permission from Nugbienyo et al. 2017)

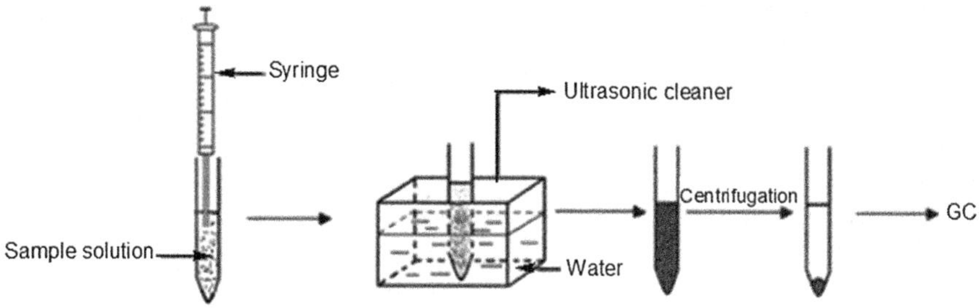

Fig. 6.7 Ultrasound microextraction method. Deep eutectic solvent is added to the powdered sample; the mixture is subjected to ultrasound in a bath followed by centrifugation, and then the deep eutectic solvent-rich phase is analyzed by gas chromatography (GC). (Figure modified from Yan et al. 2011)

of the application of microwave extraction technique with deep eutectic solvents as extraction solvents are listed in Table 6.6.

Comparing to other conventional techniques, microwave extraction was faster and allows the automation of the extraction, but it is more costly and demand cleanup steps (Chen et al. 2016). Also, it shows higher extraction efficiency using deep eutectic solvent than other conventional solvents. The maximum extraction yield of baicalin using deep eutectic solvent-based microwave extraction was slightly higher than the extraction by 70 vol% ethanol-based hot reflux-assisted extraction and higher than ultrasound extraction (Cvjetko Bubalo et al. 2016; Chanioti and Tzia 2018; Wang et al. 2018).

6.2.4 Vortex Extraction

This technique is a rapid microextraction technique. The suspension is subjected to mechanical stirring by a vortex stirrer, which further favors the dispersion of the donor phase in the aqueous phase. In this way, the analytes are extracted in the tiny droplets formed. Subsequently, the suspension is centrifuged to separate the two phases (González et al. 2018; Ojeda and Rojas 2018). This leads to a greater efficiency of the extraction procedure. Table 6.7 displays some examples of the application of vortex extraction technique with deep eutectic solvents as extraction solvents.

This process was pursued by García et al. (2016) for the extraction of phenolic compounds. An increase in the extraction yield of oleacein and oleocanthal of 20–33% and approximately 68%, respectively, was proved by extraction with the deep eutectic solvent with respect to the conventional solvent (80% methanol:water (v/v)) (García et al. 2016). Wang et al. (2017) explored the potency of deep eutectic solvent to extract and quantify rhodamine B. Its recovery value using deep eutectic solvent was 75% higher than in control experiments using water (10% recovery). Compared to methanol, deep eutectic solvent showed high selectivity for rhodamine

Table 6.5 Examples of the application of the ultrasound microextraction technique with deep eutectic solvents as extraction solvents

Extraction medium/ target compounds	DES:HBA/HBD (molar ratio)	Selected DES	References
Leave plants of *Chamaecyparis obtusa*/ flavonoids: myricetin and amentoflavone	Choline chloride/ethylene glycol (1:2), choline chloride/glycerol (1:2), choline chloride/1,2-butanediol (1:2), choline chloride/1,3-butanediol (1:2), choline chloride/1,4-butanediol (1:2), choline chloride/2,3-butanediol (1:2), and choline chloride/1,6-hexanediol (1:2)	Choline chloride/1,4-butanediol (1:5) with 35 vol % of water	Bi et al. (2013)
Sheep, bovine, and chicken liver samples/ iron	Choline chloride/oxalic acid (1:1), choline chloride/ethylene glycol (1:1), choline chloride/ glycerol (1:1), and choline chloride/lactic acid (1:1, 1:2, 1:1.5, 2:1)	Choline chloride/lactic acid (1:1) DES led to the best recovery (>95%) compared to the other DES.	Yilmaz and Soylak (2015)
Vegetable oil samples: olive, almond, sesame, and cinnamon/phenolic acids: ferulic, caffeic, and cinnamic acids	Choline chloride/ethylene glycol (1:2) and choline chloride/glycerol (1:2)	Choline chloride/ ethylene glycol (1:2)	Khezeli et al. (2016)
Greek medicinal plants (dittany, fennel, marjoram, mint, and sage)/polyphenols	Choline chloride/L-lactic acid (1:3), sodium acetate/L-lactic acid (1:3), ammonium acetate/L-lactic acid (1:3), glycine/L-lactic acid (1:3), and water/ glycine/L-lactic acid (3:1:3)	Natural deep eutectic solvent with 80% (v/v) water was better than natural deep eutectic solvent without water	Bakirtzi et al. (2016)
Agri-food waste biomass (lemon peels, olive leaves, onion solid wastes, red grape pomace, spent filter coffee, and wheat bran)/ polyphenols	Choline chloride/glycerol (1:3), sodium acetate/ glycerol (1:3), and sodium potassium tartrate/glycerol/ water (4:1:5 and 3:1:5)	Natural deep eutectic solvent tested was as 90% (v/v) in water	Mouratoglou et al. (2016)

(continued)

Table 6.5 (continued)

Extraction medium/ target compounds	DES:HBA/HBD (molar ratio)	Selected DES	References
Wine lees/anthocyanins	Choline chloride/citric acid, choline chloride/ oxalic acid, choline chloride/malic acid, choline chloride/glucose, choline chloride/fructose, choline chloride/xylose, and choline chloride/ glycerol (n.m.)	Choline chloride/malic acid. Also, it provided more effective extraction than ethanol	Bosiljkov et al. (2017)
Tartary buckwheat hull/ rutin	Choline chloride/1,2-propanediol (1:1), choline chloride/glycerol (1:1), choline chloride/glucose (2:5), choline chloride/ sucrose (1:1), choline chloride/xylitol (1:2), choline chloride/sorbitol (2:5), glycerol/L-proline (3:1), glycerol/L-alanine (3:1), glycerol/glycine (3:1), glycerol/L-histidine (3:1), glycerol/L-threonine (3:1), glycerol/L-lysine (4.5:1), and glycerol/L-arginine (4.5:1)	Choline chloride/ glycerol (1:1) (highest extraction efficiency 95%). Natural deep eutectic solvent exhibited 660–1577 times higher solubilization efficiency for rutin compared to water; the solubility of rutin in choline chloride-based natural deep eutectic solvent was higher than in glycerol-based natural deep eutectic solvent.	Huang et al. (2017)
Platycladi Cacumen: Chinese herbal medicine/flavonoids glycosides (myricetin and quercetin), and aglycones (amentoflavone and hinokiflavone)	Choline chloride/levulinic acid (1:2), choline chloride/ethylene glycol (1:2), choline chloride/ dimethylurea (1:1), choline chloride/glucose (1:1), betaine/levulinic acid (1:2), betaine/ethylene glycol (1:2), betaine/methylurea (1:1), betaine/glucose (1:1), proline/levulinic acid (1:2), proline/glycerol (1:2.5), proline/acetamide (1:1), and proline/glucose (1:1)	Choline chloride/ levulinic acid in 75% water	Zhuang et al. (2017)
Dioscorea opposita Thunb (Chinese yam) / polysaccharides	Choline chloride/ethylene glycol, choline chloride/ glycerol, choline chloride/1,4-butanediol, and choline chloride/1,6-hexanediol (n.m.)	Choline chloride/1,4-butanediol (1:4 molar ratio) with 30% (v/v) water	Zhang and Wang (2017)

(continued)

Table 6.5 (continued)

Extraction medium/ target compounds	DES:HBA/HBD (molar ratio)	Selected DES	References
Soy-food products (soybeans, flour, pasta, breakfast cereals, cutlets, tripe, soy drink, soy nuts, soy cubes, and three different dietary supplements)/ isoflavones: daidzin, genistin, genistein, and daidzein	Choline chloride/D(+)-glucose (2:1), choline chloride/L(+)-tartaric acid (1:1), choline chloride/ citric acid (1:1, 2:1, and 1:2), choline chloride/ saccharose (2:1), choline chloride/glycerine (1:1 and 2:1), choline chloride/D(+)-xylose (2:1), choline chloride/urea (1:1), citric acid/urea (1:2), L(+)-tartaric acid/urea (1:2), L(+)-tartaric acid/ glycerine (1:2), D(+)-glucose/glycerine (1:2), citric acid/glycerine (1:2), and choline chloride/citric acid/glycerine (1:1:1 and 2:2:1) with 10, 15, 20, 25, 30, 40, 50, and 75% of deionized water	Choline chloride/citric acid (1:1) in 30% (w/w) water	Bajkacz and Adamek (2017)

(continued)

Table 6.5 (continued)

Extraction medium/ target compounds	DES:HBA/HBD (molar ratio)	Selected DES	References
Green tea (*Camellia sinensis*)/catechins: epigallocatechin-3-gallate (EGCG)	Betaine/sucrose (4:1), citric acid/sucrose (1:1), glycerol/sucrose (3:1), betaine/glycerol/maltitol (4:4:1), betaine/glycerol/glucose (4:4:1, 4: 5:1, 4:10:1, 4:15:1, and 4:20:1), betaine/sorbitol (2:1), citric acid/sorbitol (1:1), glycerol/sorbitol (2:1), betaine/glycerol/maltose (4:4:1), betaine/maltose (4:1), citric acid/maltose (2:1), glycerol/maltose (3:1), betaine/glycerol/urea (1:1:2), betaine/glucose (4:1), citric acid/glucose (1:1), glycerol/glucose (3:1), betaine/glycerol/citric acid (1:1:1), betaine/maltitol (4:1), citric acid/maltitol (2:1), glycerol/maltitol (3:1), betaine/xylitol (4:1), citric acid/xylitol (1:1), glycerol/xylitol (2:1), citric acid/glycerol/maltitol (2:4:1), betaine/urea (1:2), citric acid/fructose (1:1), glycerol/fructose (3:1), citric acid/glycerol/maltose (2:4:1), betaine/glycerol (1:1), citric acid/glycerol (1:2), glycerol/galactose (3:1), citric acid/glycerol/glucose (1:2:1), betaine/citric acid (1:1), glycerol/urea (1:1), urea/glycerol/maltose (3:3:1), urea/glycerol/maltitol (3:3:1), and urea/glycerol/glucose (2:2:1)	81% betaine/glycerol/D-glucose (4:20:1) is the best as an extraction and stabilizer solvent.	Jeong et al. (2017)

(continued)

Table 6.5 (continued)

Extraction medium/ target compounds	DES:HBA/HBD (molar ratio)	Selected DES	References
Morus alba leaves/ phenolic compounds	Choline chloride/urea (1:2), choline chloride/ ethylene glycol (1:2), choline chloride/glycerol (1:2), choline chloride/ citric acid (2:1), choline chloride/DL-malic acid (1:1), betaine/levulinic acid (1:2), betaine/DL-lactic acid (1:1), betaine/glycerol (1:2), L-proline/DL-malic acid (1:1), L-proline/ glycerol (2:5), L-proline/ levulinic acid (1:2), and L-proline/lactic acid (1:1)	Choline chloride/citric acid (2:1)	Zhou et al. (2018)
Edible oils/ tert-butylhydroquinone	Choline chloride/ascorbic acid (1.2:1, 2:1, 2.5:1, 1:1.6, 1:2, and 1:2.5)	Choline chloride/ ascorbic acid (1:2) was chosen because it has the highest thermal stability.	Liu et al. (2018)

DES deep eutectic solvent, *HBA* hydrogen bond acceptor, *HBD* hydrogen bond donor; *n.m.* not mentioned

B; however, other dyes were extracted also with methanol (Wang et al. 2017). Cao et al. (2018) applied a vortex extraction method for the extraction of proanthocyanidin. Sixteen different deep eutectic solvents were tested, and it was concluded that organic acid-based deep eutectic solvents were better that alcohol-based ones, and this was related to the higher polarity of the organic acid-based ones suitable for the extraction of the hydrophilic proanthocyanidin. Extraction yield with deep eutectic solvent (22.19 ± 0.71 mg/g) was much higher than those with conventional organic solvents (70% methanol (7.87 ± 0.21 mg/g), 70% ethanol (7.84 ± 0.10 mg/g), and 70% acetone (13.26 ± 0.54 mg/g)) (Cao et al. 2018).

6.2.5 *Heating and Stirring*

The heating and stirring method is the simplest one. It is based on the solubilization of the target analytes in the deep eutectic solvent under heating and stirring conditions (Guo et al. 2013; Das et al. 2016; Bağda et al. 2017; Saravana et al. 2018a, b). The latter are optimized depending on the target compound (Ribeiro et al. 2013; Helalat–Nezhad et al. 2015; Athanasiadis et al. 2018). Centrifugation and filtration steps can be used to separate the deep eutectic solvent phase from the sample solution. Table 6.8 presents examples of the application of heating and stirring extraction technique with deep eutectic solvents as extraction solvents.

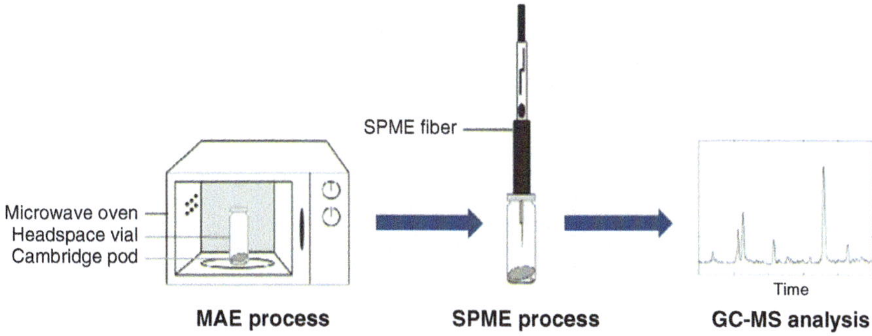

Fig. 6.8 Microwave extraction (MAE) method. Deep eutectic solvent is added to the sample in a test tube which is immersed in a microwave flask and then in a microwave extractor. The solutions are then filtrated via a solid-phase microextraction (SPME) process and then analyzed by gas chromatography-mass spectrometry (GC-MS). (Figure reprinted with permission from Sha et al. 2013)

Dai et al. (2016) used natural deep eutectic solvent to extract anthocyanins from purple and orange petals of *Catharanthus roseus*. This method was compared with ultrasound extraction and ultrasound extraction with heating. Among these three methods, the least efficient one was the ultrasound extraction method at 25 °C. But with ultrasound extraction with heating, an improvement of the yield of 2–20% was observed. The best extraction yield was obtained by heating and stirring at 40 °C, and it was 35–55% greater than that obtained with sonication at 25 °C (Dai et al. 2016). Also, Peng et al. (2018) obtained an improved yield of extracted rutin using deep eutectic solvent and water (18.1%) compared with the methanol-water solution or ethanol-water solution (60% water). The yellow rutin powder was obtained with a yield of 62.7% with purity higher than 95%. The extracted rutin has an excellent antioxidant activity (Peng et al. 2018).

6.2.6 Solid-Phase Microextraction

Solid-phase microextraction is based on the partition of the analytes between a liquid solution (sample matrix or extract) and a viscous liquid immobilized on a solid support (sorbent phase). Different types of sorbents are used for the extraction of a wide range of volatile compounds, via absorption/adsorption mechanism. The choice of the sorbent depends on the target analytes and the type of interactions. It should have large surface area, high purity, and chemical stability in the conditions of extraction. This technique is rapid, simple, safe, efficient with good recoveries, and reproducible, and it combines the extraction and the preconcentration steps into a single step (Karimi et al. 2017). It can be automated and the solvent can be reused. The solid-phase microextraction is limited for a specific number of commercially available cartridges, used as sorbents, that are expensive and demand a

Table 6.6 Examples of the application of microwave extraction technique with deep eutectic solvents as extraction solvents

Extraction medium	Target compounds	DES:HBA/HBD (molar ratio)	Selected DES	References
Pigeon pea roots	Genistin, genistein, and apigenin	Choline chloride/sucrose (1:1), choline chloride/1,2-propanediol (1:1), choline chloride/glucose (1:1), choline chloride/sorbitol (1:1), choline chloride/glycol (1:1), choline chloride/glycerol (1:1), choline chloride/1,3-butanediol (1:1), choline chloride/1,4-butanediol (1:1), glucose/L-proline (1:1), glucose/lactic acid (1:1), and choline chloride/1,6- hexanediol (1:1, 2:1, 3:1, 4:1, 5:1, 6:1, 7:1, 8:1)	Choline chloride/1,6-hexanediol (7:1)	Cui et al. (2015)
Lonicerae japonicae Flos	Phenolic acids: chlorogenic acid, caffeic acid, 3,5-dicaffeoylquinic acid, 3,4-dicaffeoylquinic acid, and 4.5-dicaffeoylquinic acid	Choline chloride/1,2-propanediol (1:2), glycerol (1:2), choline chloride/ethylene glycol (1:2), choline chloride/1-3 butanediol (1:2), choline chloride/1-4 butanediol (1:2), choline chloride/urea (1:2), propanedioic acid (1:2), choline chloride/glucose (1:2), choline chloride/sorbitol (1:2), oxalic acid/sucrose (1:2), lactic acid/sucrose (1:2), and lactic acid/glucose (1:2)	Choline chloride/1,3-butanediol (6:1) proved the highest extraction ability.	Peng et al. (2016)
Radix Salviae Miltiorrhizae	Hydrophilic and hydrophobic components: rosmarinic acid, lithospermic acid, salvianolic acid B, salvianolic acid A, and tanshinone 2 A	Choline chloride/ethylene glycol (1:2), choline chloride/1,2-propanediol (1:1, 1:2, and 1:3), choline chloride/glycerol (1:1, 1:2, 1:3, and 1:4), choline chloride/1,4-butanediol (1:2 and 1:4), choline chloride/oxalic acid (2:1 and 1:1), choline chloride/succinic acid (2:1 and 1:1), choline chloride/lactic acid (1:1), choline chloride/malonic acid (1:1 and 1:2), choline chloride/malic acid (1:1), choline chloride/citric acid (2:1), choline chloride/maltose (1:1), choline chloride/fructose (2:1), choline chloride/glucose (2:1), and choline chloride/urea (1:1, 2:1, and 1:2)	Choline chloride/1,2-propanediol (1:1) in 20 vol% of water	Chen et al. (2016)
Red grape skin	Phenolic compounds	Choline chloride/glycerol (1:2), choline chloride/oxalic acid (1:1), choline chloride/malic acid (1.5:1), choline chloride/sorbose (1:1), and choline chloride/proline/malic acid (1:1:1). 10%, 25%, and 50% w/w DES/water were tested.	Choline chloride/oxalic acid (1:1) in 25% water	Cvjetko et al. (2016)

Olive pomace	Phenolic compounds	Choline chloride/citric acid (1:2), choline chloride/lactic acid (1:2), choline chloride/maltose (1:2), and choline chloride/glycerol (1:2)	Choline chloride-based organic acids natural deep eutectic solvent (citric acid and lactic acid) with the addition of 20% v/v water	Chanioti and Tzia (2018)
Scutellaria baicalensis Georgi (Chinese herbal medicine)	Baicalin	Tetrabutylammonium chloride/decanoic acid (2:1, 1:1, and 1:2), tetraoctylammonium bromide/decanoic acid (1:2), methyltrioctylammonium chloride/decanoic acid (2:1, 1:1, and 1:2), and methyltrioctylammonium bromide/decanoic acid (1:2 and 1:1)	Tetrabutylammonium chloride/decanoic acid (2:1) in 30% vol/vol water. It increased the baicalin solubility (240 and 10 times in comparison with water and 70% ethanol, respectively). This was mainly related to the polarity and the physical properties of the solvent.	Wang et al. (2018)

DES deep eutectic solvent, *HBA* hydrogen bond acceptor, *HBD* hydrogen bond donor

Table 6.7 Examples of the application of vortex extraction technique with deep eutectic solvents as extraction solvents

Extraction medium	Target compounds	DES:HBA/HBD (molar ratio)	Selected DES	References
Virgin olive oil	Phenolic compounds	Choline chloride/glycerol (1:2), choline chloride/lactic acid (1:2), choline chloride/urea (1:2), choline chloride/sucrose (1:1 and 4:1), choline chloride/1,4-butanediol (1:5), choline chloride/xylitol (2:1), choline chloride/1,2-propanediol (1:1), choline chloride/malonic acid (1:1), choline chloride/urea/glycerol (1:1:1), and D-fructose/D-glucose/sucrose (1:1:1)	Choline chloride/xylitol (2:1) and choline chloride/1,2-propanediol (1:1)	García et al. (2016)
Chili oil	Rhodamine B	Choline chloride/ethylene glycol (1:3)	Choline chloride/ethylene glycol (1:3): the extraction process requires only 1 min of vortexing.	Wang et al. (2017)
Ginkgo biloba leaves	Polymer polyphenol compound: proanthocyanidin	Choline chloride/glycerol (1:2), choline chloride/ethylene glycol (1:2), choline chloride/propylene glycol (1:2), choline chloride/1,3-butanediol (1:3), choline chloride/D-sorbitol (1:1), choline chloride/xylitol (1:1), choline chloride/1,5-pentanedioic acid (1:1), choline chloride/glycolic acid (1:1), choline chloride/malonic acid (1:1), choline chloride/DL-malic acid (1:1), choline chloride/levulinic acid (1:1), choline chloride/lactic acid (1:2), choline chloride/citric acid (1:1), choline chloride/L-(+)-tartaric acid (1:1), choline chloride/urea (2:1), and choline chloride/oxalic (1:1)	Choline chloride/malonic acid (1:2) with 55% (w/w) water	Cao et al. (2018)

(continued)

Table 6.7 (continued)

Extraction medium	Target compounds	DES:HBA/HBD (molar ratio)	Selected DES	References
Vanilla pods	Vanillin	Choline chloride/malic acid/water (1:1:6), choline chloride/citric acid/water (1:1:6), choline chloride/glycerol (1:1), malic acid/glucose/fructose/water (1:1:1:9), citric acid/fructose/glucose/water (1:1:1:9), fructose/glucose/water (1:1:6), malic acid/glucose/water (1:1:6), betaine/sucrose/water (2:1:6), betaine/malic acid/glucose/water (1:1:1:9), betaine/citric acid/water (1:1:6), β-alanine/citric acid/water (1:1:6), L-serine/malic acid/water (1:1:6), lactic acid/1,2-propanediol (1:1), and lactic acid/fructose (5:1)		González et al. (2018)

DES deep eutectic solvent, *HBA* hydrogen bond acceptor, *HBD* hydrogen bond donor

time-consuming pretreatment (Mohebbi et al. 2018). Several types of sorbents are used such as carbon nanotube, SiO_2, C_{18}, and NH_2. However, these sorbents present some limitations such as single adsorption mechanism and low special selectivity, which limit their further application (Li et al. 2017). Hence, scientists are working on the development of new selective sorbents via the applications of deep eutectic solvent to increase their selectivity and capacity (Gan et al. 2016). All upcoming examples of the application of solid-phase microextraction techniques with deep eutectic solvents as extraction solvents are listed in Table 6.9.

Deep eutectic solvent-modified sorbents exhibited higher extraction efficiency and recovery values than tested sorbents using conventional organic solvents with a remarkable affinity and selectivity toward the target molecule; this was observed in all presented cases.

Ball Mill-Assisted Extraction

This method was developed by Wang et al. in 2016. Compared to other extraction methods (methanol-based ultrasound extraction and heat reflux extraction), the ball mill-assisted extraction was faster with higher extraction capacity of "tanshinones" and lower consumption of solvents. Also, in comparison with conventional solvents (n-hexane, ethanol, methanol, acetonitrile, acetone, and ethyl acetate), it was demonstrated that with deep eutectic solvent (six different choline chloride-based deep eutectic solvents), higher extraction efficiencies were obtained likely because of

Table 6.8 Examples of the application of heating and stirring extraction technique with deep eutectic solvents as extraction solvents

Extraction medium	Target compounds	DES:HBA/HBD (molar ratio)	References
Toluene oil	Phenol	The quaternary ammonium salts studied include tetramethylammonium chloride, tetramethylammonium bromide, tetraethylammonium chloride, methyltriethylammonium chloride, tetraethylammonium bromide, tetrapropylammonium chloride, tetrabutylammonium chloride, ammonium chloride, and choline bromide; the HBD was phenol.	Guo et al. (2013)
Plant species: Juá (*Ziziphus joazeiro*) and sisal (*Agave sisalana*)	Saponins, triterpenic, or steroidal glycosides	Choline chloride/urea (1:2), choline chloride/glycerol (1:2), choline chloride/ethylene glycol (1:2), choline chloride/acetic acid (1:2), choline chloride/propionic acid (1:2), choline chloride/lactic acid (1:2), choline chloride/malonic acid (1:2), choline chloride/oxalic acid (1:2), and choline chloride/phenylacetic acid (1:2)	Ribeiro et al. (2013)
Safflower: *Carthamus tinctorius L.* (Asteraceae)	Phenolic compounds	Choline chloride/glucose (5:2), choline chloride/sucrose (4:1), malic acid/proline (1:1); glucose/lactic acid (1:5), and choline chloride/xylitol (4:1)	Dai et al. (2014)
Fish and macroalgae samples and an uncontaminated Hamour fish sample	PAHs: phenanthrene, anthracene, fluoranthene, pyrene, benz[a]anthracene, chrysene, benzo[e]pyrene, and benzo[a]pyrene	Choline chloride/oxalic acid (1:1, 1:2, and 1:3)	Helalat-Nezhad et al. (2015)
Seaweed: *Kappaphycus alvarezii*	k-Carrageenan	Choline chloride/urea (1:2), choline chloride/ethylene glycol (1:2), and choline chloride/glycerol (1:2)	Das et al. (2016)
Purple and orange petals of the flower *Catharanthus roseus*	Anthocyanins	Choline chloride/1, 2-propanediol/water (1:1:1), glucose/lactic acid/water (1:5:3), malic acid/proline/water (1:1:3), choline chloride/malic acid/water (1:1:2), choline chloride/glucose/water (5:2:5), and malic acid/β-alanine/water (1:1:3)	Dai et al. (2016)

(continued)

Table 6.8 (continued)

Extraction medium	Target compounds	DES:HBA/HBD (molar ratio)	References
Vegetable samples: basil herb, spinach, dill, and cucumber barks	Manganese	Choline chloride/tartaric acid (1:1), choline chloride/oxalic acid (1:1), and choline chloride/citric acid (1:1)	Bağda et al. (2017)
Sophora japonica bud	Rutin	Choline chloride/triethylene glycol (1:4) + 18.1% water	Peng et al. (2018)
Shrimp shells (*Marsupenaeus japonicus*)	Chitin and chitin films	Choline chloride/lactic acid (1:2), choline chloride/1,4-butanediol (1:2), choline chloride/ethylene glycol (1:2), choline chloride/urea (1:2), choline chloride/1,6-hexanediol (1:2), choline chloride/glycerol (1:2), choline chloride/oxalic acid (1:2), choline chloride/malonic acid (1:2), choline chloride/citric acid (1:2), choline chloride/malic acid (1:2), choline chloride/propylene glycol (1:2), choline chloride/L-tartaric acid (1:2), choline chloride/maleic anhydride (1:2), and choline chloride/thiourea (1:2). Choline chloride/malonic acid (1:2) was the best extraction solvent.	Saravana et al. (2018a, b)
Olea europaea Leaves	Polyphenols	Glycerol/glycine/water (7:1:3)	Athanasiadis et al. (2018)

DES deep eutectic solvent, *HBA* hydrogen bond acceptor, *HBD* hydrogen bond donor

multiple interactions, such as hydrogen bonding, electrostatic, and Van der Waals interactions, between deep eutectic solvent and the target molecules. Extracted compounds were quite stable in deep eutectic solvent (Wang et al. 2016). This method is described by a scheme in Fig. 6.9.

Magnetic Solid-Phase Extraction

This method gained much attention in recent years. It has the same principle as the classical solid-phase microextraction, but it is mainly related to the magnetic sorbents facilitating the extraction process by applying an external magnetic field (Oller-Ruiz et al. 2018). Furthermore, these sorbents do not necessarily need to be packed into cartridge and can be recycled and reused; also, the centrifugation and the filtration could be avoided (Huang et al. 2015; Xu et al. 2016a, b).

Huang et al. (2015) synthesized magnetic graphene oxide impregnated with deep eutectic solvent (Fe_3O_4@GO-DES); the latter showed increased water solubility and improved proteins extraction efficiency than graphene oxide alone. Figure 6.10 illustrates the extraction process (Huang et al. 2015).

The same method was used by Xu et al. (2016a, b) to extract proteins. The adsorbed proteins on the Fe_3O_4–NH_2@GO@DES nanoparticles were eluted with no modification of their conformation. Also, Xu et al. (2016a, b) introduced a new type of polymer-immobilized magnetic silica materials with high thermal stability, long lifetime, and good durability for the extraction of trypsin. This sorbent can be recycled six times without significant loss of its extraction capacity and retained a high extraction capacity after eight cycles (Xu et al. 2016a, b). The new synthesized Fe_3O_4–NH_2@GO@DES sorbent gave the best extraction efficiency compared to Fe_3O_4–NH_2@graphene oxide and Fe_3O_4–NH_2 sorbents. This was linked to the numerous oxygen-containing functional groups existing on the surface of graphene oxide in addition to the hydroxyl groups of deep eutectic solvent that strengthen the interactions between proteins and deep eutectic solvent. In addition, the extraction efficiency was the best at a certain pH where opposite charges between magnetic microspheres and proteins exist. Another parameter related to the extraction capacity was the molecular weight of the proteins: the smaller the protein, the easier is its extraction (Xu et al. 2016a, b).

Li et al. (2017) developed a new synthesized sorbent C_8-amino-bifunctionalized ordered mesoporous organosilica for the extraction of triazine herbicides from watermelon samples. The synthesis of these new sorbents is illustrated in Fig. 6.11. Deep eutectic solvent composed of choline chloride/ethylene glycol 1:2 molar ratio was used as the extraction solvent. This method demonstrated lower limits of detection and relative standard deviations values in addition to high recoveries than other methods (pressurized liquid extraction, nonaqueous cavitation extraction, molecular-imprinted polymer-based solid-phase microextraction, cloud point extraction, matrix solid-phase dispersion). These sorbents present many advantages such as large surface area, regular and uniform pore size, and hydrothermal stability. In addition, these sorbents have two functional groups: C_8 and amino (the octyl chains gave the hydrophobic character, while the amino groups gave the hydrophilic one simultaneously), which improved the adsorption selectivity of C_8-amino-bifunctionalized ordered mesoporous organosilica. The latter has a good stability and reusability.

Online-Flow-Injection-Flame Atomic Absorption Spectrometry Solid-Phase Extraction

Karimi et al. (2017) extracted copper and nickel metal ions from water and biological samples (human serum and urine) by synthesizing deep eutectic solvent of choline chloride/urea (1:2 molar ratio) immobilized on cotton fibers. Desorption of the ions on deep eutectic solvent was due to the formation of a complex between nitrogen donor moieties of urea in the deep eutectic solvent and the analyte ions. This method is characterized by its rapidity, high sampling rate, simplicity, low reagent consumption, high accuracy, and freedom from contamination. Additionally, it presents a high advantage because it includes a natural cheap abundant renewable

Table 6.9 Examples of the application of solid-phase microextraction techniques with deep eutectic solvents as extraction solvents

Extraction medium	Target compounds	DES:HBA/HBD (molar ratio)	Sorbent type	References
Grape juice and mineralized drinking water	Organic acids: benzoic acid, *p*-anisic acid, salicylic acid, and cinnamic acid	Choline chloride/urea (1:2)	Silica cartridge was prepared in DES to obtain nitro-substituted tris(indolyl) methane-modified silica	Wang et al. (2016)
Phyllanthus flexuosus (dried plant materials)	Cleistanthol	Choline chloride/ glycerol (1:2)	DES-modified anion exchange resin. Positive choline groups of DES can make electrostatic interactions with the negatively charged analyte.	Gan et al. (2016)
Salvia miltiorrhiza Bunge	Tanshinones: cryptotanshinone, tanshinone I, and tanshinone IIA	Choline chloride/ ethylene glycol (1:3), choline chloride/ glycerol (1:3), choline chloride/1,2-butanediol (1:4), choline chloride/1,3-butanediol (1:4), choline chloride/1,4-butanediol (1:4), and choline chloride/2,3-butanediol (1:4)	Beads	Wang et al. (2016)
Proteins	Bovine serum albumin, ovalbumin, lysozyme, and bovine hemoglobin	Choline chloride/urea (1:1), choline chloride/D (+)-glucose (1:1), choline chloride/ ethylene glycol (1:2), and choline chloride/ glycerol (1:1)	Magnetic graphene oxide impregnated with DES (Fe_3O_4@ GO-DES); optimal DES: choline chloride/D-(+)-glucose (1:1)	Huang et al. (2015)
Bovine whole blood	Bovine serum albumin, ovalbumin, lysozyme, and bovine hemoglobin	Choline chloride/ glycerol (1:1), choline chloride/ethylene glycol (1:2), choline chloride/D-glucose (2:1), and choline chloride/D-sorbitol (1:1)	Fe_3O_4-NH_2@ GO@DES magnetic nanoparticles using choline chloride/glycerol (1:1) as the best DES	Xu et al. (2016a, b)

(continued)

Table 6.9 (continued)

Extraction medium	Target compounds	DES:HBA/HBD (molar ratio)	Sorbent type	References
Crude bovine pancreas extract	Trypsin	Choline chloride/ itaconic acid (n.m.)	Poly(DES)-grafted silica-coated magnetic microspheres (Fe_3O_4@ SiO_2-MPS@ PDES)	Xu et al. (2016a, b)
Watermelon	Triazine herbicides	Choline chloride/ ethylene glycol (1:2)	C8-amino-bifuctionalized ordered mesoporous organosilica in DES	Li et al. (2017)
Water (spring water, tap water, and seawater) and biological samples (human serum samples and urine samples)	Metal ions: copper and nickel	Choline chloride/urea (1:2)	DES immobilized on cotton fibers	Karimi et al. (2017)
Milk samples from China	Chloramphenicol	Choline chloride/ ethylene glycol (1:1), choline chloride/ glycerol (1:1), and choline chloride/ propylene glycol (1:1)	Ternary DES molecularly imprinted polymer	Tang et al. (2017)
Ilex chinensis Sims leaves	3,4-dihydroxybenzoic acid	Ternary DES of choline chloride/3, 4-dihydroxybenzoic acid and ethylene glycol (1:1:1, 1:1:2, and 1:1:3)	Ternary DES molecularly imprinted polymer	Li and Row (2018)

DES deep eutectic solvent, *HBA* hydrogen bond acceptor, *HBD* hydrogen bond donor; *n.m.* not mentioned

biopolymer sorbent, with increased selectivity by immobilization of deep eutectic solvent (Karimi et al. 2017).

Fig. 6.9 Ball mill-assisted deep eutectic solvent-based extraction of tanshinones from *Salvia miltiorrhiza Bunge*. This method can induce an optimized motion to disrupt cells through the multidirectional, simultaneous beating of specialized beads on the sample. This leads to a full release of intracellular products into the deep eutectic solvent (DES). (Figure reprinted with permission from Wang et al. 2016)

Mini-Solid-Phase Microextraction or Pipette-Tip Solid-Phase Extraction Method

The needle of the syringe system is suitable to be used as a meticulous mini-solid-phase microextraction cartridge because of its special miniconical shape with different diameters in the two ends where a certain amount of adsorbent can be packed, for example, molecular-imprinted polymer adsorbent. Compared to the conventional solid-phase extraction cartridges, extraction, using this technique, can be carried out faster and with facilitated sample-loading procedure (Li and Row 2018). Molecular-imprinted polymers are used as sorbents for purifying complex samples with numerous advantages such as high-recognition ability, mechanical stability, and resistance to a wide range of pH (Hu et al. 2012; Tang et al. 2017).

This technique was also applied by Li and Row in 2018 for the extraction of 3,4-dihydroxybenzoic acid from *Ilex chinensis Sims* leaves as shown in Fig. 6.12. Ternary deep eutectic solvent was used by mixing choline chloride, ethylene glycol, and 3, 4-dihydroxybenzoic acid. Compared to deep eutectic solvent-modified non-imprinted polymers (without template), a molecular-imprinted polymer (without deep eutectic solvent), and NIP (without deep eutectic solvent and without template), ternary deep eutectic solvent molecularly imprinted polymer demonstrated the best extraction efficiency. This technique presents an additional advantage than the other solid-phase extraction techniques since lower amounts of sorbent mass are used (Li and Row 2018).

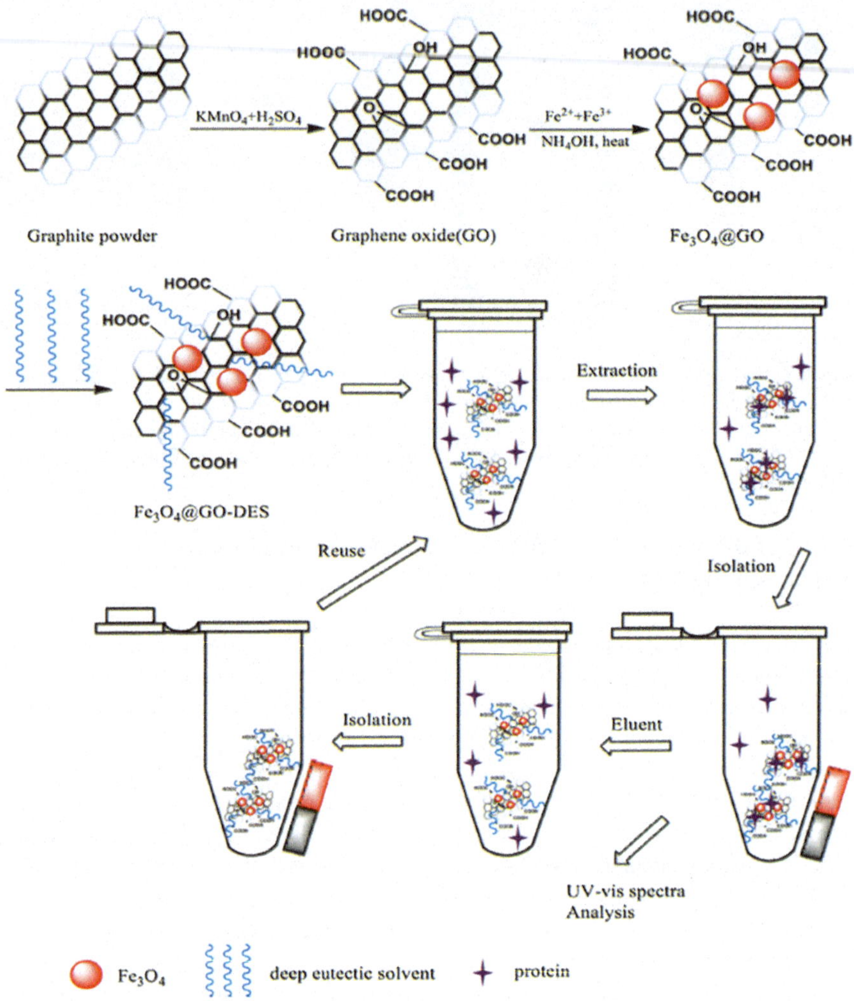

Fig. 6.10 Preparation of Fe_3O_4@GO magnetic sorbent impregnated with deep eutectic solvent and its application for the magnetic solid-phase extraction of protein. These sorbents are recovered via an external magnetic field. (Figure reprinted with permission from Yanhua Huang et al. 2015)

6.2.7 Combined Extraction Techniques

Extraction of target molecules has been also studied via application of combined techniques. Examples of these applications with deep eutectic solvents as extraction solvents are explained in this paragraph and presented in Table 6.10.

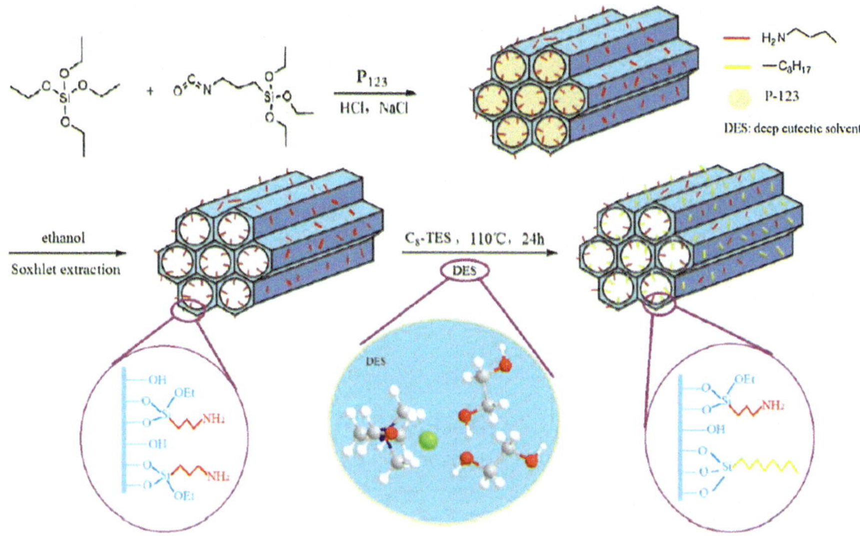

Fig. 6.11 Synthetic protocol of C_8-amino-bifunctionalized ordered mesoporous organosilica. Coating the sorbent with the deep eutectic solvent (DES) is also presented. (Figure reprinted with permission from Li et al. 2017)

Deep Eutectic Solvent-Based Negative-Pressure Cavitation-Assisted Extraction Method Combined with Macroporous Resin Enrichment

In this method (Fig. 6.13), millions of tiny vapor bubbles (voids) are formed in the liquid by the help of a machine that induces pressure (for example, pumps, turbines, and propellers). Firstly, the heating system of the negative-pressure cavitation-assisted extraction is turned on, and the water is heated. The sample is then introduced. After adding the deep eutectic solvent, this device is connected to the vacuum pump during all the extraction time. The deep eutectic solvent extraction solution obtained under optimized extraction conditions flowed through the column packed with macroporous resins at a constant flow rate. The adsorbed analytes were washed with deionized water and then eluted with 95% aqueous ethanol (v/v). The ethanolic fraction was collected and analyzed. Deep eutectic solvent showed better extractability than tested 80% ethanol solvent with the same technique. Moreover, deep eutectic solvent negative-pressure cavitation-assisted extraction yields were higher than that of deep eutectic solvent-based ultrasound extraction method (Qi et al. 2015). This method can be highly used for the thermolabile compounds because negative-pressure cavitation-assisted extraction is performed at room temperature. Additionally, the oxidation of these compounds is avoided as air is excluded in the extraction process (Liu et al. 2009).

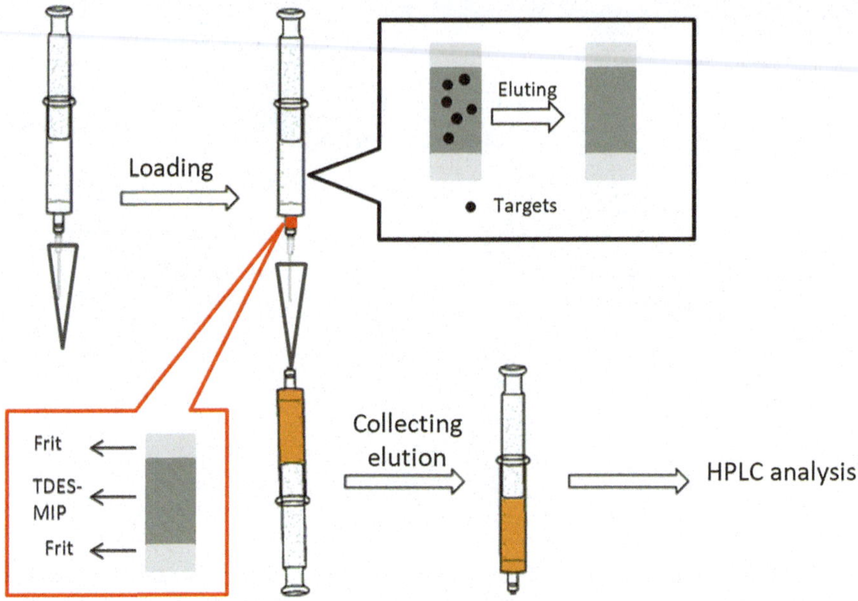

Fig. 6.12 Pipette-tip solid-phase extraction method where the sorbent is packed inside the end of the syringe (in this case, the sorbent is the ternary deep eutectic solvent molecularly imprinted polymer (TDES-MIP)). Each time the solution was sucked up into and out of micropipette type of the needle containing the adsorbent by pulling and pushing the syringe, respectively. The extraction complex retained on the adsorbent was washed, eluted, and subjected to analysis

Microwave Extraction and Solid-Phase Extraction

Wei et al. (2015) used the microwave extraction method for the extraction of four flavonoids from *Radix Scutellariae,* and their extraction yields were 33, 9, 9, and 2 mg/g, respectively. Thirteen natural deep eutectic solvents were tested; choline chloride/lactic acid (1:2 molar ratio) + 20% water (v/v) was selected because it was the most suitable one for the simultaneous extraction of these compounds. Aqueous ethanol 60% (v/v) was selected as reference extraction solvent. A solid-phase extraction method was also applied where the natural deep eutectic solvent extraction solution was flowed through the column packed with ME-2 macroporous resin for the separation of the flavonoids from the natural deep eutectic solvent solution. Recovery yields of baicalin, wogonoside, baicalein, and wogonin were 84%, 80%, 85%, and 82%, respectively (Wei et al. 2015).

Table 6.10 Examples of the application of combined extraction techniques with deep eutectic solvents as extraction solvents

Extraction medium	Target compounds	DES:HBA/HBD (molar ratio)	Extraction method	References
Equisetum palustre L.	Flavonoids	Choline chloride/glycerol, betaine hydrochloride/glycerol, choline chloride/1,4 butanediol, betaine hydrochloride/1,4 butanediol, choline chloride/1,3-butanediol, betaine hydrochloride/1,3-butanediol, choline chloride/ethylene glycol, betaine hydrochloride/ethylene glycol, and choline chloride/betaine hydrochloride/ethylene glycol (n.m.). Choline chloride/betaine hydrochloride/ethylene glycol (1:1:2) was the optimal DES.	DES-based negative-pressure cavitation-assisted extraction method coupled to macroporous resin enrichment	Qi et al. (2015)
Radix Scutellariae	Flavonoids: baicalin, wogonoside, baicalein, and wogonin	Choline chloride/1,4-butanediol (1:2), choline chloride/glycerol (1:2), choline chloride/ethylene glycol (1:2), choline chloride/citric acid (1:2), choline chloride/malic acid (1:2), choline chloride/lactic acid (1:2, 3:1, 2:1, 1:1, 1:3, 1:4), choline chloride/glucose (1:2), choline chloride/sorbitol (1:2), choline chloride/sucrose (1:2), choline chloride/maltose (1:2), citric acid/sucrose (1:2), citric acid/glucose (1:2), and lactic acid/sucrose (1:2)	Natural deep eutectic solvent-based microwave extraction coupled to direct macroporous resin adsorption and desorption process	Wei et al. (2015)
Bamboo shoot shell	Lignin and xylan	Choline chloride/organic acid (1:2), choline chloride/ethylene glycol (1:2), and choline chloride/glycerol (1:2)	Biological treatment based on microorganisms coupled to DES pretreatment	Dai et al. (2017)

(continued)

Table 6.10 (continued)

Extraction medium	Target compounds	DES:HBA/HBD (molar ratio)	Extraction method	References
Food and herbal tea samples: turmeric liquid supplement, herbal tea, turmeric drug or powder, and root herbal tea	Curcumin	Choline chloride/phenol (1:4)	DES-based vortex extraction combined to emulsification liquid-liquid microextraction	Aydin et al. (2018)
Human urine and plasma samples	Tricyclic antidepressant drugs: amitriptyline, nortriptyline, desipramine, clomipramine, and imipramine	Choline chloride/4-chlorophenol DES formed by mixing 1.39 g choline chloride with 2.58 g 4-chlorophenol	Dispersive solid-phase extraction coupled to DES-based air-assisted liquid-liquid microextraction	Mohebbi et al. (2018)
Ginkgo biloba and *Panax ginseng*	Phenolics, phenolic acids, and terpenoids from *G. biloba* leaves and ginsenosides from *P. ginseng* leaves	Choline chloride/malic acid/(1:1), glucose/malic acid (1:1), choline chloride/glucose (5:2), proline/malic acid (1:1), sucrose/fructose/glucose (1:1:1), and sucrose/proline/glycerol (1:4:9). All of the prepared NADES were mixed with 10% (w/w) water.	Ultrasound extraction combined with solid-phase extraction	Liu et al. (2018)

DES deep eutectic solvent, *HBA* hydrogen bond acceptor, *HBD* hydrogen bond donor; *n.m.* not mentioned

Biological and Deep Eutectic Solvent Pretreatments

Dai et al. (2017) highlighted the importance of biodiesel production from microbial lipids obtained from the fermentation of low-cost carbon sources by microorganisms. Waste biomass could be treated and used as one of those carbon sources (hydrolysates). Bamboo shoot shell, highly agricultural waste produced in China, is treated; it is mainly composed of cellulose, hemicellulose, and lignin. The combination of the biological treatment and deep eutectic solvent was tested for the removal of lignin and hemicellulose from bamboo shoot shell in order to enhance the biomass conversion into fermentable sugars for producing biofuel.

In this study, untreated bamboo shoot shell was subjected to biological pretreatment with Galactomyces sp. CCZU11-1 and used to produce cellulases. After 3 days, maximum amount of xylan and lignin were removed and compared to untreated bamboo shoot shell. After fermentation, the residues of bamboo shoot shell were pretreated by deep eutectic solvent. Notably, choline chloride/oxalic acid

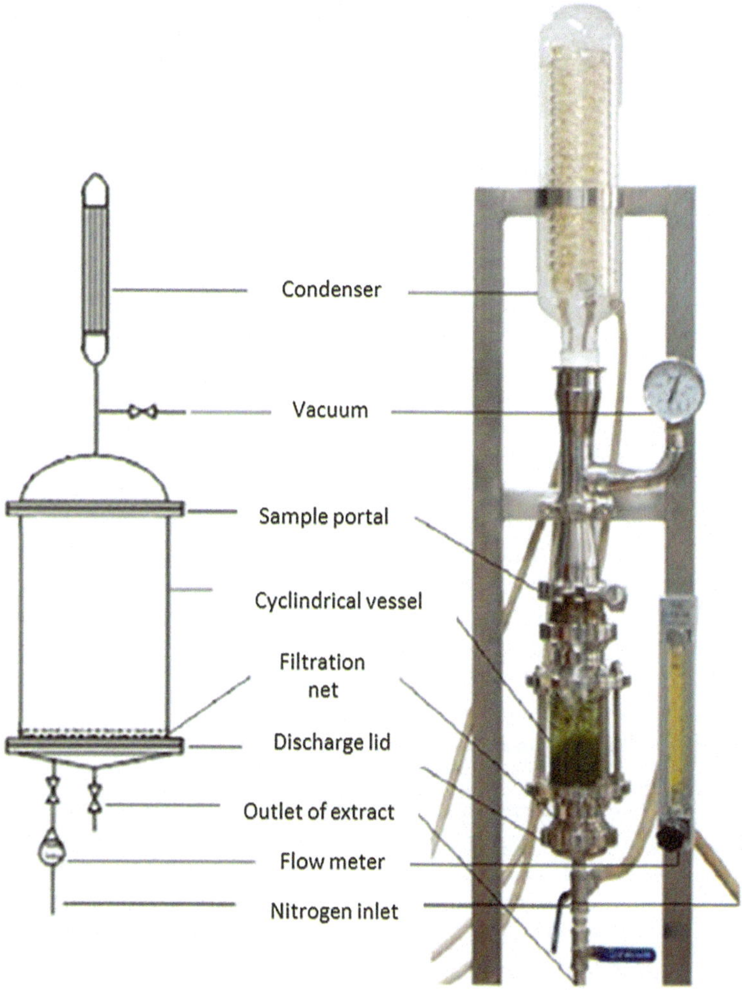

Fig. 6.13 Schematic representation of the negative-pressure cavitation-assisted extraction device and the actual negative-pressure cavitation-assisted extraction device. (Figure reprinted with permission from Liu et al. 2009)

(1:2 molar ratio) was found to be the best solvent for pretreating bamboo shoot shell with the highest xylan removal (53%) and delignification (48%). After deep eutectic solvent addition, more surface areas in the pretreated bamboo shoot shell were formed which allowed cellulases to further attack cellulose and residual hemicellulose. Thus, enzymatic saccharification of bamboo shoot shell was increased after deep eutectic solvent pretreatment. The reducing sugars yield from the enzymatic hydrolysis of 50 g/L deep eutectic solvent bamboo shoot shell was maximal (90%). Furthermore, it was found that combination pretreatment was better than one-step pretreatment, and it effectively removed xylan and lignin after treatment with deep eutectic solvent (Dai et al. 2017).

Deep Eutectic Solvent-Based Vortex Extraction Combined with Emulsification Liquid-Liquid Microextraction

Aydin et al. (2018) developed a new technique using deep eutectic solvent (choline chloride/phenol 1:4 molar ratio) as a water-miscible extraction solvent for the extraction of curcumin (Fig. 6.14). Recoveries of curcumin using different deep eutectic solvent were above 96% (Aydin et al. 2018).

Dispersive Solid-Phase Extraction in Combination with Deep Eutectic Solvent-Based Air-Assisted Liquid-Liquid Microextraction

The proposed method is represented in Fig. 6.15. The process started with dispersive solid-phase extraction method. Then air-assisted liquid-liquid microextraction method was applied. In the proposed method, the synthesized deep eutectic solvent was used as an elution/extraction solvent. Results showed many advantages such as good repeatability, high enrichment factors and extraction recoveries, low limits of detection and limits of quantification, and simplicity of operation. According to Mohebbi et al. (2018), this method can be used for the routine analysis of many drugs in the pharmaceutical and clinical laboratories with no harm on human health and environment (Mohebbi et al. 2018).

Ultrasound Extraction and Solid-Phase Extraction

Liu et al. (2018) proposed a combined technique for the extraction of different classes of natural products (phenolics, terpenoids, and phenolic acids) from *G. biloba* leaves and ginsenosides from *P. ginseng* leaves. Six different natural deep eutectic solvents were tested (Table 6.10). The presence of natural deep eutectic solvent caused severe tailing of spots in high-performance thin-layer chromatography analysis; therefore, it was important to recover the analytes from natural deep eutectic solvent before analysis. Hence, solid-phase extraction method was employed using polymeric reversed-phase sorbent cartridges. Of the natural deep eutectic solvents, choline chloride/malic acid (1:1 molar ratio) and glycerol/proline/sucrose (1:1:1 molar ratio) were the best for *G. biloba* leaves, and choline chloride/malic acid (1:1 molar ratio) and glucose/malic acid (1:1 molar ratio) for *P. ginseng* leaves showing the highest yields of the target compounds. The addition of water to natural deep eutectic solvent affected the extraction and maximum yields. The latter were obtained with approximately 20% water (w/w). Results showed that the yield of analytes obtained with the natural deep eutectic solvent is similar to that of methanol. A high advantage of the usage of natural deep eutectic solvent is their incapability to extract ginkgolic acids (considered very toxic to human) due to their low polarity and low dissolution in natural deep eutectic solvents. This method proved to be able to deliver reproducible chemical profiles from the natural deep eutectic solvent extracts (Liu et al. 2018).

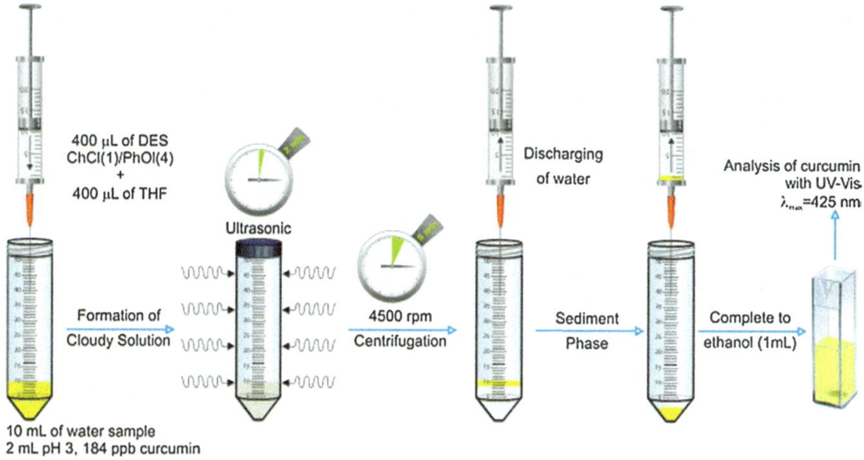

Fig. 6.14 Deep eutectic solvent-based vortex extraction combined with emulsification liquid-liquid microextraction. Deep eutectic solvent (DES) was injected rapidly to the sample solution with tetrahydrofuran as an emulsifier agent. This leads to the formation of a cloudy solution due to insoluble self-aggregation of DES droplets. The cloudy solution was subjected to ultrasonication. Then centrifugation was applied to separate the DES phase from the aqueous solution. The DES phase containing curcumin was subjected to UV-Vis analysis. (Figure reprinted with permission from Aydin et al. 2018)

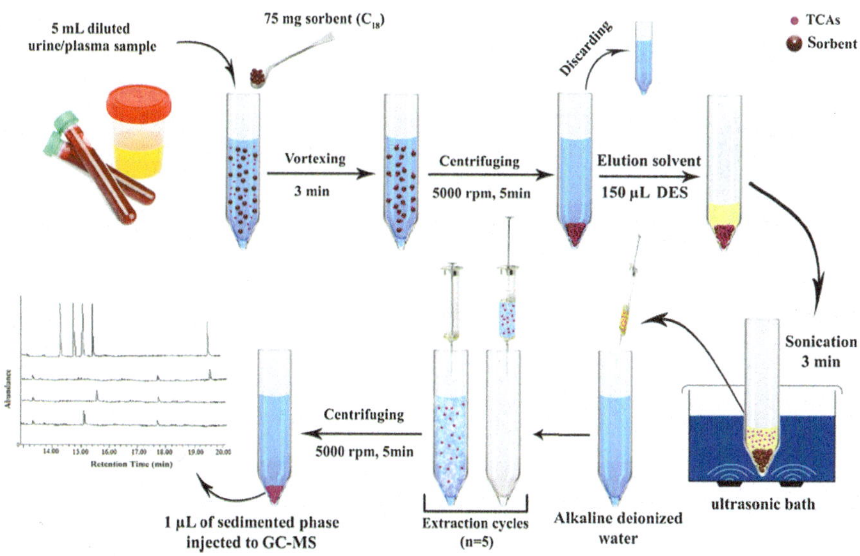

Fig. 6.15 Dispersive solid-phase extraction in combination with deep eutectic solvent (DES)-based air-assisted liquid-liquid microextraction preconcentration procedure. The process started with diluted urine or plasma samples transferred into a glass test tube in the presence of sorbent. The tube was then vortexed. Then, the adsorbed analytes were subsequently eluted with DES under sonication. The supernatant solution obtained from the previous step was removed and then subjected to five aspiration/dispersion cycles. A cloudy solution was formed. After centrifugation, DES phase was analyzed. (Figure reprinted with permission from Mohebbi et al. 2018)

6.3 Influence of Deep Eutectic Solvent Properties on the Extraction Efficiency

The properties of deep eutectic solvent highly influence the extraction efficiency of the analytes (Tang et al. 2014). These properties depend both on the physicochemical properties of each component and on the interactions between the two constituents of the deep eutectic solvent. While deep eutectic solvent melting point, density, viscosity, and compatibility with instrumental detection systems may control the extraction process feasibility, the extraction efficiency is mainly influenced by the deep eutectic solvent solubilizing capacities and by its cell structures disruption ability. Since the majority of the bioactive compounds are being extracted from a medium containing large amount of water, deep eutectic solvent properties can be influenced by the presence of water. Accordingly, El Achkar et al. (2019) discussed, in a recent review, the effect of water on the main physicochemical properties of deep eutectic solvents; the water content should be considered in the analysis of the extraction efficiency by deep eutectic solvent (El Achkar et al. 2019).

6.3.1 Melting Point

As it was mentioned above, deep eutectic solvents are characterized by a lower melting point than that of any of its individual components. For example, choline chloride has a melting point of 302 °C and urea 113 °C; when mixing these two compounds, a eutectic mixture with a melting point of 12 °C is obtained (Zhang et al. 2012). This melting point depression is due to an interaction between the halide anion Cl^- and urea via hydrogen bonds, and consequently charge delocalization occurs. This indicates that the melting point is dependent on the nature of hydrogen bond acceptor and hydrogen bond donor and is highly in correlation with the hydrogen bond strength. The molar ratio of the deep eutectic solvent components also affects the melting point (Zhang et al. 2012). So it is considered when choosing the temperature in the extraction procedure.

6.3.2 Density

Most of deep eutectic solvents possess higher density values than water, with levels ranging from 1.041 to 1.63 $g.cm^{-3}$. Density differences between deep eutectic solvents are explained by their molecular organization and packing (Zhang et al. 2012). Deep eutectic solvent density is higher than the individual components densities because of the "hole theory": the radius of the existing holes decreases after mixing. Deep eutectic solvent's density is a highly important property when separating phases in the extraction process especially in dispersive liquid-liquid microextraction techniques because most of the separation processes are based on the density differences between the two phases. Deep eutectic solvent's density decreases linearly when the temperature is increased as it was shown by Florindo et al. (2014) and as presented in Fig. 6.16. Also, it is highly influenced by the deep eutectic solvent components molar ratio. For example, it was seen by Wahaibi et al. (2019) that the densities of the deep eutectic solvent choline chloride/malonic acid with 1:0.5 molar ratio are slightly higher than the same solvent with 1:1 molar ratio at all temperatures. As it was explained by the authors, that this is due to the presence of high amount of choline chloride in the first solvent because, as a general rule, the bulkier the cation is, the lower the density is (Wahaibi et al. 2019). Another reference proved also a similar result that is presented in Fig. 6.17 (Abbott et al. 2011).

6.3.3 Viscosity

The range of deep eutectic solvent's viscosity values are around 20–1000 times higher than that of water at room temperature (Dai et al. 2013a, b). Most of the deep eutectic solvents possess high viscosity values (>100 Cp) at room temperature (Zhang et al. 2012). The choline chloride/ethylene glycol (1:4 molar ratio) deep eutectic solvent possesses the lowest viscosity (Zhang et al. 2012). Mainly, the viscosity depends on the ion size, the void volume, the temperature, and the water content (Smith et al. 2014; Tang et al. 2014). The possible high viscosity of deep eutectic solvents is attributed to the presence of an extensive hydrogen-bonding network, Van der Waals, and/or electrostatic interactions between the compounds that restrict the mobility of the free species inside deep eutectic solvent and restrict the dispersion of deep eutectic solvent in the extraction medium during the extraction process (Zhang et al. 2012; Habibollahi et al. 2018). The high viscosity is good for the prevention of entrapped agents from vaporization, but it hampers the mass transfer and thus produces lower extraction efficiency. Therefore, this problem can be solved by increasing the temperature, thus leading to a better penetration of the solvents in the sample; this is known as Arrhenius-like behavior (Bubalo et al. 2018). Also, adding a certain percentage of water to a deep eutectic solvent, at which the deep eutectic solvent's network is still maintained, is another way to decrease deep eutectic solvent's viscosity (Qi et al. 2015; Fernández et al. 2018). However, Dai et al. (2015) showed that the addition of water above 50% can dissolve the deep eutectic solvent components in water (Dai et al. 2015). Also, the addition of water can alter the hydrogen bonds between deep eutectic solvent and the target compound (Bajkacz and Adamek 2017). Water-added deep eutectic solvents are more suitable for the extraction of polar compounds, whereas deep eutectic solvents with low water content are better for the extraction of nonpolar compounds (Dai et al. 2013a, b). Other parameters that may influence the viscosity of the deep eutectic solvent are the type of its components and the molar ratio. Cao et al. (2018) showed that the extraction yield of proanthocyanidin was decreased

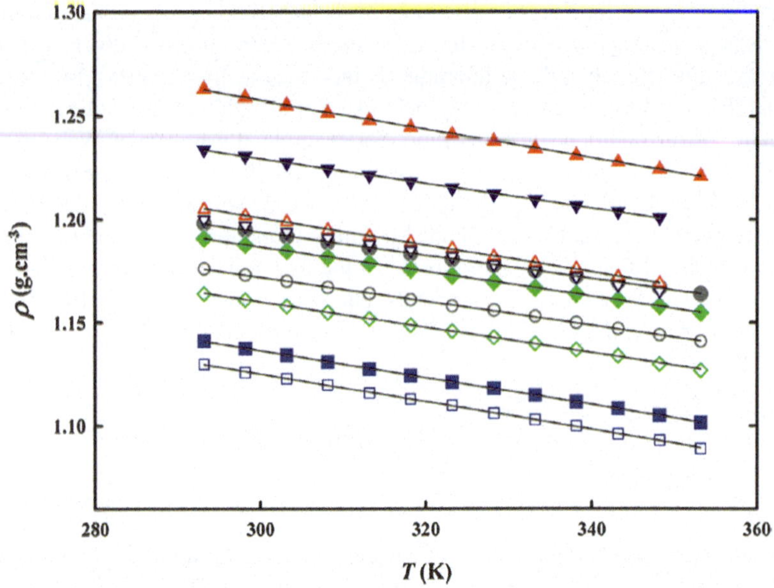

Fig. 6.16 Experimental densities (ρ) of some dried and water-saturated deep eutectic solvents as a function of temperature: choline chloride/oxalic acid (▲), choline chloride/malonic acid (▼), choline chloride/glycerol (●), choline chloride/glutaric acid (◆), and choline chloride/levulinic acid (■). The filled symbols correspond to the dried solvents, and the empty symbols correspond to the water-saturated solvents. When comparing the values obtained for the densities of the dried and the water-saturated solvents, it can be observed that the latter are lower, as expected. As observed, the density decreases linearly with temperature for all deep eutectic solvents (dried and water saturated) in the whole temperature range studied. (Figure reprinted with permission from Florindo et al. 2014)

when using deep eutectic solvent containing xylitol compared to those containing other alcohols because of the high viscosity of the first one. According to the authors, viscosities of deep eutectic solvents prepared with solid hydrogen bond donors were higher than those prepared with liquid hydrogen bond donors (Cao et al. 2018). Bi et al. (2013) showed that the amounts of the flavonoids extracted increased with decreasing choline chloride/hydrogen bond donor ratio from 1/1 to 1/5 (mol/mol) due to a decrease in viscosity (Bi et al. 2013). That means that the viscosity decreases when decreasing choline chloride concentration. However, Abbott et al. (2011) found the opposite result with choline chloride/glycerol deep eutectic solvent (Fig. 6.18). For example, at 20 °C, viscosities of this solvent with a choline chloride/glycerol molar ratio of 1:4, 1:3, and 1:2 were 503, 450, and 376 cP, respectively. This drastic decrease of the glycerol viscosity upon addition of choline chloride was attributed to the partial rupture of the intermolecular hydrogen bond network of glycerol (Abbott et al. 2011). Deep eutectic solvents with a high viscosity are good for the single-drop microextraction techniques because they facilitate the suspension of the drop at the end of the needle of a microsyringe.

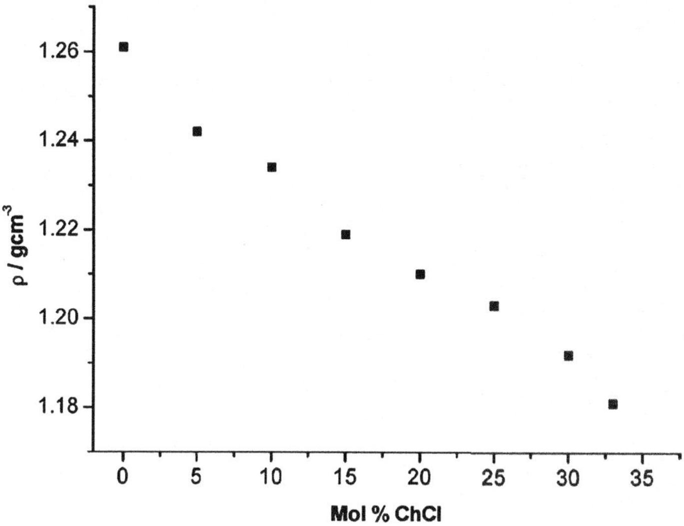

Fig. 6.17 Density (ρ) of the choline chloride/glycerol deep eutectic solvent as a function of choline chloride (ChCl) percentage. Addition of ChCl to glycerol results in a decrease of the deep eutectic solvent density. (Figure reprinted with permission from Abbott et al. 2011)

6.3.4 *Compatibility with Instrumental Detection Systems*

An extraction technique is useless without an appropriate detection method for the target compound. Therefore, a high advantage of using deep eutectic solvent in the extraction techniques is the absence of interference generally observed for these solvents with the detection methods. They showed a high compatibility with various instrumental detection systems such as mass spectrometry, flame ionization detector, UV-Vis spectroscopy, inductively coupled plasma-optical emission spectrometry, diode array detector, graphite furnace atomic absorption spectroscopy, Fourier transform infrared spectroscopy, and fluorescence; therefore, new combinations with advanced separation techniques like high-performance liquid chromatography, gas chromatography, and thin-layer chromatography are ongoing, for example, by using a certain percentage of deep eutectic solvent in the mobile phase (Tan et al. 2016).

6.3.5 *Surface Tension*

Surface tension depends on the strength of intermolecular interaction of deep eutectic solvent and on temperature (Tang et al. 2014). Therefore, it is positively dependent on the viscosity. Some deep eutectic solvent (for example, choline chloride/glycerol) showed a linear correlation between the surface tension and the

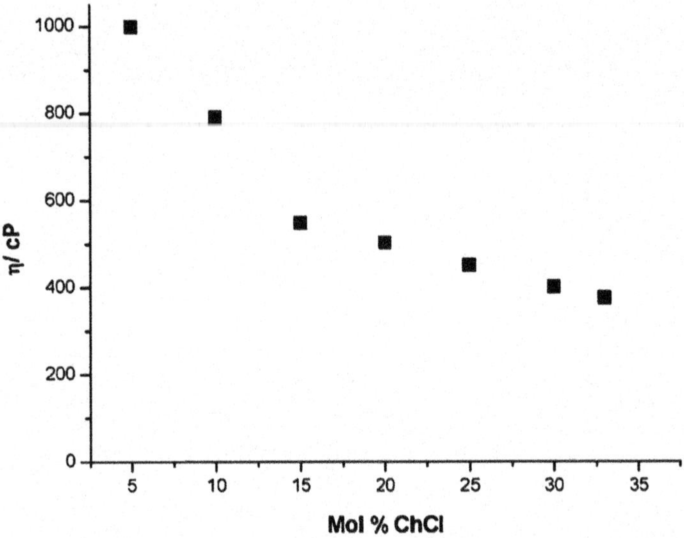

Fig. 6.18 Plot of the viscosity (η) of the choline chloride/glycerol deep eutectic solvent as a function of the choline chloride (ChCl) molar composition at 298 K. It can be seen that the viscosity decreases as the salt concentration increases. (Figure reprinted with permission from Abbott et al. 2011)

temperature but reciprocally with the choline chloride concentration. Generally, most of the deep eutectic solvents have higher surface tension than conventional solvents. Deep eutectic solvent with low surface tension can be applied as an adhesive or wetting agent in the extraction techniques (Li and Row 2016).

6.3.6 Polarity and Solubilizing Properties

Deep eutectic solvent's polarity is an important parameter regarding the miscibility of the deep eutectic solvent in other solvents or either in the sample solution (Dai et al. 2013a, b). Increasing the temperature leads to a decrease in the polarity of deep eutectic solvent because of the reduction of the hydrogen bond donating acidity of deep eutectic solvent, as observed with choline chloride/glycerol deep eutectic solvent (Tang et al. 2014). Also, a linear correlation exists between the polarity of deep eutectic solvent and the choline chloride concentration.

Deep eutectic solvents are known to have good solubilizing properties for polar and weak polar compounds, including drugs, pharmaceutical ingredients, metal oxides, carbon dioxide, and elemental species such as lead, mercury, cadmium, etc. (Aroso et al. 2016). Also, deep eutectic solvent proved better solubilizing effect, therefore better extractability compared to conventional solvents (Fernández et al. 2018). These properties may be rationalized, at least partially, by the "like dissolves like" theory, thus helping the choice of the adequate deep eutectic solvent regarding its polarity. Due to the high number of starting components, different combinations can lead to various deep eutectic solvent with different properties. Recently, deep eutectic solvents have been tailored to be target-specific via the selection of specific individual components based on the analytes via in silico methods (Fernández et al. 2018). Unique interactions between the deep eutectic solvents with target analytes make it possible to selectively separate trace of this analyte from complex matrices (Fernández et al. 2018).

6.3.7 pH

Since this parameter determines the state of the analytes, the pH of deep eutectic solvent can affect the structures of the bioactive compounds subjected to extraction and thus the extraction efficiency. In their neutral forms, analytes are generally much easier to be extracted by weakly polar solvents. However, in their ionic forms, they have fewer tendencies to be extracted. Therefore, the pH of the extraction procedure should be higher than or near to the pKa values of the studied analytes (Mohebbi et al. 2018). The pH of different deep eutectic solvent changes differently with temperature, and the acidity of deep eutectic solvent is highly affected by the type of hydrogen bond donor (Tang et al. 2014). The highly polar molecules demand an acidic environment for a better extraction; therefore, organic acid-based deep eutectic solvents or natural deep eutectic solvents are the best choice. This was observed by organic acid-based natural deep eutectic solvent that showed the best extraction results for anthocyanin (polar compounds), while sugar-based natural deep eutectic solvents were a better choice for other phenolic compounds (Radošević et al. 2016). Finally, due to their amphoteric properties, the extraction of proteins is highly influenced by the pH (Li et al. 2016). The isoelectric point of the protein should be taken into consideration (Li et al. 2016). pH contributes as well for the reduction of the matrix interference (Karimi et al. 2017).

6.3.8 Cell Disruption Ability

Cell disruption plays an important role in the extraction of analytes; therefore, it is considered as a revealed parameter for the extraction efficiency. There is no doubt that some extraction techniques cause cell disruption such as ultrasonication, microwave extraction, as well as hot reflux extraction and others. However, the solvent used have an additional impact on the cell disruption. Using different extraction techniques followed by scanning electron microscope analysis, it was proved that deep eutectic solvents cause cell rupture more efficiently than water or other conventional extraction solvents (Table 6.11). This leads to the full release of the target analyte and its subsequent dissolution in deep eutectic solvent.

6.4 Optimization of Extraction Parameters

An extraction method can be selective for a specific analyte by monitoring the process parameters like temperature, pressure, flow, or power.

6.4.1 Type of Extraction Solvent

Being applied in the extraction methods as alternatives to conventional solvents, deep eutectic solvent should have high affinity for the target analytes (Yousefi et al. 2018). They should possess low solubility in supported liquid membrane solvent and high stability in the lumen of the used membrane in the case of hollow-fiber liquid-phase microextraction method (Khataei et al. 2018). Extraction process is affected not only by the type of hydrogen bond donor but also by the proportion of functional groups. For example, in contrast to the aqueous two-phase system extraction method, Yanhua et al. (2015) declared that hydrogen bonding interactions could be the main driving force in protein extraction; the OH groups can form hydrogen bonds with proteins more than the NH_2 groups because the electronegativity of oxygen is greater than that of nitrogen (Yanhua Huang et al. 2015).

6.4.2 Sample to Deep Eutectic Solvent Volume Ratio (V/V) Optimization

The volume of deep eutectic solvent should be sufficient for the solubilization of the analytes and for a fast extraction but as low as possible in order to avoid waste and environmental toxicity and to get a final volume of deep eutectic solvent adequate for the chromatographic analysis (Khataei et al. 2018; Mohebbi et al. 2018). On the other hand, a high volume of deep eutectic solvent leads to decreasing the extraction efficiency, and this is related to the dilution effect when freeze-drying cannot be performed (Mohebbi et al. 2018).

6.4.3 Matrix Ions

Coexisting ions can enter in competition with the target analytes, thus reducing the recoveries values, and the extraction efficiency of the method (Karimi et al. 2017). This parameter reveals the selectivity of the method (Aydin et al. 2018). Salt can act as an effective dehydrating agent for more hydrophilic analytes, also called salting-out effect, in which the solubility of the analytes in the aqueous solution decreases, leading to a better extraction and, subsequently, a good-phase separation. However,

Table 6.11 Extraction techniques that proved the cell disruption ability of deep eutectic solvents

Extraction medium	Target compounds	Selected DES HBA/HBD (molar ratio)	Extraction technique	Disruption degree order	References
Radix Salviae Miltiorrhizae	Rosmarinic acid, lithospermic acid, salvianolic acid B, salvianolic acid A, tanshinone 2 A	Choline chloride/1,2-propanediol (1:1) + 20 vol% of water	Microwave extraction	DES > 75% methanol > water	Chen et al. (2016)
Green tea (*Camellia sinensis*)	Catechins: epigallocatechin-3-gallate	81% betaine/glycerol/D-glucose (4:20:1)	Ultrasound microextraction	81% DES > methanol or 70% methanol	Jeong et al. (2017)
Morus alba leaves	Phenolic compounds	Choline chloride/citric acid (2:1)	Ultrasound microextraction	DES > 75% methanol > water	Zhou et al. (2018)
Scutellaria baicalensis Georgi (Chinese herbal medicine)	Baicalin	Decanoic acid/tetrabutylammonium chloride (1: 2) + 30% v/v water	Microwave extraction or ultrasound extraction or hot reflux extraction	Microwave extraction > ultrasound extraction > hot reflux extraction. Also, DES > 70 vol% ethanol	Wang et al. (2018)

DES deep eutectic solvent, *HBA* hydrogen bond acceptor, *HBD* hydrogen bond donor

at higher concentrations, the solution viscosity can significantly increase due to the salting-in phenomena, leading to a noticeable reduction in the diffusion rate of the target analytes into the extraction liquid phase, thus reducing the extraction efficiency (Khataei et al. 2018; Moghadam et al. 2018; Mohebbi et al. 2018).

6.4.4 Type of Emulsifier Agent

Some microextraction techniques require a nonmiscibility of the deep eutectic solvent in the aqueous solution, for example, emulsification microextraction. Therefore, it is important to use an emulsifier agent or an aprotic solvent in order to release the bonding of water-miscible deep eutectic solvent from the water molecules. This leads to a self-aggregation of deep eutectic solvent molecules. Many aprotic solvents are used: tetrahydrofuran, acetone, 1,4-dioxane, dichloromethane, and others (Moghadam et al. 2018).

6.4.5 Temperature

As mentioned earlier, this parameter positively affects the viscosity of deep eutectic solvent, and therefore, it can affect the diffusivity of analytes to deep eutectic solvent; however, it should be optimized in the extraction of thermolabile analytes in order to prevent their decomposition (Li et al. 2016). When using sorbents coated by deep eutectic solvent, increasing the temperature can desorb the deep eutectic solvent molecules from the sorbents, thus diminishing the extraction capacity (Huang et al. 2015). For the extraction of volatile compounds, temperature is a critical parameter that should be controlled. At high temperature, they are rapidly volatilized which leads to a higher extraction efficiency (Tang et al. 2014). However, increasing temperature for the extraction of nonvolatile compounds enhances their desorption from the matrix and their solubilization in deep eutectic solvent (Qi et al. 2015; Huang et al. 2015). In the case of proteins, attention should be paid to the temperature for it won't cause proteins denaturation (Li et al. 2016).

6.4.6 Time

An optimal time should be chosen for the best diffusion of the analytes from the original sample to the deep eutectic solvent. This parameter should be well controlled especially when having an adsorption/desorption process because in liquid-phase microextraction methods, the extraction should reach an equilibrium (Tang et al. 2014). Also, long extraction time can cause high risk of losing solvents and formation of air bubbles (Khataei et al. 2018).

6.4.7 *Agitation or Vortexing*

It is important to accelerate the extraction; however, it is necessary to disperse the best media in order to get a good contact between the analytes and the extraction solvent. So the vortex apparatus type and the vortex time are very important parameters (Aydin et al. 2018). Ultrasonication can be also applied. It increases the mass transfer of target compounds, and it decreases the extraction time (Gu et al. 2014).

6.5 Conclusion

Extensive studies have been conducted on the different applications of deep eutectic solvents. Their low cost, their ease of synthesis, and their eco-friendly behavior make them suitable extraction solvents in the newly developed microextraction techniques. This review confirmed that the use of deep eutectic solvent in many extraction techniques, either alone or mixed in a certain percentage of water, has given many advantages compared to the highly toxic conventional organic solvents. Mainly, this was proved by the enhanced extraction efficiency for different kind of molecules and macromolecules especially drugs, proteins, and plants secondary metabolites. In addition, deep eutectic solvent properties, such as melting point, density, viscosity, compatibility with instrumental detection systems, surface tension, polarity and solubilizing properties, pH, and cell disruption ability, should be considered for a maximum extraction efficiency. Extraction method parameters including type of deep eutectic solvent, sample to deep eutectic solvent volume ratio (v/v), effect of matrix ions, type of emulsifier agent, temperature, time, and agitation should be optimized. Future studies should be carried for the extraction of other important bio-based compounds and for the development of new extraction techniques that could be even greener. Also, the application of these solvents in the industrial large-scale extraction techniques should be investigated later on.

Acknowledgments Lamia Nakhle has received a scholarship from Université du Littoral Côte d'Opale (ULCO) and from the Lebanese National Council for Scientific Research (CNRS-L).

References

Abbott AP, Capper G, Davies DL, Rasheed RK, Tambyrajah V (2003) Novel solvent properties of choline chloride/urea mixtures Electronic supplementary information (ESI) available: spectroscopic data. See http://www.rsc.org/suppdata/cc/b2/b210714g/. Chem Commun 1:70–71. https://doi.org/10.1039/b210714g

Abbott AP, Boothby D, Capper G, Davies DL, Rasheed RK (2004) Deep eutectic solvents formed between choline chloride and carboxylic acids: versatile alternatives to ionic liquids. J Am Chem Soc 126(29):9142–9147. https://doi.org/10.1021/ja048266j

Abbott AP, Harris RC, Ryder KS, D'Agostino C, Gladden LF, Mantle MD (2011) Glycerol eutectics as sustainable solvent systems. Green Chem 13(1):82–90. https://doi.org/10.1039/C0GC00395F

Alfonsi K, Colberg J, Dunn PJ, Fevig T, Jennings S, Johnson TA, Kleine HP, Knight C, Nagy MA, Perry DA, Stefaniak M (2008) Green chemistry tools to influence a medicinal chemistry and research chemistry based organisation. Green Chem 10(1):31–36. https://doi.org/10.1039/B711717E

Anastas P, Eghbali N (2010) Green chemistry: principles and practice. Chem Soc Rev 39(1):301–312. https://doi.org/10.1039/B918763B

Aroso IM, Silva JC, Mano F, Ferreira ASD, Dionísio M, Sá-Nogueira I, Barreiros S, Reis RL, Paiva A, Duarte ARC (2016) Dissolution enhancement of active pharmaceutical ingredients by therapeutic deep eutectic systems. Eur J Pharm Biopharm 98:57–66. https://doi.org/10.1016/j.ejpb.2015.11.002

Athanasiadis V, Grigorakis S, Lalas S, Makris DP (2018) Highly efficient extraction of antioxidant polyphenols from Olea europaea leaves using an eco-friendly glycerol/glycine deep eutectic solvent. Waste Biomass Valoriz 9(11):1985–1992. https://doi.org/10.1007/s12649-017-9997-7

Aydin F, Yilmaz E, Soylak M (2018) Vortex assisted deep eutectic solvent (DES)-emulsification liquid-liquid microextraction of trace curcumin in food and herbal tea samples. Food Chem 243:442–447. https://doi.org/10.1016/j.foodchem.2017.09.154

Azmir J, Zaidul ISM, Rahman MM, Sharif KM, Mohamed A, Sahena F, Jahurul MHA, Ghafoor K, Norulaini NAN, Omar AKM (2013) Techniques for extraction of bioactive compounds from plant materials: a review. J Food Eng 117(4):426–436. https://doi.org/10.1016/j.jfoodeng.2013.01.014

Bağda E, Altundağ H, Soylak M (2017) Highly simple deep eutectic solvent extraction of manganese in vegetable samples prior to its ICP-OES analysis. Biol Trace Elem Res 179(2):334–339. https://doi.org/10.1007/s12011-017-0967-5

Bajkacz S, Adamek J (2017) Evaluation of new natural deep eutectic solvents for the extraction of isoflavones from soy products. Talanta 168:329–335. https://doi.org/10.1016/j.talanta.2017.02.065

Bakirtzi C, Triantafyllidou K, Makris DP (2016) Novel lactic acid-based natural deep eutectic solvents: efficiency in the ultrasound-assisted extraction of antioxidant polyphenols from common native Greek medicinal plants. J Appl Res Med Aromat Plants 3(3):120–127. https://doi.org/10.1016/j.jarmap.2016.03.003

Bi W, Tian M, Row KH (2013) Evaluation of alcohol-based deep eutectic solvent in extraction and determination of flavonoids with response surface methodology optimization. J Chromatogr A 1285:22–30. https://doi.org/10.1016/j.chroma.2013.02.041

Bosiljkov T, Dujmić F, Cvjetko Bubalo M, Hribar J, Vidrih R, Brnčić M, Zlatic E, Radojčić Redovniković I, Jokić S (2017) Natural deep eutectic solvents and ultrasound-assisted extraction: green approaches for extraction of wine lees anthocyanins. Food Bioprod Process 102:195–203. https://doi.org/10.1016/j.fbp.2016.12.005

Cao J, Chen L, Li M, Cao F, Zhao L, Su E (2018) Efficient extraction of proanthocyanidin from Ginkgo biloba leaves employing rationally designed deep eutectic solvent-water mixture and evaluation of the antioxidant activity. J Pharm Biomed Anal 158:317–326. https://doi.org/10.1016/j.jpba.2018.06.007

Chanioti S, Tzia C (2018) Extraction of phenolic compounds from olive pomace by using natural deep eutectic solvents and innovative extraction techniques. Innovative Food Sci Emerg Technol 48:228–239. https://doi.org/10.1016/j.ifset.2018.07.001

Chemat F, Vian MA, Cravotto G (2012) Green extraction of natural products: concept and principles. Int J Mol Sci 13(7):8615–8627. https://doi.org/10.3390/ijms13078615

Chen J, Liu M, Wang Q, Du H, Zhang L (2016) Deep eutectic solvent-based microwave-assisted method for extraction of hydrophilic and hydrophobic components from radix salviae miltiorrhizae. Molecules 21(10):1383. https://doi.org/10.3390/molecules21101383

Choi YH, van Spronsen J, Dai Y, Verberne M, Hollmann F, Arends IWCE, Witkamp G-J, Verpoorte R (2011) Are natural deep eutectic solvents the missing link in understanding cellular metabolism and physiology? Plant Physiol 156(4):1701–1705. https://doi.org/10.1104/pp.111.178426

Cui Q, Peng X, Yao X-H, Wei Z-F, Luo M, Wang W, Zhao C-J, Fu Y-J, Zu Y-G (2015) Deep eutectic solvent-based microwave-assisted extraction of genistin, genistein and apigenin from pigeon pea roots. Sep Purif Technol 150:63–72. https://doi.org/10.1016/j.seppur.2015.06.026

Cvjetko Bubalo M, Ćurko N, Tomašević M, Kovačević Ganić K, Radojčić Redovniković I (2016) Green extraction of grape skin phenolics by using deep eutectic solvents. Food Chem 200:159–166. https://doi.org/10.1016/j.foodchem.2016.01.040

Cvjetko Bubalo M, Vidović S, Radojčić Redovniković I, Jokić S (2018) New perspective in extraction of plant biologically active compounds by green solvents. Food Bioprod Process 109:52–73. https://doi.org/10.1016/j.fbp.2018.03.001

Dai Y, Witkamp G-J, Verpoorte R, Choi YH (2013a) Natural deep eutectic solvents as a new extraction media for phenolic metabolites in Carthamus tinctorius L. Anal Chem 85(13):6272–6278. https://doi.org/10.1021/ac400432p

Dai Y, van Spronsen J, Witkamp G-J, Verpoorte R, Choi YH (2013b) Ionic liquids and deep eutectic solvents in natural products research: mixtures of solids as extraction solvents. J Nat Prod 76(11):2162–2173. https://doi.org/10.1021/np400051w

Dai Y, Verpoorte R, Choi YH (2014) Natural deep eutectic solvents providing enhanced stability of natural colorants from safflower (Carthamus tinctorius). Food Chem 159:116–121. https://doi.org/10.1016/j.foodchem.2014.02.155

Dai Y, Witkamp G-J, Verpoorte R, Choi YH (2015) Tailoring properties of natural deep eutectic solvents with water to facilitate their applications. Food Chem 187:14–19. https://doi.org/10.1016/j.foodchem.2015.03.123

Dai Y, Rozema E, Verpoorte R, Choi YH (2016) Application of natural deep eutectic solvents to the extraction of anthocyanins from Catharanthus roseus with high extractability and stability replacing conventional organic solvents. J Chromatogr A 1434:50–56. https://doi.org/10.1016/j.chroma.2016.01.037

Dai Y, Zhang H-S, Huan B, He Y (2017) Enhancing the enzymatic saccharification of bamboo shoot shell by sequential biological pretreatment with Galactomyces sp. CCZU11-1 and deep eutectic solvent extraction. Bioprocess Biosyst Eng 40(9):1427–1436. https://doi.org/10.1007/s00449-017-1800-4

Das AK, Sharma M, Mondal D, Prasad K (2016) Deep eutectic solvents as efficient solvent system for the extraction of κ-carrageenan from Kappaphycus alvarezii. Carbohydr Polym 136:930–935. https://doi.org/10.1016/j.carbpol.2015.09.114

de Cássia da Silveira e Sá R, Andrade L, dos Reis Barreto de Oliveira R, de Sousa D (2014) A review on anti-inflammatory activity of phenylpropanoids found in essential oils. Molecules 19(2):1459–1480. https://doi.org/10.3390/molecules19021459

El Achkar T, Fourmentin S, Greige-Gerges H (2019) Deep eutectic solvents: an overview on their interactions with water and biochemical compounds. J Mol Liq 288:111028. https://doi.org/10.1016/j.molliq.2019.111028

Farajzadeh MA, Sattari Dabbagh M, Yadeghari A (2017) Deep eutectic solvent based gas-assisted dispersive liquid-phase microextraction combined with gas chromatography and flame ionization detection for the determination of some pesticide residues in fruit and vegetable samples. J Sep Sci 40(10):2253–2260. https://doi.org/10.1002/jssc.201700052

Fernández M d l Á, Espino M, Gomez FJV, Silva MF (2018) Novel approaches mediated by tailor-made green solvents for the extraction of phenolic compounds from agro-food industrial by-products. Food Chem 239:671–678. https://doi.org/10.1016/j.foodchem.2017.06.150

Ferrone V, Genovese S, Carlucci M, Tiecco M, Germani R, Preziuso F, Epifano F, Carlucci G, Taddeo VA, (2018) A green deep eutectic solvent dispersive liquid-liquid micro-extraction (DES-DLLME) for the UHPLC-PDA determination of oxyprenylated phenylpropanoids in olive, soy, peanuts, corn, and sunflower oil. Food Chemistry 245:578–585

Florindo C, Oliveira FS, Rebelo LPN, Fernandes AM, Marrucho IM (2014) Insights into the synthesis and properties of deep eutectic solvents based on cholinium chloride and carboxylic acids. ACS Sustain Chem Eng 2(10):2416–2425. https://doi.org/10.1021/sc500439w

Gan K, Tang W, Zhu T, Li W, Wang H, Liu X (2016) Enhanced extraction of cleistanthol from *Phyllanthus flexuosus* by deep eutectic solvent-modified anion-exchange resin. J Liq Chromatogr Relat Technol 39(19–20):882–888. https://doi.org/10.1080/10826076.2017.1278609

García A, Rodríguez-Juan E, Rodríguez-Gutiérrez G, Rios JJ, Fernández-Bolaños J (2016) Extraction of phenolic compounds from virgin olive oil by deep eutectic solvents (DESs). Food Chem 197:554–561. https://doi.org/10.1016/j.foodchem.2015.10.131

Ge D, Zhang Y, Dai Y, Yang S (2018) Air-assisted dispersive liquid-liquid microextraction based on a new hydrophobic deep eutectic solvent for the preconcentration of benzophenone-type UV filters from aqueous samples. J Sep Sci 41(7):1635–1643. https://doi.org/10.1002/jssc.201701282

González CG, Mustafa NR, Wilson EG, Verpoorte R, Choi YH (2018) Application of natural deep eutectic solvents for the "green" extraction of vanillin from vanilla pods. Flavour Fragr J 33(1):91–96. https://doi.org/10.1002/ffj.3425

Gu T, Zhang M, Tan T, Chen J, Li Z, Zhang Q, Qiu H (2014) Deep eutectic solvents as novel extraction media for phenolic compounds from model oil. Chem Commun 50(79):11749–11752. https://doi.org/10.1039/C4CC04661G

Guo W, Hou Y, Wu W, Ren S, Tian S, Marsh KN (2013) Separation of phenol from model oils with quaternary ammonium salts via forming deep eutectic solvents. Green Chem 15(1):226–229. https://doi.org/10.1039/C2GC36602A

Habibollahi MH, Karimyan K, Arfaeinia H, Mirzaei N, Safari Y, Akramipour R, Sharafi H, Fattahi N (2018) Extraction and determination of heavy metals in soil and vegetables irrigated with treated municipal wastewater using new mode of dispersive liquid-liquid microextraction based on the solidified deep eutectic solvent followed by GFAAS: extraction of heavy metals using dispersive liquid-liquid microextraction. J Sci Food Agric 99(2):656–665. https://doi.org/10.1002/jsfa.9230

Hawthorne SB, Miller DJ, Pawliszyn J, Arthur CL (1992) Solventless determination of caffeine in beverages using solid-phase microextraction with fused-silica fibers. J Chromatogr A 603(1–2):185–191. https://doi.org/10.1016/0021-9673(92)85360-6

Helalat–Nezhad Z, Ghanemi K, Fallah–Mehrjardi M (2015) Dissolution of biological samples in deep eutectic solvents: an approach for extraction of polycyclic aromatic hydrocarbons followed by liquid chromatography-fluorescence detection. J Chromatogr A 1394:46–53. https://doi.org/10.1016/j.chroma.2015.03.053

Hu Y, Hu R, Zhu Q, Zhan J, Liu H, Yao B (2012) Improved selective extraction of 3,3′-dichlorobenzidine by molecularly imprinted polysiloxane microspheres. Environ Chem Lett 10(3):275–280. https://doi.org/10.1007/s10311-011-0350-2

Huang Y, Wang Y, Pan Q, Wang Y, Ding X, Xu K, Li N, Wen Q (2015) Magnetic graphene oxide modified with choline chloride-based deep eutectic solvent for the solid-phase extraction of protein. Anal Chim Acta 877:90–99. https://doi.org/10.1016/j.aca.2015.03.048

Huang Y, Feng F, Jiang J, Qiao Y, Wu T, Voglmeir J, Chen Z-G (2017) Green and efficient extraction of rutin from tartary buckwheat hull by using natural deep eutectic solvents. Food Chem 221:1400–1405. https://doi.org/10.1016/j.foodchem.2016.11.013

Ince M, Kaya G, Yaman M (2010) Solid phase extraction and preconcentration of cobalt in mineral waters with PAR-loaded Amberlite XAD-7 and flame atomic absorption spectrometry. Environ Chem Lett 8(3):283–288. https://doi.org/10.1007/s10311-009-0218-x

Jain R, Kumar A, Shukla Y, Mudiam MKR (2015) Dispersive liquid-liquid microextraction-injector port silylation: a viable option for the analysis of polar analytes using gas chromatography-mass spectrometry. Austin J Anal Pharm Chem 2(3):4

Jeong KM, Ko J, Zhao J, Jin Y, Yoo DE, Han SY, Lee J (2017) Multi-functioning deep eutectic solvents as extraction and storage media for bioactive natural products that are readily applicable to cosmetic products. J Clean Prod 151:87–95. https://doi.org/10.1016/j.jclepro.2017.03.038

Karimi M, Dadfarnia S, Shabani AMH (2017) Application of deep eutectic solvent modified cotton as a sorbent for online solid-phase extraction and determination of trace amounts of copper and nickel in water and biological samples. Biol Trace Elem Res 176(1):207–215. https://doi.org/10.1007/s12011-016-0814-0

Khataei MM, Yamini Y, Nazaripour A, Karimi M (2018) Novel generation of deep eutectic solvent as an acceptor phase in three-phase hollow fiber liquid phase microextraction for extraction and preconcentration of steroidal hormones from biological fluids. Talanta 178:473–480. https://doi.org/10.1016/j.talanta.2017.09.068

Khezeli T, Daneshfar A, Sahraei R (2016) A green ultrasonic-assisted liquid–liquid microextraction based on deep eutectic solvent for the HPLC-UV determination of ferulic, caffeic and cinnamic acid from olive, almond, sesame and cinnamon oil. Talanta 150:577–585. https://doi.org/10.1016/j.talanta.2015.12.077

Lamei N, Ezoddin M, Abdi K (2017) Air assisted emulsification liquid-liquid microextraction based on deep eutectic solvent for preconcentration of methadone in water and biological samples. Talanta 165:176–181. https://doi.org/10.1016/j.talanta.2016.11.036

Li X, Row KH (2016) Development of deep eutectic solvents applied in extraction and separation: liquid chromatography. J Sep Sci 39(18):3505–3520. https://doi.org/10.1002/jssc.201600633

Li G, Row KH (2018) Selective extraction of 3,4-dihydroxybenzoic acid in Ilex chinensis Sims by meticulous mini-solid-phase microextraction using ternary deep eutectic solvent-based molecularly imprinted polymers. Anal Bioanal Chem 410(30):7849–7858. https://doi.org/10.1007/s00216-018-1406-y

Li N, Wang Y, Xu K, Huang Y, Wen Q, Ding X (2016) Development of green betaine-based deep eutectic solvent aqueous two-phase system for the extraction of protein. Talanta 152:23–32. https://doi.org/10.1016/j.talanta.2016.01.042

Li Z, Jia J, Wang M, Zhang H, Yan H, Qiao F (2017) Bifunctionalized ordered mesoporous organosilica synthesized in deep eutectic solvent for extraction of triazine herbicides from watermelon. J Chromatogr A 1529:50–56. https://doi.org/10.1016/j.chroma.2017.10.074

Liu W, Fu Y, Zu Y, Kong Y, Zhang L, Zu B, Efferth T (2009) Negative-pressure cavitation extraction for the determination of flavonoids in pigeon pea leaves by liquid chromatography–tandem mass spectrometry. J Chromatogr A 1216(18):3841–3850. https://doi.org/10.1016/j.chroma.2009.02.073

Liu X, Ahlgren S, Korthout HAAJ, Salomé-Abarca LF, Bayona LM, Verpoorte R, Choi YH (2018) Broad range chemical profiling of natural deep eutectic solvent extracts using a high performance thin layer chromatography–based method. J Chromatogr A 1532:198–207. https://doi.org/10.1016/j.chroma.2017.12.009

Machmudah S, Lestari SD, Widiyastuti W, Kanda H, Winardi S, Goto M (2018) Subcritical water extraction enhancement by adding deep eutectic solvent for extracting xanthone from mangosteen pericarps. J Supercrit Fluids 133:615–624. https://doi.org/10.1016/j.supflu.2017.06.012

Moghadam AG, Rajabi M, Asghari A (2018) Efficient and relatively safe emulsification microextraction using a deep eutectic solvent for influential enrichment of trace main anti-depressant drugs from complicated samples. J Chromatogr B 1072:50–59. https://doi.org/10.1016/j.jchromb.2017.09.042

Mohebbi A, Yaripour S, Farajzadeh MA, Afshar Mogaddam MR (2018) Combination of dispersive solid phase extraction and deep eutectic solvent–based air–assisted liquid–liquid microextraction followed by gas chromatography–mass spectrometry as an efficient analytical method for the quantification of some tricyclic antidepressant drugs in biological fluids. J Chromatogr A 1571:84–93. https://doi.org/10.1016/j.chroma.2018.08.022

Moura L, Moufawad T, Ferreira M, Bricout H, Tilloy S, Monflier E, Costa Gomes MF, Landy D, Fourmentin S (2017) Deep eutectic solvents as green absorbents of volatile organic pollutants. Environ Chem Lett 15(4):747–753. https://doi.org/10.1007/s10311-017-0654-y

Mouratoglou E, Malliou V, Makris DP (2016) Novel glycerol-based natural eutectic mixtures and their efficiency in the ultrasound-assisted extraction of antioxidant polyphenols from

agri-food waste biomass. Waste Biomass Valoriz 7(6):1377–1387. https://doi.org/10.1007/s12649-016-9539-8

Nam MW, Zhao J, Lee MS, Jeong JH, Lee J (2015) Enhanced extraction of bioactive natural products using tailor-made deep eutectic solvents: application to flavonoid extraction from Flos sophorae. Green Chem 17(3):1718–1727. https://doi.org/10.1039/C4GC01556H

Nugbienyo L, Shishov A, Garmonov S, Moskvin L, Andruch V, Bulatov A (2017) Flow method based on liquid-liquid extraction using deep eutectic solvent for the spectrofluorimetric determination of procainamide in human saliva. Talanta 168:307–312. https://doi.org/10.1016/j.talanta.2017.03.057

Ojeda CB, Rojas FS (2018) Vortex-assisted liquid–liquid microextraction (VALLME): the latest applications. Chromatographia 81(1):89–103. https://doi.org/10.1007/s10337-017-3403-2

Okenicová L, Žemberyová M, Procházková S (2016) Biosorbents for solid-phase extraction of toxic elements in waters. Environ Chem Lett 14(1):67–77. https://doi.org/10.1007/s10311-015-0539-x

Oller-Ruiz A, Garrido I, Viñas P, Campillo N, Fenoll J, Hernández-Córdoba M (2018) Reliable analysis of chlorophenoxy herbicides in soil and water by magnetic solid phase extraction and liquid chromatography. Environ Chem Lett 16(3):1077–1082. https://doi.org/10.1007/s10311-018-0725-8

Pedersen-Bjergaard S, Rasmussen KE (1999) Liquid−liquid−liquid microextraction for sample preparation of biological fluids prior to capillary electrophoresis. Anal Chem 71(14):2650–2656. https://doi.org/10.1021/ac990055n

Peng X, Duan M-H, Yao X-H, Zhang Y-H, Zhao C-J, Zu Y-G, Fu Y-J (2016) Green extraction of five target phenolic acids from Lonicerae japonicae Flos with deep eutectic solvent. Sep Purif Technol 157:249–257. https://doi.org/10.1016/j.seppur.2015.10.065

Peng F, Xu P, Zhao B-Y, Zong M-H, Lou W-Y (2018) The application of deep eutectic solvent on the extraction and in vitro antioxidant activity of rutin from Sophora japonica bud. J Food Sci Technol 55(6):2326–2333. https://doi.org/10.1007/s13197-018-3151-9

Picó Y, Fernández M, Ruiz MJ, Font G (2007) Current trends in solid-phase-based extraction techniques for the determination of pesticides in food and environment. J Biochem Biophys Methods 70(2):117–131. https://doi.org/10.1016/j.jbbm.2006.10.010

Qi X-L, Peng X, Huang Y-Y, Li L, Wei Z-F, Zu Y-G, Fu Y-J (2015) Green and efficient extraction of bioactive flavonoids from Equisetum palustre L. by deep eutectic solvents-based negative pressure cavitation method combined with macroporous resin enrichment. Ind Crop Prod 70:142–148. https://doi.org/10.1016/j.indcrop.2015.03.026

Radošević K, Ćurko N, Gaurina Srček V, Cvjetko Bubalo M, Tomašević M, Kovačević Ganić K, Radojčić Redovniković I (2016) Natural deep eutectic solvents as beneficial extractants for enhancement of plant extracts bioactivity. LWT 73:45–51. https://doi.org/10.1016/j.lwt.2016.05.037

Raynie DE (2004) Modern extraction techniques. Anal Chem 76(16):4659–4664. https://doi.org/10.1021/ac040117w

Rezaee M, Assadi Y, Milani Hosseini M-R, Aghaee E, Ahmadi F, Berijani S (2006) Determination of organic compounds in water using dispersive liquid–liquid microextraction. J Chromatogr A 1116(1–2):1–9. https://doi.org/10.1016/j.chroma.2006.03.007

Ribeiro BD, Coelho MAZ, Marrucho IM (2013) Extraction of saponins from sisal (Agave sisalana) and juá (Ziziphus joazeiro) with cholinium-based ionic liquids and deep eutectic solvents. Eur Food Res Technol 237(6):965–975. https://doi.org/10.1007/s00217-013-2068-9

Ribeiro AR, Gonçalves VMF, Maia AS, Ribeiro C, Castro PML, Tiritan ME (2015) Dispersive liquid–liquid microextraction and HPLC to analyse fluoxetine and metoprolol enantiomers in wastewaters. Environ Chem Lett 13(2):203–210. https://doi.org/10.1007/s10311-015-0498-2

Ruesgas-Ramón M, Figueroa-Espinoza MC, Durand E (2017) Application of deep eutectic solvents (DES) for phenolic compounds extraction: overview, challenges, and opportunities. J Agric Food Chem 65(18):3591–3601. https://doi.org/10.1021/acs.jafc.7b01054

Saravana PS, Cho Y-N, Woo H-C, Chun B-S (2018a) Green and efficient extraction of polysaccharides from brown seaweed by adding deep eutectic solvent in subcritical water hydrolysis. J Clean Prod 198:1474–1484. https://doi.org/10.1016/j.jclepro.2018.07.151

Saravana PS, Ho TC, Chae S-J, Cho Y-J, Park J-S, Lee H-J, Chun B-S (2018b) Deep eutectic solvent-based extraction and fabrication of chitin films from crustacean waste. Carbohydr Polym 195:622–630. https://doi.org/10.1016/j.carbpol.2018.05.018

Sha Y, Huang D, Zheng S, Liu B, Deng C (2013) Development of microwave-assisted headspace solid-phase microextraction followed by gas chromatography-mass spectrometry for the analysis of phenol in a cigarette pad. Anal Methods 5(18):4655. https://doi.org/10.1039/c3ay40273h

Smith EL, Abbott AP, Ryder KS (2014) Deep eutectic solvents (DESs) and their applications. Chem Rev 114(21):11060–11082. https://doi.org/10.1021/cr300162p

Tan T, Zhang M, Wan Y, Qiu H (2016) Utilization of deep eutectic solvents as novel mobile phase additives for improving the separation of bioactive quaternary alkaloids. Talanta 149:85–90. https://doi.org/10.1016/j.talanta.2015.11.041

Tang B, Bi W, Zhang H, Row KH (2014) Deep eutectic solvent-based HS-SME coupled with GC for the analysis of bioactive terpenoids in Chamaecyparis obtusa leaves. Chromatographia 77(3–4):373–377. https://doi.org/10.1007/s10337-013-2607-3

Tang W, Gao F, Duan Y, Zhu T, Ho Row K (2017) Exploration of deep eutectic solvent-based molecularly imprinted polymers as solid-phase extraction sorbents for screening chloramphenicol in milk. J Chromatogr Sci 55(6):654–661. https://doi.org/10.1093/chromsci/bmx011

van Osch DJGP, Zubeir LF, van den Bruinhorst A, Rocha MAA, Kroon MC (2015) Hydrophobic deep eutectic solvents as water-immiscible extractants. Green Chem 17(9):4518–4521. https://doi.org/10.1039/C5GC01451D

Vanda H, Dai Y, Wilson EG, Verpoorte R, Choi YH (2018) Green solvents from ionic liquids and deep eutectic solvents to natural deep eutectic solvents. C R Chim 21(6):628–638. https://doi.org/10.1016/j.crci.2018.04.002

Wahaibi IA, Wahaibi YA, Hajri RA, Jibril B, Shuwa S (2019) The novel use of malonic acid-based deep eutectic solvents for enhancing heavy oil recovery. Int J Oil Gas Coal Technol 20(1):31. https://doi.org/10.1504/IJOGCT.2019.096493

Wang M, Wang J, Zhang Y, Xia Q, Bi W, Yang X, Chen DDY (2016) Fast environment-friendly ball mill-assisted deep eutectic solvent-based extraction of natural products. J Chromatogr A 1443:262–266. https://doi.org/10.1016/j.chroma.2016.03.061

Wang W, Du Y, Xiao Z, Li Y, Li B, Yang G (2017) Determination of trace rhodamine B in chili oil by deep eutectic solvent extraction and an ultra high-performance liquid chromatograph equipped with a fluorescence detector. Anal Sci 33(6):715–717. https://doi.org/10.2116/analsci.33.715

Wang H, Ma X, Cheng Q, Xi X, Zhang L (2018) Deep eutectic solvent-based microwave-assisted extraction of baicalin from *Scutellaria baicalensis* Georgi. J Chem 2018:1–10. https://doi.org/10.1155/2018/9579872

Wei Z-F, Wang X-Q, Peng X, Wang W, Zhao C-J, Zu Y-G, Fu Y-J (2015) Fast and green extraction and separation of main bioactive flavonoids from Radix Scutellariae. Ind Crop Prod 63:175–181. https://doi.org/10.1016/j.indcrop.2014.10.013

Xu K, Wang Y, Huang Y, Li N, Wen Q (2015) A green deep eutectic solvent-based aqueous two-phase system for protein extracting. Anal Chim Acta 864:9–20. https://doi.org/10.1016/j.aca.2015.01.026

Xu K, Wang Y, Ding X, Huang Y, Li N, Wen Q (2016a) Magnetic solid-phase extraction of protein with deep eutectic solvent immobilized magnetic graphene oxide nanoparticles. Talanta 148:153–162. https://doi.org/10.1016/j.talanta.2015.10.079

Xu K, Wang Y, Li Y, Lin Y, Zhang H, Zhou Y (2016b) A novel poly(deep eutectic solvent)-based magnetic silica composite for solid-phase extraction of trypsin. Anal Chim Acta 946:64–72. https://doi.org/10.1016/j.aca.2016.10.021

Yan H, Wang H, Qin X, Liu B, Du J (2011) Ultrasound-assisted dispersive liquid–liquid microextraction for determination of fluoroquinolones in pharmaceutical wastewater. J Pharm Biomed Anal 54(1):53–57. https://doi.org/10.1016/j.jpba.2010.08.007

Yang M, Xi X, Yang X, Bai L, Lu R, Zhou W, Zhang S, Gao H (2015) Determination of benzoylurea insecticides in environmental water and honey samples using ionic-liquid-mingled air-assisted liquid–liquid microextraction based on solidification of floating organic droplets. RSC Adv 5(32):25572–25580. https://doi.org/10.1039/C5RA00140D

Yilmaz E, Soylak M (2015) Ultrasound assisted-deep eutectic solvent extraction of iron from sheep, bovine and chicken liver samples11This study is a part of PhD thesis of Erkan Yilmaz. Talanta 136:170–173. https://doi.org/10.1016/j.talanta.2014.12.034

Yilmaz E, Soylak M (2018) A novel and simple deep eutectic solvent based liquid phase microextraction method for rhodamine B in cosmetic products and water samples prior to its spectrophotometric determination. Spectrochim Acta A Mol Biomol Spectrosc 202:81–86. https://doi.org/10.1016/j.saa.2018.04.073

Yousefi SM, Shemirani F, Ghorbanian SA (2018) Enhanced headspace single drop microextraction method using deep eutectic solvent based magnetic bucky gels: application to the determination of volatile aromatic hydrocarbons in water and urine samples. J Sep Sci 41(4):966–974. https://doi.org/10.1002/jssc.201700807

Zeng Q, Wang Y, Huang Y, Ding X, Chen J, Xu K (2014) Deep eutectic solvents as novel extraction media for protein partitioning. Analyst 139(10):2565. https://doi.org/10.1039/c3an02235h

Zeng H, Qiao K, Li X, Yang M, Zhang S, Lu R, Li J, Gao H, Zhou W (2017) Dispersive liquid-liquid microextraction based on the solidification of deep eutectic solvent for the determination of benzoylureas in environmental water samples. J Sep Sci 40(23):4563–4570. https://doi.org/10.1002/jssc.201700890

Zhang L, Wang M (2017) Optimization of deep eutectic solvent-based ultrasound-assisted extraction of polysaccharides from Dioscorea opposita Thunb. Int J Biol Macromol 95:675–681. https://doi.org/10.1016/j.ijbiomac.2016.11.096

Zhang Q, De Oliveira Vigier K, Royer S, Jérôme F (2012) Deep eutectic solvents: syntheses, properties and applications. Chem Soc Rev 41(21):7108. https://doi.org/10.1039/c2cs35178a

Zhang H, Wang Y, Zhou Y, Chen J, Wei X, Xu P (2018) Aqueous biphasic systems formed by deep eutectic solvent and new-type salts for the high-performance extraction of pigments. Talanta 181:210–216. https://doi.org/10.1016/j.talanta.2018.01.014

Zhou P, Wang X, Liu P, Huang J, Wang C, Pan M, Kuang Z (2018) Enhanced phenolic compounds extraction from Morus alba L. leaves by deep eutectic solvents combined with ultrasonic-assisted extraction. Ind Crop Prod 120:147–154. https://doi.org/10.1016/j.indcrop.2018.04.071

Zhuang B, Dou L-L, Li P, Liu E-H (2017) Deep eutectic solvents as green media for extraction of flavonoid glycosides and aglycones from Platycladi Cacumen. J Pharm Biomed Anal 134:214–219. https://doi.org/10.1016/j.jpba.2016.11.049

Chapter 7
Extraction of Plant and Algal Polyphenols Using Eutectic Solvents

Lucie Percevault, Emmanuelle Limanton, Fabienne Gauffre, Corinne Lagrost, and Ludovic Paquin

Contents

L. Percevault · E. Limanton · F. Gauffre · C. Lagrost (✉) · L. Paquin
Univ Rennes, ISCR, CNRS UMR 6226, Rennes, France
e-mail: lucie.percevault@univ-rennes1.fr; emmanuelle.limanton@univ-rennes1.fr; fabienne.gauffre@univ-rennes1.fr; corinne.lagrost@univ-rennes1.fr; ludovic.paquin@univ-rennes1.fr

S. Fourmentin et al. (eds.), *Deep Eutectic Solvents for Medicine, Gas Solubilization and Extraction of Natural Substances*, Environmental Chemistry for a Sustainable World 56, https://doi.org/10.1007/978-3-030-53069-3_7

Abstract Natural ingredients are currently attracting growing interest for industrial applications such as functional food, nutraceuticals, cosmetics, and pharmaceuticals. A key point is the societal acceptance for the naturally occurring compounds which are believed to be safer to humans and environment than artificial compounds. Polyphenols are one of the most interesting plant substances which are carrying many biological activities. Thousands of phenolic compounds have been identified, isolated, and selected from the plant kingdom including terrestrial and marine sources. A challenge concerns the development of efficient and green technologies for the extraction of plant and algal polyphenols. Ionic liquids have been successfully evaluated, but they are facing strong issues related to their cost, sourcing, and green nature, precluding any industrial development. Being cheaper and eco-friendly, deep eutectic and natural deep eutectic solvents have a great potential as solvents for the extraction of polyphenols. The ability of deep eutectics to form hydrogen bonds is particularly of interest for the selective extraction of polyphenols since these compounds are generally rich in hydrogen bond donor groups.

Here we propose a comprehensive review of the use of deep eutectic solvent for polyphenol extraction processes, with more than 100 references published over the last decade. Despite the large possible combination of components for preparing deep eutectics with tunable properties, choline chloride-based eutectics are the most largely used extractant media. Important parameters including sampling, sample preparation, water content in deep eutectic solvent, extraction duration, and temperature along with methodology to account with all these interacting parameters are examined. These parameters have been put in perspective with extraction technologies. All are specifically considered in terms of extraction performances when using deep eutectic solvents. Here again, it turns out that only a few of the available technologies have been indeed evaluated with deep eutectic solvents, leaving ample perspectives for improvement in the field of polyphenol extraction in the near future. Finally, identification and quantification technologies are also discussed since they are also a key aspect in the process. The possibility of using deep eutectic solvents for a given application without having to extract the polyphenols also appears to be a very promising prospect, strengthened by the longer preservation of their antioxidant properties in deep eutectics.

Keywords Polyphenols · Deep eutectic solvents · Natural deep eutectic solvents · Hydrogen bond · Extraction · Maceration · Selectivity · Ultrasound · Microwave · Antioxidant properties

Abbreviations

ABTS	2,2′-Azino-bis(3-ethyl-benzothiazoline-6-sulfonic) acid
BBD	Box-Behnken design
DES	Deep eutectic solvents
DPPH	2,2-Diphenyl-1-picrylhydrazyl hydrate
FRAP	Ferric reducing-antioxidant power

ORAC	Oxygen radical absorbance capacity
HBA	Hydrogen bond acceptor
HBD	Hydrogen bond donor
HRE	High reflux extraction
MAE	Microwave-assisted extraction
RSM	Response surface methodology
SPE	Solid phase extraction
UAE	Ultrasound-assisted extraction

7.1 Introduction and Positioning of the Review

Ionic liquids have been used for more than 20 years as media for the extraction of natural products from various matrices (Tang et al. 2012; Ventura et al. 2017). Some comparative studies have shown that they have the potential to outperform conventional solvent extraction at the industrial level, for example, for artemisinin extraction (Lapkin et al. 2006). However, industrial development of ionic liquids as extractants for natural products is facing strong issues including their cost, sourcing, and "green" nature that remains somewhat questionable (Plechkova and Seddon 2008; Romero et al. 2008). As early as the mid-2000s, new ionic liquids started being developed to meet those requirements (Pena-Pereira and Namieśnik 2014) while keeping their modular character (Abbott et al. 2001; Dai et al. 2013a). In this context, biosourced ionic liquids appeared, for example, based on natural amino acids (Fukumoto et al. 2005; Gao et al. 2005). With the aim of employing always greener solvents along with cheaper processes, a new class of tunable solvents has emerged over the last 10 years, the deep eutectic solvents (DES) (Abbott et al. 2004), also called low-transition temperature mixture (LTTM). A further step has been taken, thanks to the development of the even more eco-friendly natural deep eutectic solvents (NADES), only prepared from natural products (Vanda et al. 2018). To date (2000–2019), more than 2000 publications deal with the extraction of natural products in deep eutectic solvent.

Considering the extraction of polyphenols themselves, a hundred of publications focusing on the use of deep eutectic solvent and natural deep eutectic solvent can be found between 2012 and 2019, showing the growing interest of these solvents in the field (Huang et al. 2019b). Almost all types of plant and different extraction processes have been evaluated (Zainal-Abidin et al. 2017; Cvjetko Bubalo et al. 2018; Cunha and Fernandes 2018). Because polyphenols are widely present in the vegetable kingdom, in its broad sense, i.e., not only plants but also algae, they are frequently considered in extraction studies beyond their economic potential in many industrial fields such as cosmetics, pharmaceuticals, and agrifood. The rich variety of polyphenols and their sources allow variation of both targets and matrices (Ruesgas-Ramón et al. 2017). While all polyphenol families have been studied, flavonols and flavones are by far the most investigated (see Sect. 7.2). They represent the largest part of publications dealing with the extraction of polyphenols by eutectic solvents, whatever the associated extraction technology (Fig. 7.1).

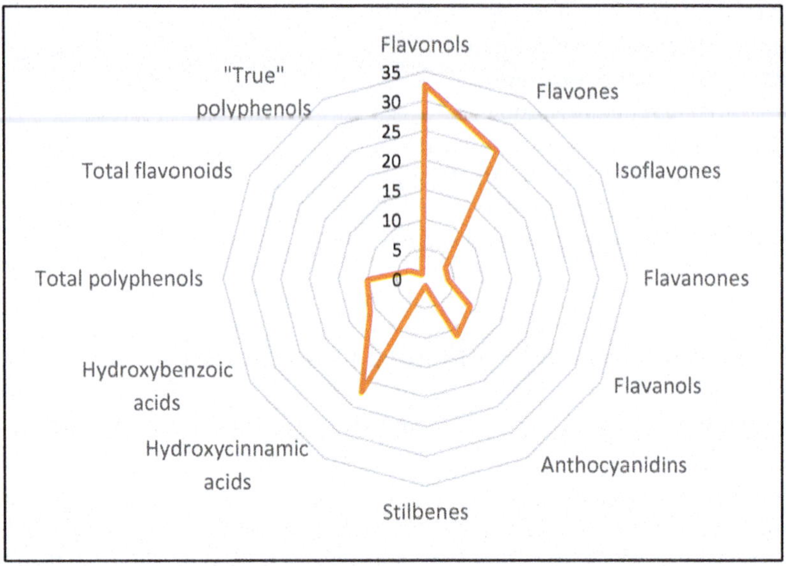

Fig. 7.1 Number of articles as a function of types of polyphenols targeted in extraction studies using deep eutectic solvents and natural deep eutectic solvents. "True" polyphenols denominate a category of polyphenol polymers like condensed or hydrolyzable tannins. Note that flavones and flavonols are the most investigated polyphenol families

Herein, the recent literature data about the extraction of polyphenols in deep eutectic solvent and natural deep eutectic solvent is thoroughly reviewed. First, a section describes the structures and properties of plant and algal polyphenols. Then definition, structure, and main physicochemical properties of deep eutectic solvent are discussed, highlighting their expected potential as extractant media toward polyphenols.

The extraction processes are implemented according to general steps which are illustrated in Fig. 7.2. For each of these steps, the nature and impact of deep eutectics is considered, ranging from the selection of plant matrix, mixing with deep eutectic solvent, extraction methodology to identification and purification of targeted bioactive compounds. All the deep eutectic solvents commonly used for polyphenol extraction are reported, showing that they are mainly based on choline chloride. Key parameters involved in the extraction processes are reviewed, namely, sampling, sample preparation, temperature, water content, duration, etc. A focus is made on the extraction technologies, demonstrating that only few of the available technologies are indeed employed. Finally, a section is dedicated to the polyphenol recovery in the deep eutectic solvent extracts, showing a promising prospect of keeping deep eutectic solvent as a part of polyphenol formulation. A last point of importance regarding the quantification and the identification of polyphenols is examined. We will then end the review with a critical conclusion with perspectives.

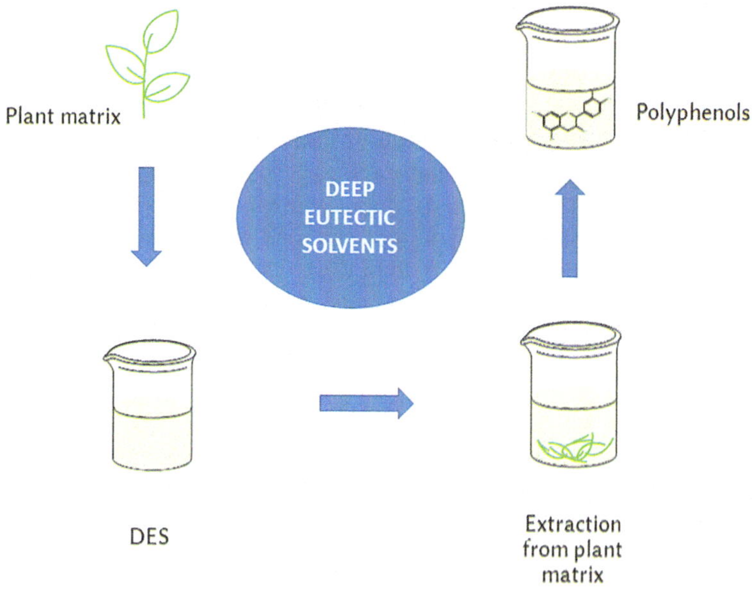

Fig. 7.2 Scheme of general steps illustrating the implementation of extraction process. In each step, deep eutectic solvent may have an impact that contributes to the overall performance of the extraction of polyphenols

7.2 Plant and Algal Polyphenols

Polyphenols are one of the most widespread secondary metabolites in the plant kingdom, with several thousand structures (Cheynier 2005; Tsao 2010; Quideau et al. 2011; Belščak-Cvitanović et al. 2018). Plant polyphenols can be found in almost all organs of edible and nonedible plants and have a pivotal role in their growth and development. Fruits, vegetables, seeds, cereals, bark, coffee, and tea are known to be a rich source of polyphenols, not only to the scientific community but also to the general public because of the current promotion of the beneficial effects of polyphenols for health. Thus, crude extracts of fruits, herbs, vegetables, cereals, etc. have been attracted growing interests in food and cosmetic industries, because they are rich in phenolic compounds, especially flavonoids. Algae, notably marine algae (seaweeds) or microalgae (mainly *Chlorella* sp.), are also an interesting source of polyphenols although much less explored than their plant counterparts (Hajimahmoodi et al. 2010; Goiris et al. 2012; Joana Gil-Chávez et al. 2013; Wang et al. 2014; Michalak and Chojnacka 2015; Zakaria et al. 2017). The reason might be that carotenoid pigments have been the most usual antioxidant target considered in algal extracts (Crampon et al. 2011; Goiris et al. 2012; Kadam et al. 2013). However, polyphenolic compounds, including catechins and flavonols (aglycone and glucosides), have been extracted in some red, brown, and green algae (Yoshie et al. 2000; Santoso et al. 2004; Shoubaky et al. 2016; Tanna and Mishra 2018). Phlorotannins which are the dominant polyphenols in brown algae (Alariaceae, Fucaceae,

Sargassaceae) have recently gained strong interest because of their biological properties (Creis n.d.; Wang et al. 2014), hence contributing to the recent focus toward marine algae as a source of polyphenols. In contrast to marine algae, very little is known about the phenolic composition of microalgae extracts, with hydroxycinnamic acids or hydroxybenzoic acids being mostly identified. But some works have shown that a variety of phenolic compounds are present in *Scenedesmus* microalgae (Kováčik et al. 2010; Klejdus et al. 2010), meaning that further work is required to better understand the phenolic composition and consequently the difference in properties of microalgae species. Starting from these few introductive words, it may be easily inferred the wide diversity of the family of polyphenols, having diverse structures, from the simplest one to much more complex systems (Fig. 7.3).

7.2.1 Is there a Ubiquitous "Chemical" Definition of Polyphenols?

Chemically speaking, polyphenols are a class of natural compounds featuring phenolic moieties. A primary point is their exact definition, which has been disputed in many publications. Quideau and coworkers have recently described this aspect in a

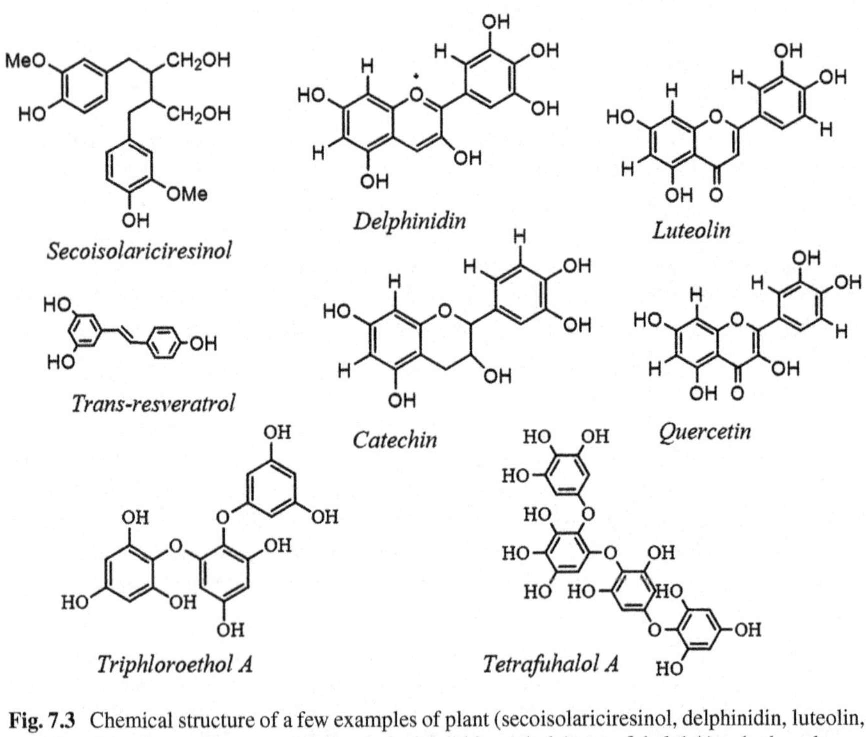

Fig. 7.3 Chemical structure of a few examples of plant (secoisolariciresinol, delphinidin, luteolin, trans-resveratrol, catechin, quercetin) and algal (triphloroethol A, tetrafuhalol A) polyphenols

very nice verbatim review (Quideau et al. 2011). Considering that the original WBSSH (from White, Bate-Smith, Swain, Haslam) definition of polyphenols (Swain and Bate-Smith 1962; Haslam and Cai 1994) is (too) restrictive, they propose a new definition: "The term polyphenol should be used to define plant secondary metabolites derived exclusively from the shikimate-derived phenylpropanoid and/or polyketide pathway(s), featuring more than one phenolic ring and being devoid of any nitrogen-based functional group in their most basic structural expression" (Quideau et al. 2011). This definition takes into account structural features and biosynthetic routes, while the former WBSSH one is based on the capacity of "polyphenols" to engage interaction with other biomolecules, thus practically considering only water-soluble compounds that can be divided into only three classes. Hence, this new definition excludes monophenolic structures (i.e., **one** phenyl ring bearing one or more hydroxyl groups such as phenolic alcohol and hydroxycinnamic or hydroxybenzoic acids) as well as their naturally occurring derivatives (methyl phenyl esters and O-phenyl glycosides) (Fig. 7.4). These compounds should be better regarded as polyphenol sub-units than "true" polyphenols, being precursors or metabolites. However, they are widely used and/or studied in polyphenol-related works and are commonly equated to polyphenols. Following Quideau and coworkers (Quideau et al. 2011), we acknowledge that these monophenolic compounds have their place in works dealing with polyphenols but cannot be strictly defined as polyphenols.

7.2.2 *Structural Diversity and Classification*

Two metabolic pathways, namely, phenylpropanoid and polyketide, lead to plant polyphenols. Most of the plant polyphenols are synthesized through the phenylpropanoid pathway (Hollman 2001). The combination of both pathways produces flavonoids, which are probably the largest class of polyphenols with more than 8000 identified structures (Cheynier 2005; Tsao 2010; Quideau et al. 2011; Belščak-Cvitanović et al. 2018). Interestingly, flavonoids exhibit several subclasses (flavonols, flavones, isoflavones, flavanones, anthocyanidins, flavanols) of different

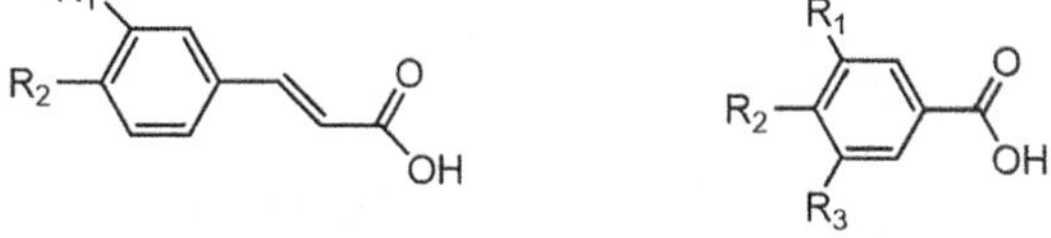

$R_1 = R_2 = OH$, caffeic acid
$R_1 = H$, $R_2 = OH$, *p*-coumaric acid
$R_1 = OMe$, $R_2 = OH$, ferulic acid

$R_1 = R_2 = R_3 = OH$, gallic acid

Fig. 7.4 Structure of common monophenolic compounds to be compared to polyphenol structure in Fig. 7.3

structures despite this common biosynthetic origin. The hybrid phenylpropanoid/polyketide pathway also produces another important class of polyphenols, the polyhydroxystilbenes. The *trans*-resveratrol is an archetypal member of this family. The biosynthetic route to flavonoids is suspected to form condensed tannins (proanthocyanidins) and theatannins with flavanols as precursors, but the condensation and polymerization phases are not yet fully elucidated (Quideau et al. 2011; Belščak-Cvitanović et al. 2018). Hydrolyzable tannins are produced from gallic acid or hexahydroxydiphenic acid (Quideau et al. 2011). In algae, phlorotannins are essentially derived from oligomerization of phloroglucinol (dehydrogenative coupling). In addition to this already large chemical diversity, polyphenols could be associated with various carbohydrates and/or organic acids. Thus, many polyphenols exist as glycosides with different sugar units, sometimes acylated sugars, at various positions onto the polyphenolic backbone.

Such a diversity and wide distribution of polyphenols have led to different ways of classification, according to their chemical structure, natural distribution, source, or biological functions (Belščak-Cvitanović et al. 2018). As chemists, we were logically more interested by the chemical classification. The most popular one follows the chemical structure of aglycone polyphenols. However, this could lead to several different ways, depending on consideration of the skeleton (giving 16 major classes) or on consideration of the number of phenyl rings along with their structural connection to each other. This latter classification leads generally to five major classes for plant polyphenols: phenolic acids, flavonoids, stilbenes, lignans, and others (Manach et al. 2004; Belščak-Cvitanović et al. 2018). Note that this classification includes monophenolic acids that cannot be really considered as true polyphenols while the "others" category involves polymers like condensed or hydrolyzable tannins. On the other hand, phlorotannins are systematically classified into six subclasses: phlorethols, fucols, fuhalols, fucophlorethols, isofuhalols, and eckols (Wang et al. 2014). Fucols incorporate phloroglucinol units that are linked with C-C (aryl-aryl) bonds and have linear and branched structure. In phlorethols, the phloroglucinol units are only coupled with aryl-ether bonds (C-O-C). Fucophlorethols are hybrid structures having both aryl-aryl and aryl-ether linkages. Fuhalols contain only ether bonds, arranged in a regular sequence of *para*- and *ortho*-bridges and exhibit additional hydroxyl groups on some units. Eckols exhibit at least one three-ring moiety with a dibenzo-1,4-dioxin element substituted by a phenoxy group at C4-position. Eckols are quite specific, being isolated from some specific *genera* of brown algae (*Ecklonia*, *Eisenia*, and *Alarieae*) (Wang et al. 2014). Similarly, isofuhalols are a specialized group isolated from *Chorda filum* and are generally of low molecular weight (Wang et al. 2014). Moreover, phlorotannins can be also sulfated, and such sulfated polyphenols are widely distributed among brown algae. Chlorinated or bromated phlorethols and fucophlorethols have been also detected, but the origin of some of these halogenated polyphenols is suspected to be an artifact (Wang et al. 2014). By taking into account the generally admitted chemical classification, we have finally summarized the classification of natural polyphenols from algae or plants in some large classes (Scheme 7.1).

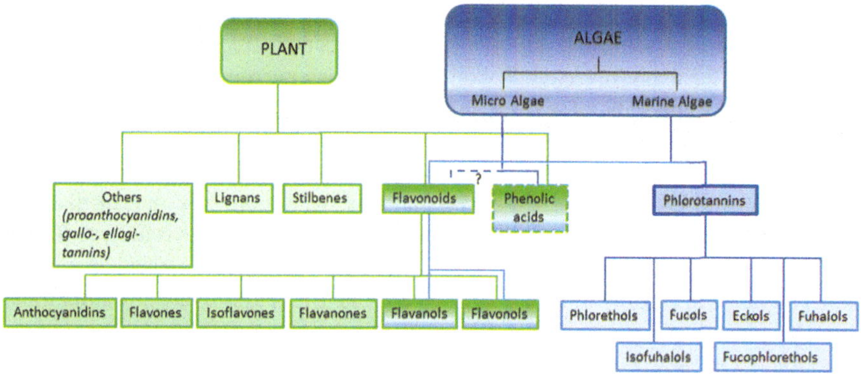

Scheme 7.1 General classes of polyphenols in plant (green), algae, and microalgae (blue). Monophenolic acids, which belong to the "phenolic acids" group, are not polyphenols but are generally present in the extracts

From the broad structural diversity of polyphenols, a wide variability of physico-chemical properties can be expected. Therefore, the presentation of a unique and general method for extracting and isolating these compounds is almost impossible. This is also the reasons why polyphenol extraction still generates strong interest, searching for efficient technologies and extractant media.

7.2.3 Antioxidant Properties

Polyphenols display a wide range of bio-physicochemical properties that derive from their multi-phenolic structures. These properties have been thoroughly reviewed in many publications (Manach et al. 2004; Cheynier 2005; Belščak-Cvitanović et al. 2018; Tanase et al. 2019). Polyphenols are essential for plant growth and development. They have important roles in controlling their size and their pigmentation, with a pivotal contribution in photosynthesis. They also provide resistance to microbial pathogens (Treutter 2006) and herbivorous animals, especially insects (Boulogne et al. 2012). The bitterness due to polyphenols also discourages plant consumption. Polyphenols act as signaling molecules to distinguish symbionts, as UVB screens for protecting against solar radiation, hence preventing DNA damages. Finally, among many other properties, they promote some beneficial interactions with other organisms.

However, their most acknowledged property is unquestionably antioxidative action. Polyphenols are natural antioxidants, which are capable to directly scavenge reactive oxygen species (ROS, i.e., $O_2^{\bullet-}$, $OH^\bullet$, H_2O^-, $H_2O^\bullet$, etc.), nitrogen reactive species (RNS, i.e., $NO^\bullet$, $ONOO^-$, $NO_2^\bullet$, etc.), and other free radicals ($RO^\bullet$ and $ROO^\bullet$) that are derived from oxidation of low-density lipoproteins, proteins, and oligonucleic acids (DNA and RNA) (Li et al. 2000; Shi et al. 2000; Vaya et al. 2003; Manach et al. 2005; Neudörffer et al. 2006; Giftson et al. 2010; Velmurugan et al. 2018). The strong free radical scavenging activities of polyphenols are well-documented and have been extensively reviewed. Two main mechanisms are generally considered, based on the capability of phenol groups to donate hydrogen atom or electron (Wright et al. 2001).

The hydrogen atom transfer between phenol group and radical is especially involved in the deactivation of peroxyl radicals, which are key intermediates as chain propagators, thereby terminating the chain reaction. Peroxyl radicals (LOO$^•$) are issued from the auto-oxidation of lipids (LH). A propagating radical chain reaction can occur according to LOO$^•$ + LH → LOOH + L$^•$. After hydrogen atom transfer, the phenol group becomes a phenoxy radical (ArO$^•$). The efficiency of the reaction (LOO$^•$ + ArOH → LOOH + ArO$^•$) depends on both the kinetics of the H transfer to LOO$^•$ and on the stability of phenoxy radical. The phenoxy radical is mainly stabilized through electron delocalization, becoming less reactive and hence unable to react back neither with LOOH nor LH. A key point is the structural features of ArOH, more precisely the position and number of hydroxyl groups, allowing electron delocalization over the molecule along with the formation of intramolecular hydrogen bonds (Quideau et al. 2011). This aspect can be quantified using O-H bond dissociation enthalpy values as a parameter (de Heer et al. 1999). The lower the bond dissociation enthalpies, the more efficient is the hydrogen atom transfer reaction.

The second mechanism involves a single electron transfer from phenol group to free radical, yielding a radical cation (Cren-Olivé et al. 2002). Similar to hydrogen atom transfer process, the structure of phenols plays a key role, and the same determining factors apply. The quantitative physicochemical parameter is the redox potentials or ionization potentials. The easier the oxidation of the polyphenols, i.e., lower redox potentials, the more efficient is the electronic transfer reaction.

Deciphering the nature of mechanisms is then strongly related to polyphenol structures, and many predictive works based on structure-activity relationship have been undertaken, mainly on plant polyphenols, while algae remain still poorly investigated. Moreover, works for establishing a correlation between antioxidant activity and phlorotannins have even given contradictory conclusions (Wang et al. 2014). However, many works combining density functional theory calculations with experimental determinations of O-H bond dissociation enthalpies and redox potentials/ionization potentials of plant polyphenols allow a fairly good rationalization of their antioxidant properties and underlying mechanism (Quideau et al. 2011). A very interesting point with plant polyphenols is that the products formed during ROS scavenging process are phenols themselves, which may keep a further ROS scavenging activity. As a consequence, they are long-lasting antioxidant, then able to potentially reduce much more ROS than the most efficient antioxidant, i.e., α-tocopherol (Roche et al. 2005).

In addition to direct radical scavenging, other modulation effects have been suggested to explain the antioxidant activities of polyphenols, for instance, synergistic effect with other potent antioxidants like α-tocopherol (Zhou et al. 2005; Dai et al. 2008; Achat et al. 2016) or in cells through modulations of the protein kinase and lipid kinase signaling pathways (Williams et al. 2004). Inhibition of xanthine oxidase and elevation of endogenous antioxidants have been also mentioned as important mechanisms (Tsao 2010). Polyphenols can induce antioxidant enzymes such as glutathione peroxidase, catalase, and superoxide dismutase that decompose hydroperoxides, hydrogen peroxide, and superoxide anions, respectively, and inhibit the expression of enzymes such as xanthine oxidase (Du et al. 2007). These emerging views of polyphenols' actions question the actual role of polyphenols in vivo, which is based on the idea of a low cellular level of concentration of polyphenols and metabolites. The low in vivo availability should indeed preclude any direct radical scavenging action.

Polyphenols are also known as metal chelators, able to make complexes with trace metals such as Al^{3+}, Fe^{3+}, and Cu^{+}. These metals play an important role in oxygen metabolism and free radical formation. Chelation of metals like Fe^{2+} or Cu^{+} can limit Fenton reaction, thus preventing oxidation caused by highly reactive hydroxyl radicals ($OH^{\bullet}$) (Pietta 2000; Perron and Brumaghim 2009).

7.3 Deep Eutectic Solvent and Natural Deep Eutectic Solvent, Definition and General Properties

7.3.1 What Are Deep Eutectic Solvents?

The term *eutectic* is used to design a mixture whose melting temperature is lower than these of its pure components. As a result, two solids can liquefy upon mixing. Although this observation might seem intriguing, it is a quite general phenomenon, since a compound with an impurity solidifies with more difficulty than the pure compound. A well-known example of eutectic mixture is the mixture of NaCl and iced water, freezing below 0°C at atmospheric pressure. *Deep* eutectics are characterized by a "large" temperature depression, which is obviously a subjective feature, but can be defined as a temperature depression which is significantly deeper than that of the ideal liquid mixture (Martins et al. 2019). The gap to ideality generally results from strong intermolecular interactions between the different components of the mixture. The different types of interaction will be discussed in more detail below. There is a single composition, called the "eutectic composition," where the pure liquid state reaches the lowest temperature. It is important to mention that contrary to what is often reported in the literature, the eutectic composition is not related to the formation of a stoichiometric complex, but rather to the intersection of two independent melting curves. An emblematic preparation of deep eutectics is the mixture of hydroxyethyltrimethyl ammonium (i.e., choline chloride) (melting point T_m = 302 °C) with urea (T_m = 133 °C), which results in a eutectic of very low melting point (T_m = 12 °C) (Abbott et al. 2003). The recent discovery of many room temperature eutectics and of their ability to solubilize both organic and inorganic compounds has launched an intense research activity about their uses as solvents, and the term *deep eutectic solvents* (DES) is now widely used in this field.

The formation of a deep eutectic solvent requires strong and multiple interactions between its components. Deep eutectic solvents generally contain an organic salt Cat^+X^-, where Cat^+ is an organic bulky cation (ammonium, sulfonium, phosphonium) and X^- is a Lewis base, in most cases a halide anion. This organic salt is combined with a Lewis or an organic hydrogen bond donor Y, which interacts with the anionic species X^- through electrostatic interactions or hydrogen bonds, respectively. The role of the bulky (and generally of low symmetry) Cat^+ is to lower the lattice energy of the salt Cat^+X^- and to reduce the freezing temperature of the mixture. The first successful achievement of a room temperature deep eutectic solvent was observed in a mixture of choline chloride as the Cat^+X^- salt and urea as the hydrogen bond donor, Y (Abbott et al. 2003). Today, deep eutectic solvents are classified into five types (Table 7.1).

Types I-IV involve ionic species and can be conveniently described by the general formula Cat^+X^-zY, z being the number of interacting Y molecules, which defines the so-called "stoichiometry" of the deep eutectic solvent (Abbott et al. 2001). Type I corresponds to deep eutectic solvent formed from metal halides such as $ZnCl_2$ associated with an organic salt. The number of type I deep eutectic solvents being liquid at room temperature is relatively limited. The development of type II deep eutectic solvent based on hydrated metal halides has allowed to overcome this limitation and widely opened the scope of deep eutectic solvent. Particularly, type II deep eutectic solvent has attracted much interest in many industrial applications due to their insensitivity to moisture.

Alternatively, organic salts such as choline chloride have been successfully combined to a variety of hydrogen bond donors (HBD) including amide, carboxylic acids, and ketones. These all-organic deep eutectic solvents form the type III family. Among all deep eutectic solvents, type III deep eutectic solvents present the lowest freezing point and are relatively easy to prepare and unreactive to water. Their physical properties can be tailored for specific applications through a selection of the hydrogen bond donor, and they can be prepared from biodegradable and/or low-cost materials if necessary. For these reasons, type III deep eutectic solvents are the ones

Table 7.1 The five types of deep eutectic solvents. (HBD, hydrogen bond donor; HBA, hydrogen bond acceptor). Note that types III and V are all-organic deep eutectic solvents. Type V is nonionic

Type	Hydrogen bond donor/ Lewis acid	Hydrogen bond acceptor	Example
Type I	**Y :** Metal salt	**Cat^+X^- :** Organic salt	$ZnCl_2$:choline chloride
Type II	**Y :** Metal salt hydrate	**Cat^+X^- :** Organic salt	$CoCl_2$,$6H_2O$:choline chloride
Type III	**Y :** Organic HBD	**Cat^+X^- :** Organic salt	Urea:choline chloride
Type IV	Organic HBD	Metal halide	$ZnCl_2$ in urea ((urea • $ZnCl^+$), ($ZnCl_3^-$))
Type V	Nonionic HBD (phenols)	Nonionic HBA	Thymol:menthol

that have been used for natural product extraction. Interestingly, they also present a relatively wide electrochemical window, which has been used for metal deposition.

Type IV is a specific and relatively limited class of deep eutectic solvents that do not contain an organic salt. Although non-hydrated inorganic salts generally do not form deep eutectic solvents at low temperature, deep eutectic solvents have been prepared at ambient temperature from transition metal chlorides such as $ZnCl_2$ associated with an organic HBD, such as urea or ethylene glycol. Metal halides are not normally expected to dissociate in nonaqueous media; however, the structure $MCl_{(x-1)}^{+}.Y, MCl_x^{-}$, where M is the metal atom and Y is the organic HBD, was proposed to account for the observed ionization and deep eutectic solvent formation.

The existence of a significant number of eutectics based solely on nonionic species (e.g., thymol:lidocaine or thymol:menthol) has very recently led Coutinho et al. to define a fifth type of deep eutectic solvents (Abranches et al. 2019). The authors show that phenols play a central role in type V deep eutectics, since their combination with any hydrogen bond acceptor is susceptible to produce a deep eutectic. This is due to resonant character of phenols which make them good HBD and poor HBA, then they prefer to interact with other HBA than with themselves.

In 2011, natural deep eutectic solvents were introduced as a particular class of deep eutectic solvent, mainly of type III, prepared from biomolecules, such as choline chloride and betaine as the organic salt, and urea, organic acids, amino acids, or sugars as the HBD (Hayyan et al. 2013; Dai et al. 2013b; Paiva et al. 2014; Aroso et al. 2017; Silva et al. 2018). It can be also expected that many type V deep eutectic solvents will be based on natural compounds. Remarkably, natural deep eutectic solvents are ubiquitously present in living organisms both in the intra- and extracellular media where they may play a role in the synthesis and solubilization of poorly soluble metabolites such as flavonoids, in enzymatic reactivity, and also in drought tolerance (Dai et al. 2013b; Paiva et al. 2014). In this paradigm, natural deep eutectic solvents constitute a third type of natural liquid, separated from water and lipids (Choi et al. 2011). Choi et al. investigated a series of natural deep eutectic solvents formed from abundant biomolecules, including combinations of choline chloride with citric, malic, maleic, and aconic acids; proline with citric acid; malic acid with glucose; and mixtures of sugars (fructose:glucose, fructose:sucrose, glucose:sucrose, etc.) (Choi et al. 2011). Natural deep eutectic solvents are regarded as an environmentally friendly alternative solvent for the extraction of biomolecules. Thus, the formation of a glycerol-based natural deep eutectic solvent was evaluated for the separation of glycerol from biodiesel (Abbott et al. 2007). Most type V deep eutectic solvents are also formulated from natural molecules and as such belong to the family of natural deep eutectic solvents.

Despite the significant increase in the number and variety of deep eutectic solvents, research has mainly focused on the design of hydrophilic deep eutectic solvents. The first report of hydrophobic deep eutectic solvents, sometimes referred to as HDES or HES, appeared only in 2015 (Florindo et al. 2019), but their major interest as water-immiscible solvents for liquid-liquid extraction of natural compounds was rapidly identified, provided that their viscosity is sufficiently low and their density far enough from that of water. Hydrophobic deep eutectic solvents can meet the requirement of green solvents, and therefore they constitute a promising alternative to organic solvents. They also show major advantages over ionic liquids, such as lower prices and easy preparation. Interestingly, hydrophobic deep eutectic solvents enable the extraction of metal ions from water (van Osch et al. 2016), and membranes impregnated with hydrophobic deep eutectic solvents can be used in separative devices. An increased CO_2 capture in hydrophobic deep eutectic solvents compared to hydrophilic ones is also reported (Zubeir et al. 2018). Hydrophobic deep eutectic solvents often make use of fatty acids as HBD; the hydrophobicity can also be brought about by constituents such as terpenes. A comprehensive list of hydrophobic deep eutectic solvents, can be found in a recent review (Dwamena 2019).

It is important to note that the term deep eutectic solvent and associated concepts are commonly used well beyond their strict definition, as stated in the beginning of this section. Two issues in particular should be discussed, which concern type III deep eutectic solvents principally. First, water can be part of the components of a deep eutectic solvent as hydrogen bond donor (HBD) added in stoichiometric amount. In this case, water is an integral part of the deep eutectic solvent and is extremely difficult or even impossible to remove (Dai et al. 2013b; Aroso et al. 2017). However, in many extraction studies, relatively large amounts of water are added to reduce the viscosity of the solvent or to facilitate the preparation of the mixture. Although spectroscopic analyses (NMR, FTIR) show that a moderate excess of water does not destroy the deep eutectic solvent structure, this point is rarely controlled (Dai et al. 2015). The other issue concerns the use of liquid hydrogen bond donor, such as glycerol, in type III deep eutectic solvent. Although liquid HBD may interact with organic salts in the same way as their solid analogues, in the absence of the visual criterion of liquefaction, it is difficult to distinguish a simple dissolution from the formation of a eutectic. As an alternative, Abbott et al. have used the lowest freezing point to define the "eutectic composition" of mixtures of glycerol with various quaternary ammonium salts (Abbott et al. 2007). Nevertheless, even if the thermodynamic notion of eutectic can be questioned in these two situations, it remains that excellent solvation properties can be observed and used.

Deep eutectic solvents present favorable features as solvents, such as low vapor pressure, nonflammability, and low or negligible toxicity. Compared to ionic liquids, deep eutectic solvents are generally more environmentally friendly, easier to prepare, and radically less expensive. Until recently, the two main applications of deep eutectic solvents have been metal deposition and synthesis media. However, with the emergence of natural deep eutectic solvents, applications as extraction media for natural products are becoming increasingly important. Indeed, natural deep eutectic solvents cover a large range of polarities from more polar than water to equivalent to methanol (Ruesgas-Ramón et al. 2017). They proved to be excellent solvents for metabolites of medium of low polarity that are poorly soluble in water, particularly phenolic compounds (Dai et al. 2013c). This is due to the composition of the media, which are rich both in ions and in HBD moieties. Their physicochemical and solubilizing properties can be tailored by adjusting the molecular structures and molar proportions of the components, including water (Dai et al. 2013b). Interestingly deep eutectic solvents can be combined with most of the extraction-assisted techniques such as microwave, ultrasound, high pressure, etc. Their main limitation is their viscosity, which is generally higher than those of most conventional solvents. Viscosity is an important parameter when considering a solvent for extraction, since it controls diffusion and mixing (Ruesgas-Ramón et al. 2017). However, the viscosity can be considerably lowered by adding small amounts of water or increasing the temperature.

7.3.2 Hydrogen Bonds

Hydrogen bonds (H bonds) play a major role both in the formation and in the solubilization properties of organic molecules in type III deep eutectic solvent. For this reason, we will recall here their main features. A H bond is an electrostatic force of attraction between a hydrogen atom covalently bound to a more electronegative atom (particularly N, O, or F) and another electronegative atom bearing a lone pair of electrons, the former being the hydrogen bond *donor* and the latter the hydrogen bond *acceptor*. H bonds are generally considered as electrostatic dipole-dipole interactions. However, they also share some features with covalent bonds. They are directional and the interatomic distances are shorter than the sum of the van der Waals *radii*. Their strength varies from a few $kJ.mol^{-1}$ to more than 100 $kJ.mol^{-1}$, which places them in between the van der Waals and covalent interactions. Thanks to their directionality, H bonds enable molecular recognition, which is ubiquitous in biomolecules. Thus, it strongly contributes to the structure of DNA and proteins and to substrate-protein association. The existence of hydrogen bonds in type III deep eutectic solvents was revealed by NMR spectroscopy (Abbott et al. 2003; Dai et al. 2013b). More recently, the variety of hydrogen bonds that can be formed between the different components was evidenced to play an important role in the freeze point depression (Migliorati et al. 2019). The ability of the solvent to form hydrogen bonds is particularly of interest for the selective extraction of polyphenols since these compounds are generally rich in hydrogen bond donor groups (Choi et al. 2011; Ruesgas-Ramón et al. 2017). In addition, ternary deep eutectic solvents containing polyphenols in their formulation such as the poorly soluble caffeic acid (e.g., choline chloride:caffeic acid:ethylene glycol) were also prepared as a basis of molecularly imprinted materials for the recovery of the target polyphenol (Fu et al. 2017a, b).

7.3.3 Deep Eutectic Solvent and Extraction of Polyphenols

The polarity of polyphenols ranges from LogP ~ −2.5 for the most polar ones (catechin) to LogP = 5 for amentoflavone (Bohn 2014), while LogP is equal to 10.5 for α-tocopherol (vitamin E), which bears a long alkyl chain. Moreover, polyphenols may also exhibit a polymeric structure (e.g., tannins or marine algae polyphenols) as seen in Sect. 7.2. They may also form complex with proteins, sometimes highly insoluble (e.g., tannins). Polar polyphenols have been traditionally extracted using aqueous mixtures of methanol, ethanol, and acetone in conventional solid/liquid extraction (SLE) (Ruesgas-Ramón et al. 2017). Ethanol and methanol are also used as cosolvent in supercritical carbon dioxide (Sc-CO_2) extraction of polyphenol from plant materials such as grape pomaces and seeds, rosemary and olive leaves, and pistachio hulls. The Sc-CO_2 extracts mainly contain gallic acid, catechin, and epicatechin, whereas the SLE extracts are richer in proanthocyanidin. This underlines the importance of the choice of the solvent and extraction method for extracting selectively the compounds of interest, which logically depend on their structures.

Thanks to their ability to form hydrogen bonds, polyphenols are potentially good targets for extraction using deep eutectic solvent. As mentioned previously, hydroxyl groups on polyphenols are good hydrogen donors but poor acceptors and preferably form bonds with good hydrogen bond acceptors (Burghoff et al. 2008; Martins et al. 2019). Thus, amines and phosphine oxides are excellent phenol extractants. It might be also interesting to introduce aromatic rings in the deep eutectic solvent to take benefit of π-π interactions. Depending on their constituents, deep eutectic solvents and especially natural deep eutectic solvents could be tuned with a wide range of polarities to accommodate from the less to the most polar polyphenols. In addition, they could interact *via* numerous hydrogen bonds with the hydroxyl groups of polyphenols, thus promoting their solubilization (Dai et al. 2013c). For these reasons, natural deep eutectic solvents have been considered as promising solvents for the selective solubilization and extraction of polyphenols (Faggian et al. 2016). Even if (natural) deep eutectic solvents do not overtake the extraction yields obtained with benchmark solvents, an advantage is expected in terms of selectivity in addition to environmental acceptability.

Interestingly, it is also suggested that natural deep eutectic solvents also enhance the biological activity of polyphenols (Faggian et al. 2016) and, thanks to the biocompatibility of the natural deep eutectic solvents, extracts may be used directly in food, cosmetic, and pharmaceutical formulations.

In front of the abundance of possible (natural) deep eutectic solvent formulations, the in silico methods could provide a valuable guidance to select the most promising solvents for the selective extraction of a desired compound or class of compounds (Jeliński and Cysewski 2018; Silva et al. 2018). Particularly, the COSMO-RS (quantum conductor-like screening model for real solvent) has been successfully applied to ionic liquids and deep eutectic solvents to rank solvent candidates for the extraction of polyphenols (Burghoff et al. 2008; Jeliński and Cysewski 2018) and other compounds of interest. The COSMO-RS model enables to compute the chemical potential of molecules in liquid solutions, from which other parameters such as activity and solubility can be derived. One limitation of the use of COSMO-RS for (natural) deep eutectic solvents is that the degree of dissociation of weak acids and bases in the mixture, which are the main components of natural deep eutectic solvent, is generally unknown; therefore, rough assumptions have to be made. Their accuracy can be validated by comparison with experimental solubility data sets.

7.4 Extraction Methodologies of Polyphenols in (Natural) Deep Eutectic Solvent

7.4.1 Commonly Used Deep Eutectic Solvent for Polyphenol Extraction

As mentioned previously, the (natural) deep eutectic solvents have very interesting solubilizing properties for polyphenols. For example, the solubility of rutin, a well-known flavonoid, is 50–100 times higher in various natural deep eutectic solvents than in water (Choi et al. 2011; Huang et al. 2017). Dai et al. (2013c) have investigated the extraction of polyphenols from safflower with seven natural deep eutectic solvents of various polarities and obtained higher extraction yields with the deep eutectic solvents than with conventional solvents. The same authors have reported the effect of water on the solubility of quercetin and carthamin in natural deep eutectic solvents (Dai et al. 2015). These two polyphenols are poorly soluble in water. In both cases, the solubility of the polyphenols in a weakly polar deep eutectic solvent decreases dramatically with the amount of water from 25 to 50% w/w, which correlates with the rupture of the network of hydrogen bonds as observed by measuring the viscosity. As expected, more polar polyphenols require more water for an optimal solubilization. Interestingly, the optimal water content to solubilize a given polyphenol in different natural deep eutectic solvents also seems to correlate with the viscosity, the more viscous being less prone to accommodate the solute.

By scrutinizing the literature data, it turns out that more than 60% of the publications dealing with the extraction of polyphenols concern choline chloride-based (natural) deep eutectic solvents (Fig. 7.5). Actually, choline chloride was also employed for designing “cheap” ionic liquids (Abbott et al. 2001), and we could consider that the choline chloride is the keystone from ionic liquids to deep eutectic solvents. Originally, choline chloride was mixed with conventional solvents like ethylene glycol in order to reduce the viscosity of the latter. The resulting mixtures were cheap and not very viscous, therefore attracting attention toward deep eutectic solvents (Zainal-Abidin et al. 2017).

Choline chloride is then frequently found, associated with organic acids, e.g., citric or acetic acid, or with alcohols, e.g., 1,4-butanediol (Table 7.2). The equimolar mixture of sodium acetate and glycerol is probably the only deep eutectic solvent without choline chloride that is used in the extraction of polyphenols. Note that this mixture is one of the less viscous deep eutectic solvents. As mentioned in Sect. 7.3, the properties of deep eutectic solvent can be adjusted through the variation of the HBA (hydrogen bond acceptor) or HBD (hydrogen bond acceptor), the molar ratio between components, and the water content. Changing the hydrogen bond donors allows the tuning of physical properties (viscosity, pH, color, melting point, etc.) of the solvents but also the modification of the interactions between solvent and polyphenol compounds (Zainal-Abidin et al. 2017). Often, solvents are empirically selected through researchers’ know-how and knowledge. Interestingly, the very large majority of deep eutectic solvents used for the extraction of polyphenols are

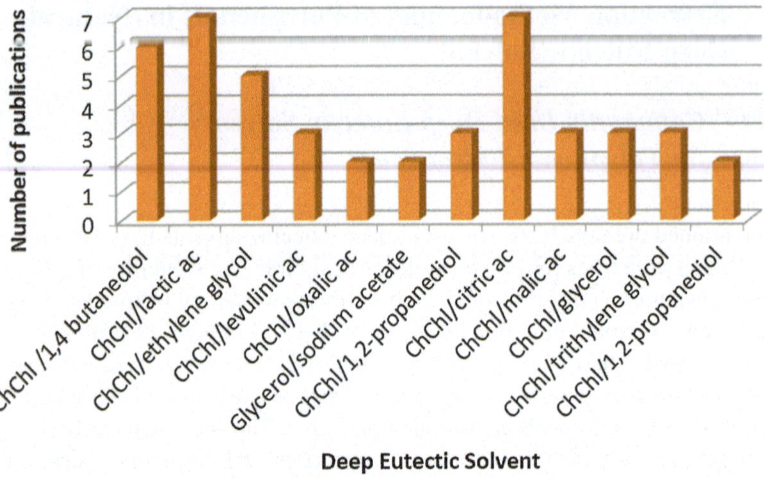

Fig. 7.5 Number of publications reported for each type of deep eutectic solvent used in extraction of polyphenols. Only deep eutectic solvent with a minimum of two citations is quoted. *ChCl* choline chloride, *ac* acid. Note that most reported deep eutectics are mainly based on choline chloride; only one example, i.e., glycerol/sodium acetate, is not composed of choline chloride

hydrophilic mixtures, containing various water contents (Table 7.2). Taking into account that many polyphenols are poorly soluble or even insoluble in water, this point is worth to mention.

7.4.2 Extraction

Figure 7.6 shows the general principle of polyphenol extraction processes, whatever the extractive methods applied and the kind of deep eutectic solvents used (Benvenutti et al. 2019). Four major stages are involved as follows: (1) the sample preparation, (2) the extraction itself, (3) the purification (separation of the residual matrix with the elimination or not of the deep eutectic solvent), and finally (4) the analytical part for the characterization and quantification of extracted compounds. In this section, we will focus on the three first stages; the last one will be developed in Sect. 7.5.

Sampling and Sample Preparation

Sampling and sample preparation are generally a key aspect in extraction processes, whatever the solvents or technologies used, but generally regarded as trivial. However, worth is outlining some critical points for extraction of polyphenols.

Table 7.2 Deep eutectic solvent mixtures used for polyphenol extraction

Deep eutectic solvent composition	References
Betaine/glycerol/D-(+)-glucose	Chanioti and Tzia (2018)
Choline chloride/1,4-butanediol	Bi et al. (2013), Cui et al. (2018), Wan Mahmood et al. (2019)
Choline chloride/butyric acid	Vieira et al. (2018)
Choline chloride/citric acid	Molnar et al. (2018), Panić et al. (2019a, b)
Choline chloride/D-(+)-glucose	Jeong et al. (2018)
Choline chloride/ethylene glycol	Li et al. (2016), Wan Mahmood et al. (2019)
Choline chloride/formic acid	Fu et al. (2017c)
Choline chloride/glycerol	Fraige et al. (2019)
Choline chloride/levulinic acid	Zhao et al. (2015)
Choline chloride/malic acid	Obluchinskaya et al. (2019)
Choline chloride/oxalic acid/ethylene glycol	Tang et al. (2017)
Choline chloride/triethylene glycol	Zhao et al. (2015), Peng et al. (2018), Mansur et al. (2019)
Choline chloride/zinc bromide	Duan et al. (2019)
Choline chloride/1,2-propanediol	Dai et al. (2016), García et al. (2016), Chen et al. (2016), Sang et al. (2018), Meng et al. (2018)
Choline chloride/1,3-butanediol	Peng et al. (2016)
Choline chloride/1,4-butanediol	Yao et al. (2015), Shang et al. (2018), Wang et al. (2019)
Choline chloride/1,6-hexanediol	Cui et al. (2015)
Choline chloride/betaine hydrochloride/ethylene glycol	Qi et al. (2015)
Choline chloride/citric acid	Bajkacz and Adamek (2017), Hamany Djande et al. (2018), Chanioti and Tzia (2018), Zhou et al. (2018)
Choline chloride/citric acid/glucose	Guo et al. (2019)
Choline chloride/ethylene glycol	Zhang et al. (2014), Xia et al. (2015), Ozturk et al. (2018)
Choline chloride/glycerol	(Abbott et al. (2003)
Choline chloride/lactic acid	Li et al. (2015), Wei et al. (2015b), Ma et al. (2018), Chanioti and Tzia (2018), Ivanović et al. (2018), Obluchinskaya et al. (2019)
Choline chloride/levulinic acid	Duan et al. (2016), Zhuang et al. (2017)
Choline chloride/levulinic acid/N-methyl urea	Xu et al. (2019)
Choline chloride/malic acid	Radošević et al. (2016), Bosiljkov et al. (2017)
Choline chloride/malonic acid	Cao et al. (2018a)
Choline chloride/maltose	Wei et al. (2015a)
Choline chloride/oxalic acid	Cvjetko Bubalo et al. (2016), Saha et al. (2019)
Choline chloride/*p-toluene* sulfonic acid	Ali et al. (2019)
Choline chloride/xylitol	García et al. (2016)
Choline chloride/phenylpropionic acid	Vieira et al. (2018)

(continued)

Table 7.2 (continued)

Deep eutectic solvent composition	References
Citric acid/D-(+)-maltose	Jeong et al. (2015)
Glycerol/choline chloride	Mouratoglou et al. (2016)
Glycerol/sodium acetate	Mouratoglou et al. (2016)
Glycerol/trisodium citrate	Jancheva et al. (2017)
Glycerol/glycine	Rajha et al. (2019)
Glycerol/sodium acetate	Karageorgou et al. (2017)
Glycerol/xylitol/D-(-)-fructose	Wang et al. (2017)
Lactic acid/ glucose	Fernández et al. (2018a), Dai et al. (2016), Mouratoglou et al. (2016), El Kantar et al. (2019)
Lactic acid/glycine	Bakirtzi et al. (2016)
Lactic acid/nicotinamide	Georgantzi et al. (2017)
Lactic acid/L-alanine	Georgantzi et al. (2017)
L-Proline/glycerol	Nam et al. (2015)
Malic acid/glucose/glycerol	Rajha et al. (2019)
$Me(Ph)_3PBr$/ethylene glycol	Tang et al. (2015)
n-Octanoic acid/1-nonanoic acid/ dodecanoic acid	Yang et al. (2019)
Tetrabutylammonium chloride/ decanoic acid	Kanberoglu et al. (2019)

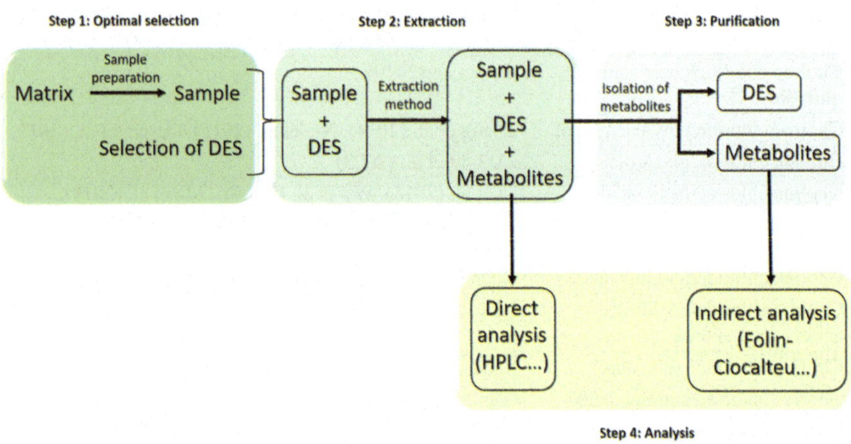

Fig. 7.6 General scheme of an extraction process, from the crude sample to quantitative analysis of the extracted metabolites

The growth period of the plant is an important parameter to consider because reproducibility in extraction results requires a homogeneous matrix. Yet, plants do not produce the same type of metabolites depending on the period of their growth. Moreover, specific stress conditions (temperature, duration and intensity of lighting, watering, etc.) could be applied to the plants for optimizing the production of metabolites of interest. This is especially relevant in the case of polyphenols because the type of produced polyphenols may be different in nature depending on the stress conditions of the plants. Indeed, polyphenols are generally powerful antioxidants protecting the plants from external attacks (insects, UV, fungi, etc.) (Naikoo et al. 2019). Their production can therefore be particularly influenced by the living conditions of plants. Our meta-analysis of *ca.* 100 publications shows that only 24% of publications dealing with the extraction of polyphenols quote the harvest period of the raw materials.

A careful selection of the part of the plants (leaves, stems, flowers, roots, etc.) is equally important for extraction. Feduraev et al. have shown that phenolic compound compositions depend on the part of plants with an accumulation in the leaves, seeds, and flowers for *Rumex crispus* and *Rumex obtusifolius* (Feduraev et al. 2019). But in some cases, the whole plant (or algae) is chosen.

For the sample preparation before extraction, the plant is usually dried by freeze-drying (Jeong et al. 2015; Dai et al. 2016; Cvjetko Bubalo et al. 2016; Radošević et al. 2016; Bosiljkov et al. 2017; Karageorgou et al. 2017; Fernández et al. 2018a; Shang et al. 2018; Panić et al. 2019a, b; Saha et al. 2019; Xu et al. 2019) or in an oven between 30 and 70 °C (Li et al. 2015; Mouratoglou et al. 2016; Tang et al. 2017; Cao et al. 2018a, b; Ozturk et al. 2018; Fraige et al. 2019; Kanberoglu et al. 2019; Panić et al. 2019b; Rajha et al. 2019; Mansur et al. 2019). Phenolic compounds could be sensitive to temperature or light; therefore, plants' drying is often performed away from the light or even at room temperature for several days (Wei et al. 2015a; Yao et al. 2015; Cui et al. 2015, 2018; Peng et al. 2016; Li et al. 2016; Jancheva et al. 2017; Vieira et al. 2018; Ivanović et al. 2018). However, there is no standard protocol. For example, Karageorgou et al. chose to wash the *Moringa oleifera* leaves prior to freeze-drying, while Hamany Djande et al. directly air-dry the leaves without prior washing (Karageorgou et al. 2017; Hamany Djande et al. 2018).

Then, the raw material is grounded. The grinding can have a strong influence on the extraction (Komaty et al. 2016). Indeed, intensive grinding can lead to heating the matter and generate degradation of the compounds. On the other hand, light grinding can prevent good penetration of the solvent in the matter, thus reducing the surface contact with the solvent and finally limiting the extraction (Azwanida 2015). The grinding can be carried out by disintegrator, domestic blender, laboratory mill, or liquid nitrogen with a mortar and pestle. The goal is to promote cell destruction and to have a better homogeneity of the mixture, namely, in the case of soft matrix (succulent leaf, some algae, etc.) (Wan Mahmood et al. 2019). The preparation of the sample is obviously different depending on the raw material. For instance, grinding virgin olive oil is not needed because it is a homogeneous liquid, while Chanioti et al. use a cutting mill to grind the olive pomace (García et al. 2016; Chanioti and Tzia 2018).

Some molecules can be hidden deeper in the plant and may require finer grinding for allowing a good extraction. This depends on the histolocalization of the targeted metabolites. A good knowledge of the selected plants is therefore particularly important for controlling the extraction of the compounds of interest (Komaty et al. 2016). In plant, phenolic compounds are present in the tissue in cellular and subcellular levels with differences between insoluble and soluble phenolics (Naczk and Shahidi 2004). Usually, when indicating, the raw material is pulverized between 0.2 and 1 mm (Bakirtzi et al. 2016; Mouratoglou et al. 2016; Georgantzi et al. 2017; Jancheva et al. 2017; Karageorgou et al. 2017; Molnar et al. 2018; Chanioti and Tzia 2018; Ozturk et al. 2018; Obluchinskaya et al. 2019). Sieves ranging from 20 to 80 mesh (841 μm to 177 μm) are used to standardize the sample size (Li et al. 2015, 2016; Wei et al. 2015a, b; Qi et al. 2015; Cui et al. 2015, 2018; Peng et al. 2016, 2018; Zhuang et al. 2017; Wang et al. 2017, 2018a, b, 2019; Ma et al. 2018; Sang et al. 2018; Cao et al. 2018a, b; Vieira et al. 2018; Shang et al. 2018; Duan et al. 2019; Saha et al. 2019; Mansur et al. 2019; Xu et al. 2019).

Storing samples before extraction is also an important step (Naczk and Shahidi 2004). The sample can be stored in the dark, at room temperature or keeping cold. Mouratoglou et al. specify that samples of lemon peel, once dried and pulverized, are stored "in the dark, at ambient temperature, for no longer than 7 days" to preserve the compounds (Mouratoglou et al. 2016).

Key Parameters for Extraction

A great number of variables may affect the extraction efficiency (Chanioti and Tzia 2018; Duan et al. 2019; Choi and Verpoorte 2019) and could be leveraged for increasing the extractive performance and optimizing the processes. The plant-to-solvent ratio, the temperature, the duration of extraction, the nature of deep eutectic solvent through variation of HBD and/or HBA, and the water content in the deep eutectic solvents are the main factors which may influence the extraction efficacy (see Table 7.3) (Cvjetko Bubalo et al. 2018). All play a critical role, with possible interactions between each other. These parameters could also strongly depend on the technologies employed, and of course technologies have a true impact on the performance (Table 7.3). Many research works use multivariate approaches to consider the interactions between these variables (Bi et al. 2013; Bosiljkov et al. 2017; Ivanović et al. 2018; Panić et al. 2019b). Ivanović et al., for example, studied the interactions of three extraction parameters including the duration of microwave irradiation, the quantity of water in the deep eutectic solvent, and the temperature. They showed the influence of these interactions on the polyphenol extraction rate (Ivanović et al. 2018). The amount of water and the temperature were found to have a major effect, while the microwave irradiation time had a minor influence.

Table 7.3 Extracted polyphenols and corresponding sources as a function of deep eutectic solvent and technologies employed. For each target compound, methodology for quantification and/or recovery is added

Sources	Target compounds	Deep eutectic solvent	Extraction process	Quantification method	Compounds' recovery from deep eutectic solvent	References
Lonicera japonica	Caffeic acid; chlorogenic acid; 3,5-dicaffeoylquinic acid; 3,4-dicaffeoylquinic acid; 4,5-dicaffeoylquinic acid	Choline chloride/1,3-butanediol 1:6	HRE UAE MAE	RP-HPLC (C18)	Macroporous resin NKA-9 from 79.25% to 86.01%	Peng et al. (2016)
Cajanus cajan leaves	Orientin; vitexin; luteolin; apigenin-6,8-di-C-α-L-arabinoside; apigenin-8-C-α-L-arabinoside luteolin; apigenin; isorhamnetin; formononetin; cajanol; pinostrobin chalcone; pinostrobin; longistyline C; cajaninstilbene acid; cajanuslactone	Choline chloride/maltose 1:2	MAE	UPLC-UV (C18)		Wei et al. (2015a)
Equisetum palustre L.	Kaempferol-3-O-β-D-glucopyranoside-7-O-β-D-glucopyranoside; kaempferol-3-O-β-D-rutinoside-7-O-β-D-glucopyranoside; luteolin-7-O-β-D-glucopyranoside; quercetin-3-O-β-D-glucopyranoside; apigenin-5-O-β-D-glucopyranoside; genkwanin-5-O-β-D-glucopyranoside; luteolin; apigenin; genkwanin	Choline chloride/betaine hydrochloride/ethylene glycol 1:1:2	Negative pressure cavitation-assisted extraction (NPCE) UAE	HPLC-DAD (Curosil-PFP column)	Macroporous resin HPD-826 from 64.52% to 89.25%	Qi et al. (2015)

(continued)

Table 7.3 (continued)

Sources	Target compounds	Deep eutectic solvent	Extraction process	Quantification method	Compounds' recovery from deep eutectic solvent	References
Pyrola incarnata Fisch.	Hyperin; 2′-O-galloyl-hyperin; quercetin; quercetin-O-rhamnoside; chimaphilin	Choline chloride/1,4-butanediol 1:4	Stirring + heating	RP-HPLC-UV (C18W)		Yao et al. (2015)
			UAE			
			MAE			
Chamaecyparis obtusa leaves	Quercitrin; myricetin; amentoflavone	$Me(Ph)_3PBr$/ethylene glycol 1:5	Heating	HPLC-UV (C18)		Tang et al. (2015)
Camellia sinensis leaves	(-)-Epigallocatechin gallate; (-)-epigallocatechin; (-)-epicatechin gallate; (-)-epicatechin	Choline chloride/lactic acid 1:2	HRE	HPLC-DAD (C18)	Macroporous resin AB-8 from 75.2% to 86.1%	Li et al. (2015)
			MAE			
Radix scutellariae	Wogonoside; baicalein; baicalin; wogonin	Choline chloride/lactic acid 1:2	MAE	HPLC-UV (C18)	Macroporous resin ME-2 from 79.5% to 85.0%	Wei et al. (2015b)
Green tea	Catechin; (+)-epicatechin gallate; (−)-epigallocatechin gallate	Choline chloride/ethylene glycol 1:5	Stirring	HPLC-UV (C18)		Zhang et al. (2014)
			Stirring + heating			
			UAE			
Prunella vulgaris	Rosmarinic acid; salviaflaside	Choline chloride/ethylene glycol 1:4	Stirring	UPLC-UV (C18)		Xia et al. (2015)
			Stirring + heating			
			UAE			

Berberidis Radix; *Epimedii Folium*; *Notoginseng Radix Rhizoma Rhei Rhizoma et Radix*; *Salviae Miltiorrhizae Radix*	Aloe-emodin; rhein; emodin; physcion; chrysophanol; ginsenoside Rg1; notoginsenoside R1; ginsenoside Rb1L; lithospermic acid; rosmarinic acid; salvianolic acid; jatrorrhizine hydrochloride; palmatine hydrochloride; berberine hydrochloride; icariin	Choline Chloride/levulinic acid 1:2	UAE	HPLC-UV (C18)		Duan et al. (2016)
Platycladi Cacumen	Myricitrin; quercitrin; amentoflavone; hinokiflavone	Choline chloride/levulinic acid 1:2	UAE	HPLC-DAD (C18)	Macroporous resin LX-38 from 77.44% to 98.92%	Zhuang et al. (2017)
Grape skin	(+)-Catechin; delphinidin-3-O-glucoside; cyanidin-3-O-glucoside; petunidin-3-O-glucoside; peonidin-3-O-glucoside; malvidin-3-O-glucoside; quercetin-3-O-glucoside	Choline chloride/oxalic acid 1:1	Stirring MAE UAE	HPLC-DAD (C18)		Cvjetko Bubalo et al. (2016)
Grape skin	Total anthocyanins; total polyphenols; delphinin; cyanidin; peonidin; malvidin glycoside	Citric acid/D-(+)-maltose 4:1	Stirring	UPLC-Q-TOF-MS (C18)		Jeong et al. (2015)
			Heating Vortexing Stirring + heating UAE	Folin-Ciocalteu, pH differential method		
Pigeon pea roots	Genistin; genistein; apigenin	Choline chloride/1,6-hexanediol 1:7	HRE UAE MAE	RP-HPLC (C18)		Cui et al. (2015)

(continued)

Table 7.3 (continued)

Sources	Target compounds	Deep eutectic solvent	Extraction process	Quantification method	Compounds' recovery from deep eutectic solvent	References
Flos sophorae	Quercetin; kaempferol; isorhamnetin glycosides	L-Proline/glycerol 1:2.5	Stirring	LC-UV (C18)	SPE C18 from 81% to 87%	Nam et al. (2015)
			Heating	UPLC-Q-TOF-MS (C18)		
			Stirring + heating			
			UAE			
Chamaecyparis obtuse leaves	Myricetin; amentoflavone	Choline chloride/1,4-butanediol 1:5	Stirring	HPLC-UV (C18)		Bi et al. (2013)
			Heating			
			UAE			
Ginkgo biloba leaves	Proanthocyanidin	Choline chloride/malonic acid 1:2	Stirring	DMAC method (spectrophotometric method)	Macroporous resin D-101	Cao et al. (2018a)
			Heating			
			Stirring + heating			
			Water-bath shaking			
			Air-bath shaking			
			UAE			

Satureja thymbra	Total polyphenols; flavone derivatives; rosmarinic acid derivatives; luteolin derivatives; rosmarinic acid derivative; quercetin rhamnoside; quercetin rutinoside; quercetin derivatives; apigenin derivatives; quercetin glucoside or galactoside; luteolin glucuronide; apigenin glucuronide; apigenin glucuronide; luteolin glucoside or galactoside; quercetin; luteolin	Glycerol/trisodium citrate 15:1	Stirring + heating	Folin-Ciocalteu		Jancheva et al. (2017)
				LC-DAD-MS (RP-18)		
Moringa oleifera leaves	Total polyphenols; chlorogenic acid; quercetin diglycoside; multiflorin B; apigenin *C*-glycoside; quercetin glycoside; quercetin glycoside derivative; quercetin malonylglycoside derivative; kaempferol glycoside derivative; quercetin rhamnoside derivative; kaempferol malonylglucoside	Glycerol/sodium acetate 6:1	Stirring + heating	Folin-Ciocalteu		Karageorgou et al. (2017)
				$AlCl_3$ reagent		
				LC-DAD-MS (RP-18)		
Flower petals of *Catharanthus roseus*	Petunidin 3-O-(6-O-p-coumaroyl) glucose; malvidin 3-O-(6-O-p-coumaroyl) glucose; hirsutidin 3-O-(6-O-p-coumaroyl) glucose; 7-O-methylcyanidin; 7,3'-O-dimethylcyanidin	Lactic acid/glucose 5:1	Stirring + heating	HPLC-DAD (C18)		Dai et al. (2016)
		1,2-Propanediol/choline chloride 1:1	UAE	UPLC-TOF-MS (C18)		
Olive cake	Apigenin; caffeic acid; catechin; gallic acid; luteolin; 3-hydroxytyrosol; naringenin; quercetin dehydrate; rutin hydrate; tyrosol; coumaric acid; trans-ferulic acid; oleuropein; cinnamic acid	Lactic acid/ glucose 5:1	UAE	HPLC-DAD (Zorbax SB-Aq column)		Fernández et al. (2018a)
Onion						
Tomato waste						
Pear waste						

(continued)

Table 7.3 (continued)

Sources	Target compounds	Deep eutectic solvent	Extraction process	Quantification method	Compounds' recovery from deep eutectic solvent	References
Virgin olive oil	Oleacein; oleocanthal; hydroxytyrosol; tyrosol; ligstroside aglycon; oleuropein aglycon; 1-acetoxypinoresinol; luteolin; apigenin; vanillic acid; 4-(acetoxyethyl)-1,2-dihydroxybenzene; p-coumaric acid; elenolic acid; 1-acetoxypinoresinol; ligstroside aglycon	Choline chloride/xylitol 2:1	Stirring + heating	HPLC-DAD (C18)	Amberlite	García et al. (2016)
		Choline chloride/1,2-propanediol 1:1		HPLC-ESI-MS (C18)	XAD-16	
Orange peel	Total polyphenols; ferulic acid; gallic acid; ρ-coumaric; caffeic acid; syringic acid; rutin; naringin; trans-cinnamic acid; thymol	Choline chloride/ ethylene glycol 1:4	Stirring + heating	Folin-Ciocalteu		Ozturk et al. (2018)
				RP-HPLC-UV (C18)		
Radix *salviae miltiorrhizae*	Rosmarinic acid; lithospermic acid, salvianolic acid A and B; tanshinone IIA	Choline chloride/1,2-propanediol 1:1	HRE	RP-HPLC (C18)		Chen et al. (2016)
			UAE			
			MAE			
Ficus carica leaves	Caffeoylmalic acid; psoralic acid-glucoside; rutin; psoralen; bergapten	Glycerol/ xylitol/D-(-)-fructose 3:3:3	UAE	HPLC-UV (C18)	Macroporous resin D-101 from 75.3% to 85.5%	Wang et al. (2017)
			MAE			

Lippia citriodora leaves	Iridoid glycosides; gardoside; ixoside; theveside; teucardoside; lippioside; durantoside I; phenylpropanoid glycosides; verbasoside; cistanoside F; β-hydroxy-verbascoside/β-hydroxy-isoverbascoside; oxoverbascoside; campneoside; verbascoside; isoverbascoside; forsythoside A; eukovoside; martynoside; flavonoid glycosides; luteolin-7-diglucuronide; apigenin-7-O-diglucuronide; chrysoeriol-7-diglucuronide; acacetin-7-diglucuronide; dimethyl quercetin	Choline chloride/lactic acid 1:2	MAE	Folin-Ciocalteu		Ivanović et al. (2018)
				HPLC-DAD-ESI-TOF-MS (C18)		
Flos Sophorae Immaturus	Rutin; nicotiflorin; narcissin; quercetin; kaempferol; isorhamnetin	Choline chloride/1,4-butanediol 1:2	HRE	RP-HPLC-UV	Macroporous resin AB-8 from 75.5% to 84.1%	Wang et al. (2019)
			UAE			
			MAE			
Olive kernel	Total polyphenols	Choline chloride/lactic acid 1:2	MAE	Folin-Ciocalteu		Chanioti and Tzia (2018)
Olive leaves						
Olive pomace	Oleuropein; hydroxytyrosol; caffeic acid; vanillin; rutin; luteolin	Choline chloride/citric acid 1:2	UAE	Folin-Ciocalteu		Chanioti and Tzia (2018)
		Choline chloride/lactic acid 1:2	MAE	HPLC-DAD (C18)		
			High hydrostatic pressure-assisted extraction (HHPAE)			
			Homogenate-assisted extraction (HAE)			

(continued)

Table 7.3 (continued)

Sources	Target compounds	Deep eutectic solvent	Extraction process	Quantification method	Compounds' recovery from deep eutectic solvent	References
Camellia sinensis leaves (green tea)	Epicatechin; epigallocatechin; epicatechin-3-gallate; epigallocatechin-3-gallate	Betaine/glycerol/D-(+)-glucose 4:20:1	Stirring	LC-UV (C18)		Jeong et al. (2017)
			Heating			
			Heating + stirring			
			UAE			
Coridothymus capitatus	Total polyphenols	Lactic acid/nicotinamide 7:1	UAE	Folin-Ciocalteu Spectrophotometric method with reagent		Georgantzi et al. (2017)
Origanum vulgare hirtum		Lactic acid/L-alanine 7:1				
Salvia fruticosa (triloba)						
Salvia officinalis						
Thymus vulgar						
Grape skin	(+)-Catechin; delphinidin-3-O-monoglucoside; cyanidin-3-O-monoglucoside; petunicin-3-O-monoglucoside; peonidin-3-O-monoglucoside; malvidin-3-O-monoglucoside; malvidin-3-O-acetylmonoglucosides; peonidin-3-(6-O-p-coumaroyl) monoglucosides; malvidin-3-(6-O-p-coumaroyl)monoglucosides; quercetin-3-O-glucoside	Choline chloride/malic acid 1:1	UAE	Folin-Ciocalteu		Radošević et al. (2016)
				Bisulfite bleaching procedure		
				HPLC-DAD (C18)		

Aegle marmelos	Apigenin; ascorbic acid; gallic acid; kaempferol; p-coumaric acid; protocatechuic acid; chlorogenic acid; caffeic acid; epicatechin; syringic acid; vanillic acid; ferulic acid; sinapinic acid; trans-cinnamic acid; (+)-catechin; rosmarinic acid; rutin; quercetin	Choline chloride/oxalic acid 1:1	UAE	Folin-Ciocalteu	Amberlite XAD-7 resin	Saha et al. (2019)
				pH differential method		
				$AlCl_3$ method		
				HPLC-DAD (C18)		
Morus alba leaves	Gallic acid; gentisic acid; chlorogenic acid; catechinic acid; vanillic acid; caffeic acid; syringic acid; epicatechin; rutin; hyperin; benzoic acid; astragalin; quercetin	Choline chloride/citric acid 2:1	UAE	HPLC-UV (C18)	Antisolvent from 60.7% to 80.8%	Zhou et al. (2018)
Origanum dictamnus	Total polyphenols; total flavonoids	Lactic acid/glycine 3:1	UAE	Folin-Ciocalteu		Bakirtzi et al. (2016)
Foeniculum vulgare				$AlCl_3$ method		
Origanum majorana						
Salvia officinalis						
Mentha spicata						
Tomato	Quercetin	Tetrabutylammonium chloride/decanoic acid 1:3	Ultrasound-assisted emulsification microextraction (USAEME)	UV-VIS spectrophotometry at 370 nm		Kanberoglu et al. (2019)
Onion						
Grape						
Pomegranate peels	Caffeic acid; kaempferol; luteolin; protocatechuic acid; ellagic acid; chlorogenic acid; hydroxybenzoic acid; gallic acid; quercetin	Glycerol/glycine 3:1	Heating	Folin-Ciocalteu		Rajha et al. (2019)
		Malic acid/glucose/glycerol 1:1:1	UAE	Spectrophotometric method with reagent		
			Ired-Irrad®	HPLC-DAD (C18)		

(continued)

Table 7.3 (continued)

Sources	Target compounds	Deep eutectic solvent	Extraction process	Quantification method	Compounds' recovery from deep eutectic solvent	References
Lycium barbarum fruits	Myricetin; morin; rutin; luteolin; hyperoside; quercitrin; apigenin	Choline chloride/*p*-toluene sulfonic acid 1:2	Heating + stirring	HPLC-UV (C18-WR)	Antisolvent	Ali et al. (2019)
			UAE			
Wine lees	Total anthocyanins; anthocyanin-3-*O*-monoglucosides; delphinidin-3-*O*-glucoside	Choline chloride/malic acid 1:1	UAE	HPLC-DAD (C18)		Bosiljkov et al. (2017)
	Petunidin-3-*O*-glucoside; malvidin-3-*O*-glucoside; anthocyanin-3-(6-*O*-*p*-coumaroyl)monoglucosides; peonidin-3-(6-*O*-*p*-coumaroyl) monoglucoside; malvidin-3-(6-*O*-*p*-coumaroyl)monoglucoside			Bisulfite bleaching procedure		
Nobis tangerine peel	Narirutin; hesperidin; sinensetin; nobiletin	Choline chloride/ levulinic acid/N-methyl urea 1:1.2:0.8	UAE	UHPLC-UV (C18)	Antisolvent	Xu et al. (2019)
	Tangeretin; kaempferol-3-O-rutinoside; isosinensetin; 3,5,6,7,8,3′,4′-pentamethoxyflavone; 5-hydroxy-6,7,8,3′,4′-pentamethoxyflavone			UHPLC-Q-Orbitrap-MS/MS		
Mulberry	Cyanidin-3-O-glucoside	Choline chloride/citric acid/glucose 1:1:1	High-speed homogenization and cavitation-burst extraction (HSH-CBE) CBE	UV-VIS spectrophotometry RP-HPLC-UV (C18W)		Guo et al. (2019)

Grapefruit peels	Naringin	Lactic acid/glucose 5:1	Stirring + heating	Folin-Ciocalteu HPLC-DAD (C18)		El Kantar et al. (2019)
			High-voltage electrical discharges (HVED)			
Ginkgo biloba leaves	Quercetin; myricetin	Choline chloride/oxalic acid/ethylene glycol 1:1:3	Immersion stirring extraction (ISE)	HPLC-UV (C18)	Antisolvent	Tang et al. (2017)
			HRE			
			UAE			
Tartary buckwheat hull	Rutin	Choline chloride/ glycerol 1:1	UAE	RP-HPLC-UV (XBD-C18)	Antisolvent 95.1%	Huang et al. (2017)
Sophora japonica bud	Rutin	Choline chloride/ triethylene glycol 1:4	Stirring + heating	RP-HPLC-UV (SB-C18)	Macroporous resin AB-8	Zhao et al. (2015), Peng et al. (2018)
		Choline chloride/ levulinic acid 1:2				
Ruta graveolens leaves	Rutin	Choline chloride/citric acid 1:1	Stirring + heating	RP-HPLC-UV (C18)		Molnar et al. (2018)
Hippophae rhamnoides leaves	Rutin; quercetin-3-O-glucoside; quercetin; kaempferol; isorhamnetin	Choline chloride/1,4-butanediol 1:3	HRE	HPLC-UV (C18)	Macroporous resin AB-8 from 72.36% to 84.99%	Cui et al. (2018)
			UAE			
			MAE			
Pollen Typhae	Quercetin; naringenin; kaempferol; isorhamnetin	Choline chloride/1,2-propanediol 1:4	UAE	HPLC-DAD (C18)		Meng et al. (2018)
Cyclocarya paliurus (Batal.) Iljinskaja leaves	Kaempferol-7-O-α-L-rhamnoside; kaempferol; quercetin; quercetin-3-O-β-D-glucuronide; kaempferol-3-O-β-D-glucuronide	Choline chloride/1,4-butanediol 1:5	UAE	LC-MS	Macroporous resin SP-270	Shang et al. (2018)

(continued)

Table 7.3 (continued)

Sources	Target compounds	Deep eutectic solvent	Extraction process	Quantification method	Compounds' recovery from deep eutectic solvent	References
Camellia oleifera flowers	Quercetin; kaempferol; quercetin-3-O-rhamnoside; kaempferol-3-O-rhamnoside	Choline chloride/lactic acid 1:2	UAE	HPLC-DAD (C18)		Ma et al. (2018)
Moringa oleifera leaves	Vicenin-2 (apigenin C-diglycoside); vitexin (apigenin hexose); isovitexin (apigenin C-hexose); quercetin dihexose; quercetin acetyl dihexose; quercetin rutinoside; quercetin hexose; quercetin malonyl hexose; quercetin acetyl hexose isomers I and II; quercetin-hydroxymethylglutaroyl hexose I and II; kaempferol acetyl dihexose; kaempferol rutinoside; kaempferol hexose; kaempferol malonyl hexose isomer; kaempferol-hydroxymethylglutaroyl hexose; kaempferol acetyl hexose; isorhamnetin rutinoside; isorhamnetin hexose; isorhamnetin acetyl hexose; isorhamnetin-hydroxymethylglutaroyl hexose; glucomoringin; acetyl 4-(α-L-rhamno-pyranosyloxy)-benzyl glucosinolate; caffeoylquinic acid; 3-O-caffeoylquinic acid; 3-p-coumaroylquinic acid; 3-p-coumaroylquinic acid; feruloylquinic acid; 4-feruloylquinic acid	Choline chloride/citric acid 1:1	UAE	UHPLC-Q-TOF-MS (C8)		Hamany Djande et al. (2018)

Soy products	Daidzin; genistin; genistein; daidzein	Choline chloride/citric acid 1:1	UAE	UHPLC-UV (EC-C18)		Bajkacz and Adamek (2017)
Dalbergia odorifera leaves	Prunetin; tectorigenin; genistein; biochanin A	Choline chloride/ ethylene glycol 1:2	NPCE	RP-HPLC	Macroporous resin AB-8 from 85.6% to 57.9%	Li et al. (2016)
			UAE			
Lycium ruthenicum fruit	Cyanidin-3-O-glucoside; petunidin-3-O-(trans-p-coumaroyl)-rutinoside-5-O-glucoside; petunidin-3-O-(cis-p-coumaroyl)-rutinoside-5-O-glucoside; petunidin-3-O-(caffeoyl)-rutinoside-5-O-glucoside; delphinidin-3-O-(p-coumaroyl)-rutinoside-5-O-glucoside; petunidin-3-O-(cis-p-coumaroyl)-rutinoside-5-O-glucoside; petunidin-3-O-(trans-p-coumaroyl)-rutinoside-5-O-glucoside; malvidin-3-O-(p-coumaroyl)-rutinoside-5-O-glucoside; petunidin-3-O-(p-coumaroyl)-rutinoside	Choline chloride/1,2-propanediol 1:2	UAE	HPLC-DAD-ESI-MS/MS (C18)	Macroporous resin LX-32	Sang et al. (2018)
Juglans regia leaves	Neochlorogenic acid; quercetin-3-O-glucoside; quercetin-O-pentoside	Choline chloride/butyric acid 1:2	Heating + stirring	HPLC-DAD (ODS-2C18)		Vieira et al. (2018)
		Choline chloride/ phenylpropionic acid 1:2				
Mentha piperita leaves	Total polyphenols; eucalyptol; menthone; isomenthone; menthol; neomenthol; pulegone; menthyl acetate	Choline chloride/D-(+)-glucose 5:2	Stirring	Folin-Ciocalteu		Jeong et al. (2018)
			Heating	$AlCl_3$ method (HS-SPME)-GC-MS UPLC-Q-TOF-MS (C18)		
			Heating + stirring			
			UAE			

(continued)

Table 7.3 (continued)

Sources	Target compounds	Deep eutectic solvent	Extraction process	Quantification method	Compounds' recovery from deep eutectic solvent	References
Olive oil Almond oil Sesame oil Cinnamon oil	Ferulic acid; caffeic acid; cinnamic acid	Choline chloride/ ethylene glycol 1:2	Ultrasonic-assisted liquid-liquid microextraction method (UALLME) + UAE	RP-HPLC-UV (C8)		Khezeli et al. (2016)
Carthamus tinctorius L.	Hydroxysafflor yellow; cartormin; carthamin; N^1,N^{10},N^5-(Z)-tri-p-coumaroylspermidine; N^1-(E)-N^5,$N1^0$-(Z)-tri-p-coumaroylspermidine; N^1,$N1^0$-(E)-N^5-(Z)-tri-p-coumaroylspermidine; N^1,N^{10},N^5-(E)-tri-p-coumaroylspermidine	Proline/malic acid 1:1	Heating + stirring	HPLC-DAD (C18) NMR ESI-TOF-MS	Macroporous resin HP-20 from 84% to 92%	Dai et al. (2013c)
Zingiber officinale Roscoe rhizome	6-Gingerol; 8-shogaol; (5R)-1,7-bis-(4'-hydroxy-3'-methoxyphenyl)-5-hydroxyheptan-3-one; (3R,5S)-3,5-dihydroxy-1,7-bis-(4'-hydroxy-3'-methoxyphenyl)-heptane; (3S,5R)-gingerdiols; (3R,5S)-gingerdiols	Proline/lactic acid 1:1	Stirring HRE UAE	UV spectroscopy HPLC-UV NMR FTIR	Macroporous resin HP-20	Rajan et al. (2015)

Lemon peels	Total polyphenols	Glycerol/choline chloride 3:1	UAE	Folin-Ciocalteu $AlCl_3$ method		Mouratoglou et al. (2016)
Olive leaves						
Onion solid wastes						
Red grape pomace						
Spent filter coffee						
Wheat bran						
Extra virgin olive oil	Total polyphenols; hydroxybenzoic acid; protocatechuic acid; vanillic acid; tyrosol; p-coumaric acid; caffeic acid; apigenin; pinoresinol	Lactic acid/glucose 6:1	Vortexing	UV spectroscopy		Paradiso et al. (2016a, b)
				Folin-Ciocalteu		
Trachycarpus fortunei bark	Protocatechuic acid; catechins; epicatechin; caffeic acid	Choline chloride/formic acid 1:1	HRE	HPLC-UV-MS (C18)	SPE C_{18}/C_8	Fu et al. (2017c)
Ginkgo biloba leaves	Total procyanidins; total flavonoids	Methyltrioctylammonium chloride/capryl alcohol/octylic acid 1:2:3 with choline chloride/levulinic acid 1:2 or choline chloride/malonic acid 1:2	Two-phase system with: Stirring	DMAC method		Cao et al. (2018b)
			UAE	Rutin method (spectrophotometric method with reagent)		
Olive oil	Ferulic acid; umbelliferone; boropinic acid; 7-isopentenyloxycoumarin; 4′-geranyloxyferulic acid; auraptene	Phenylacetic acid/betaine 2:1	Dispersive liquid-liquid microextraction (DLLME)	UHPLC-PDA (C18)		Ferrone et al. (2018)
Soy oil						
Peanut oil						
Corn oil						
Sunflower oil						

(continued)

Table 7.3 (continued)

Sources	Target compounds	Deep eutectic solvent	Extraction process	Quantification method	Compounds' recovery from deep eutectic solvent	References
Byrsonima intermedia leaves	Proanthocyanidin dimer; galloylproanthocyanidin dimer; digalloyl quinic acid; quercetin-*O*-hexoside; galloyl quercetin hexoside; quercetin-*O*-pentoside; galloyl quercetin pentoside	Choline chloride/glycerol 1:1, 1:2, or 2:1	UAE	HPLC-DAD (C18)		Fraige et al. (2019)
				HPLC-MS/MS (C18)		
Grape pomace	Gallic acid; hydroxytyrosol; tyrosol	Choline chloride/citric acid 2:1	UAE + MAE	HPLC-DAD (C18)		Panić et al. (2019b)
Olive pomace	Vanillic acid; vanillin; catechin; pinoresinol; quercetin-3-*O*-glucoside; delphinidin-3-*O*-monoglucoside; petunidin-3-*O*-monoglucoside; peonidin-3-*O*-monoglucoside					
	Malvidin-3-*O*-monoglucoside; peonidin-3-acetylmonoglucosides; malvidin-3-acetylmonoglucosides; cyanidin-3-(6-*O*-p-coumaroyl) monoglucosides; peonidin-3-(6-*O*-p-coumaroyl)monoglucosides; malvidin-3-(6-*O*-p-coumaroyl)monoglucosides					
Scutellaria baicalensis Georgi	Baicalin	Choline chloride/lactic acid 1:1	Ultrahigh pressure extraction (UPE)	HPLC-UV (XBP-C18)		Wang et al. (2018b)

Buckwheat sprouts	Orientin; isoorientin; vitexin; isovitexin; quercetin-3-O-robinobioside; rutin	Choline chloride/ triethylene glycol 1:4	UAE	Spectrophotometric method with reagent	SPE C_{18} from 97.51% to 99.34%	Mansur et al. (2019)
				RP-HPLC-PAD (C18)		
				Q-TOF-MS		
Grape pomace	Total anthocyanins; peonidin-3-O-monoglucoside; petunidin-3-O-monoglucoside; delphinidin-3-O-monoglucoside; malvidin-3-O-monoglucoside	Choline chloride/citric acid 2:1	MAE	HPLC-DAD (C18)	Macroporous resin SP207	Panić et al. (2019a)
			UAE			
			MAE + UAE		70.36%	
Fucus vesiculosus	Total polyphenols	Choline chloride/lactic acid 1:1, 1:2, or 1:3	Stirring + heating	Folin-Ciocalteu		Obluchinskaya et al. (2019)
Ascophyllum nodosum						
Flos Trollii	2''-O-Galactopyranosylorientin; orientin; vitexin	Choline chloride/zinc bromide 1:1	UAE	HPLC-DAD (C18)	Macroporous resin D101 from 75.4% to 80.5%	Duan et al. (2019)
Chlorella vulgaris	Total polyphenols; gallic acid; p-coumaric acid; ferulic acid; caffeic acid	Choline chloride/ ethylene glycol 1:4	Stirring + heating	Folin-Ciocalteu method		Wan Mahmood et al. (2019)
		Choline chloride/1,4-butanediol 1:4		HPLC-DAD		
Scutellaria baicalensis Georgi	Baicalin	Decanoic acid/ tetrabutylammonium chloride 1:2	MAE	HPLC-UV (XBP-C18)		Wang et al. (2018a)
			UAE			
			HRE			

Plant/Solvent Ratio and Water Content in Deep Eutectic Solvents

The plant material-to-solvent ratio is classically set between 2.5 and 6% in conventional extraction. The high solubility of polyphenols in deep eutectic solvent yet suggests the possibility of using higher ratio than in traditional solvents (Dai et al. 2016). However, the largest used ratio is *ca.* 10%, which is not significantly more. Cao and coworkers used a RSM study (response surface methodology; see below) for the optimization of the plant/solvent ratio on the extraction of proanthocyanidins from the leaves of *Ginkgo biloba* (Cao et al. 2018a). It is shown that beyond 10%, the rate of extraction strongly decreases. Tang and coworkers managed to obtain a solid/solvent ratio of 80% in the case of *Chamaecyparis obtusa* for flavonoid extraction (Tang et al. 2015). To our knowledge, to date, it is the only reference with such a large ratio. The high viscosity of deep eutectic solvent is usually invoked for explaining modest ratio. Indeed, a sufficient amount of solvent is necessary to fully impregnate the matrix and successfully perform the extraction. It could be noted that high plant/solvent ratio (but still 3 to 6% at the maximum) is often used with deep eutectic solvent containing rather large proportions of water (around 25% w/w).

The use of water in eutectic mixtures is reported in almost all works. The dilution of deep eutectic solvent with water allows a significant lowering of their viscosity. However, it must be emphasized that it strongly changes the structure of deep eutectic solvent mixtures, altering their composition from a binary mixture to a ternary one. Significant changes have been observed in both the nature of extracted molecules and the extraction rates according to the percentage of water (Ruesgas-Ramón et al. 2017; Zainal-Abidin et al. 2017). A very recent study from Panic et al. focuses on the impact of plant/solvent ratio and water dilution on the extraction of anthocyanins from grape pomace (Panić et al. 2019b). Several technologies combining with deep eutectic solvent were evaluated. They finally demonstrated that the maximum ratio was 3%. Likewise, they showed that the dilution of deep eutectic solvent in water had a significant impact on this ratio. In addition, the type of studied solids strongly influenced this ratio value. The ideal conditions emerging from this study are a plant/solvent ratio of 3% and an amount of water fixed at 25% (w/w). These optimal conditions are valid regardless of the techniques applied (microwave, ultrasound, or both) (Panić et al. 2019b). Surprisingly, the use of water is only considered through a pragmatic approach by comparing the extraction rates. However, a key issue is the water content limit able to preserve the properties of the deep eutectic solvent mixtures. This limit is obviously dependent on the type of solvents. However, most authors seem to agree on a maximum limit of 30% w/w of water (Gutiérrez et al. 2009; Huang et al. 2017; Liu et al. 2018; Benvenutti et al. 2019). Thus, Cui and coworkers carried out the extraction of genistin, genistein, and apigenin from pigeon pea roots by employing a deep eutectic microwave-assisted extraction (deep eutectic solvent-MAE) procedure (Cui et al. 2015). Optimal conditions for deep eutectic solvent-MAE were established through the single factor and the Box-Behnken design tests (BBD). They consisted of 30% water in 1,6-hexanediol/ChCl (7:1) as extraction solvent, a microwave power of 600 W, and a solid/solvent ratio of about 6% (Cui et al. 2015). Since water is known to play a key role in the molecular structure of the deep eutectic solvent, and therefore on extraction performances (Passos et al. 2016), fundamental studies of the effect of water in the extraction process should be taken further.

Extraction Duration

The extraction durations of polyphenols in deep eutectic solvent media vary from a few minutes (Li et al. 2015; Wei et al. 2015a; Khezeli et al. 2016; Wang et al. 2018b) to several hours (Fu et al. 2017c). The optimal extraction time will depend on various parameters according to the type of polyphenol, the type of matrix, and all the abovementioned parameters. With liquid matrices like oils, extraction times at room temperature are, for example, much shorter (Paradiso et al. 2016a, b) than in the case of grape skins (Jeong et al. 2015). Thus, Paradiso et al. carried out an effective extraction of polyphenols from olive oil at room temperature in less than 15 minutes (Paradiso et al. 2016a, b). However, it is important to note that this extraction is a liquid/liquid extraction, hence probably requiring less contact time. Maceration is generally much longer. For example, Cvjetko Bubalo and coworkers, in order to compare assisted extraction under microwave irradiation with more traditional conditions, carried out 12 h-tests on the extraction of phenolic compounds from the grape skin (Cvjetko Bubalo et al. 2016). In the case of particularly resistant matrices (typically grape skin), the extraction time of 1 h in maceration at 65 °C is generally found (Radošević et al. 2016).

Importantly, this parameter is also very dependent on the technologies employed. For example, the use of microwave irradiation drastically reduces duration, allowing effective extractions in less than 20 min. Chen et al. have performed parameter optimization on *Radix Salviae miltiorrhizae* by combining RSM and BBD (Box-Behnken design) methodologies. A maximum extraction rate was determined after 11 min under microwave irradiation at 70 °C (800 W max) (Chen et al. 2016). However, extractions without activation, at room temperature, are very rarely performed in less than 60 min (Dai et al. 2013c; Fu et al. 2017c; Xu et al. 2019). A direct correlation between temperature and extraction time has been established by Bajkacz et al. for the extraction of isoflavones from soy products (Bajkacz and Adamek 2017). In this study, the central composite design (CCD) was employed to find the optimal values between several variables, notably temperature and extraction time. Not surprisingly, the higher the temperature, the shorter is the extraction time. By using ultrasound, it is however possible to reduce the extraction time even at room temperature (25 °C). Meng and colleagues have shown that the optimal extraction time for 4 flavonoids from Pollen *Typhae* was only 35 min (Meng et al. 2018).

Extraction duration influences both the efficiency and selectivity of extraction. Depending on the affinity of the metabolites for a given solvent but also depending on their location in the plant, they are more or less rapidly extracted. Thus, if a longer extraction time may improve the extraction efficiency, it could also increase the range of extracted metabolites. It is therefore essential to find the right balance between efficiency and selectivity. However, in the case of compounds which are not very stable like most of polyphenols (Larrauri et al. 1997; Manach et al. 2004; Quideau et al. 2011; Volf et al. 2014; Deng et al. 2018), long extraction or maceration duration may lead to the degradation of compounds.

Temperature

Temperature is an important parameter in the extraction processes by both increasing the solubility of the polyphenols to be extracted and decreasing the viscosity of deep eutectic solvents. Choice of temperature is mainly ruled by thermal stability of polyphenols that is generally well below that of deep eutectic solvents (generally stable up to 120 °C). The applied temperatures usually range from 20 to 100 °C (Ali et al. 2019; Yılmaz et al. 2019; Bakirtzi et al. 2016; Fernández et al. 2018b; Panić et al. 2019b). Our meta-analysis of more than 100 publications shows that more than 70% of the extractions are carried out with temperatures between 45 °C and 55 °C. These temperatures are in the range of the thermal stability of polyphenols (Yan et al. 2010; Volf et al. 2014). They can be regarded as optimal classical conditions.

Extraction conditions at room temperature or even at 5 °C have been also reported (Nam et al. 2015; Qi et al. 2015; Jeong et al. 2017; Hamany Djande et al. 2018; Cao et al. 2018a; Meng et al. 2018; Ali et al. 2019). While such conditions prevent the thermal decomposition of polyphenols, they could drastically reduce the extraction performances, notably its kinetics. However, in some cases according to the polyphenols and/or deep eutectic solvent structure, temperature has been increased beyond the generally applicable temperatures in conventional solvents, i.e., up to 90 °C (Zhang et al. 2014; Yao et al. 2015; Cui et al. 2015; Vieira et al. 2018). For instance, a microwave-assisted extraction of baicalin from *Scutellaria baicalensis* Georgi was optimized (RMS method), and the authors have shown that good performance is obtained providing that the temperature reaches 85 °C (Wang et al. 2018a). The possibility to use such a high temperature was explained by the potential stabilization of polyphenols by deep eutectic solvent (Dai et al. 2016). Actually, the challenge is therefore to find the best compromise between the amount of extracted polyphenols and the amount of thermally degraded polyphenols.

7.4.3 Impact of Extraction Technologies

All the parameters above could be tuned for optimizing the extraction performance of a process. Nevertheless, apart from the choice of solvent, the selection of extracting technologies is of primary importance. Among all the possible extraction technologies, 6 extraction technologies have been mainly used in deep eutectic solvents (Fig. 7.7), and they mostly correspond to extraction methods that were developed in ionic liquids.

The selection of the technology has a major impact on the extraction rates observed (Table 7.3). Maceration is one of the methods commonly used in extraction. It is generally a method compatible with the majority of solvents. The maceration is also often named in publications as "heating", "heating reflux", or even "stirring heating". This traditional method is still widely employed in many publications with deep eutectic solvents. This is not for reasons of performance but more likely to study the potential benefit of deep eutectic solvent on the extraction compared to traditional organic solvents (Duan et al. 2016; Cvjetko Bubalo et al. 2016; Jancheva et al. 2017; Cui et al. 2018; Chanioti and Tzia 2018; Cunha and Fernandes 2018). In other words, maceration technique can be regarded as a basic reference technology. Thus, Cui and coworkers have carried out a detailed study comparing both the influence of deep eutectic solvents and the technologies used (Cui et al. 2015). Maceration was used as a reference technology to highlight the effect of microwaves and ultrasound activations on extraction rates. The study involved 11 different deep eutectic solvents, but mainly based on choline chloride (9 over 11). According to single factor experiments and a Box-Behnken design test, the extraction parameters have been optimized. Optimal performance was obtained with a temperature of 80 °C, an extraction time of 11 min, using 1,6-hexanediol/ChCl solvent with a ratio of 7:1 with a water content of 30%, and a solvent-to-solid ratio of 14 mL/g. The highest extraction yields of genistin, genistein, and apigenin were obtained under microwave irradiation (0.4688, 0.6410, and 0.2664 mg/g, respectively), while the extraction yields from maceration were a little bit lower (0.4085, 0.5531, and 0.2442 mg/g, respectively) (Cui et al. 2015).

The ultrasound-assisted extraction (UAE) is the most employed method (Khezeli et al. 2016; Ma et al. 2018; Marcus 2019) (Fig. 7.7). UAE is often used because it is a simple, inexpensive, and efficient alternative to conventional extraction

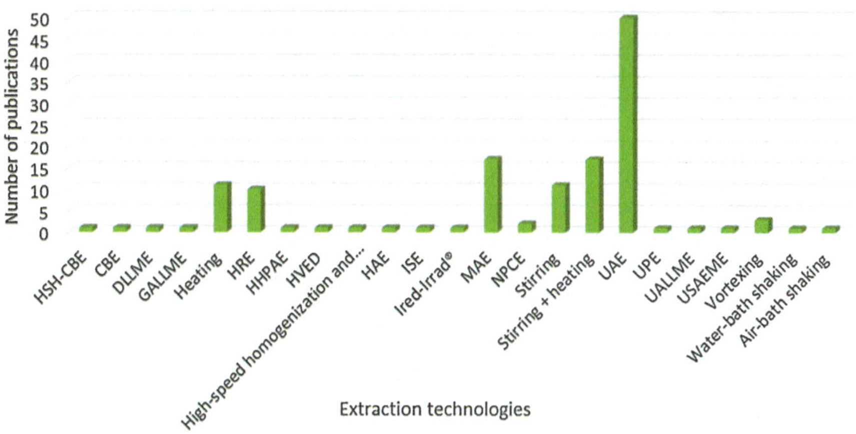

Fig. 7.7 Number of publications reporting each extraction technology used for the extraction of polyphenols using deep eutectic solvent. Note that publications have mainly focused on six extraction methods, and they correspond to technologies also evaluated in ionic liquids

techniques. It has been proven that ultra-sonication has no effect on natural deep eutectic solvent structure and may be an energy-effective combination (Singh et al. 2013). UAE in solid/liquid extraction generally exhibits faster kinetics and very good extraction yield. This technique seems particularly suitable for the extraction of polyphenol in deep eutectic solvents. Bakirtzi et al. studied the efficiency of ultrasound-assisted extraction of antioxidant polyphenols from common native Greek medicinal plants (Bakirtzi et al. 2016). Deionized water and 60% (v/v) aqueous ethanol were chosen as control solvents to compare with selected deep eutectic solvents. The amount of total polyphenols obtained by ultrasound with deep eutectic solvents is on average twice as high as those obtained in water or in ethanol (Bakirtzi et al. 2016). Actually, the technique combines moderate heating that is compatible with the thermal stabilities of polyphenols with extraction optimization due to ultrasound effect. Ultrasounds generally promote diffusion of dissolved substances from the inner part of the plant matrix to the extractant medium and penetration of the latter into the plant matrix. This is due to a mechanical effect on the plant, thanks to cavitation effects (Tiwari 2015). On the whole, cavitation results in an increase of the polarity of the system, including extractants, analytes, and matrix. Unfortunately, the viscosity of deep eutectic solvent is systematically regarded as a major limitation in this technology because the diffusion of ultrasound appears less effective in viscous media (Goula et al. 2017; Huang et al. 2019a). To overcome this issue, addition of water or a slight increase of temperature is often applied.

Extractions under microwave irradiation have also been studied. Although several publications refer to the use of activation by microwave irradiation, the overall number of related studies remains quite low compared to what is generally found in the field of extraction of natural products. Indeed, the effects of microwaves on kinetics and extraction performance have been demonstrated for a long time (Du et al. 2007; Casazza et al. 2010; Li et al. 2011; Dahmoune et al. 2015). It should be noted that it is not the combination of microwaves and deep eutectic solvent that is underrepresented but the specific extraction of polyphenols. This was already the case for polyphenol extractions with traditional solvents because microwave activation is hardly compatible with the range of thermal stability of the thermosensitive polyphenol compounds. However, worth is to mention that deep eutectic solvent is fully compatible with microwave activation, as shown by the number of recent publications combining microwave and eutectic solvents (Li et al. 2015; Wang et al. 2017; Chanioti and Tzia 2018; Ivanović et al. 2018; Panić et al. 2019a) (Fig. 7.7).

A few publications compare microwave and ultrasound technologies and point to greater efficiency of microwave compared to ultrasound (Yao et al. 2015; Cui et al. 2015, 2018; Peng et al. 2016; Chen et al. 2016; Wang et al. 2018a, 2019; Panić et al. 2019a). Only one publication reports a more positive performance in favor of ultrasounds for anthocyanin extraction (Cvjetko Bubalo et al. 2016). Generally, the microwave irradiation leads to an increase of temperature, which in turn allows an increase of the extraction rate, except in the case of anthocyanins. These latter are indeed much more thermosensitive, and a microwave irradiation typically induces temperature above 65 °C, sufficient for degrading these compounds.

A recent work reported by Panic and coworkers has shown that the extraction rate is doubled using a concomitant combination of ultrasound and microwave irradiations compared to the same experiments conducted by using the technologies separately (Fig. 7.8) (Panić et al. 2019a).

The combination of the two technologies takes advantage of both activation methods. Microwaves allow a decrease of the viscosity, thanks to an increase of temperature. This lower viscosity allows a better propagation of ultrasounds within the eutectic mixture. Ultrasound increases the penetration of the solvent into the core of the matrix and maximizes the solubility of polyphenols in the solvent at a lower temperature than that involved under microwave activation only.

Whatever the extraction technologies, viscosity is one of the most limiting factors associated to the use of eutectic mixtures. If the eutectic solvents based on choline chloride are the most frequently mentioned, it is undoubtedly because their viscosity at ambient temperature is among the lowest observed in eutectic mixtures. As explained above, viscosity is an important parameter for optimizing the contact between plant matrix and solvent. Moreover, the plant particles may be trapped in these viscous solvents and separation could be demanding if the amount of plant is too important. Filtration and centrifugation could be difficult or even impossible to achieve when the viscosity becomes too important. Filtration at higher temperatures and hot centrifugation are sometimes carried out, but the thermosensitivity of the

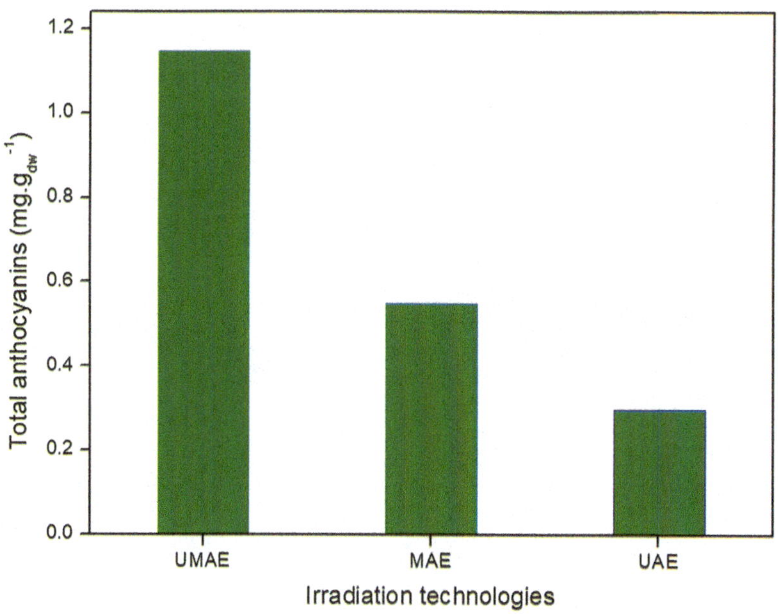

Fig. 7.8 Total anthocyanins extracted as a function of technologies used, UMAE (ultrasound + microwave), MAE (microwave), and UAE (ultrasound). Total anthocyanins (TA in mg g_{dw}^{-1}) were expressed as a sum of identified anthocyanins. Content of anthocyanins was expressed as the means (n = 3) ± S.D. (Adapted from (Panić et al. 2019a) copyright Elsevier 2019)

polyphenols is an issue. While viscosity is sometimes a complex problem to solve in the laboratory, it may constitute a real barrier to the industrial development of extraction applications.

7.4.4 Response Surface Methodology

As mentioned above, the extraction efficiency could be maximized through optimization of multiple variables which may interact with each other. Response surface methodology (RSM) is now often employed to address the different extraction conditions by considering all the parameters together (Fig. 7.9). Bi and coworkers have

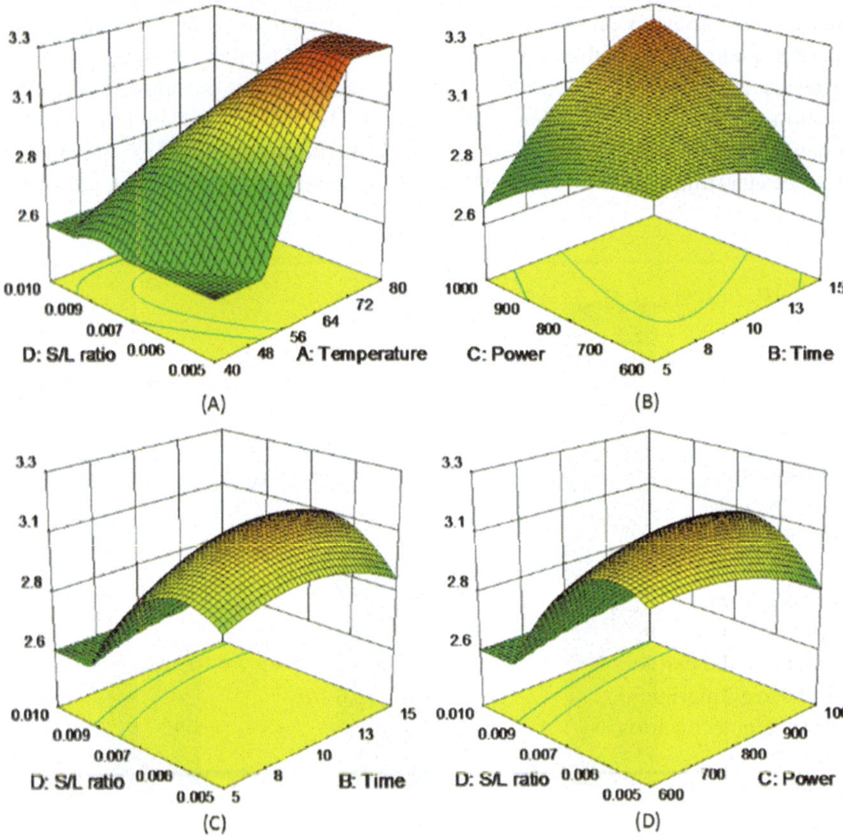

Fig. 7.9 Response surface methodology plot of the model for extraction of lithospermic acid in Radix Salviae miltiorrhizae from (Chen et al. 2016) copyright MDPI 2016. Plant/solvent ratio (A), (B), (C), power of irradiation (B), (D), temperature (A), and duration (B), (C) were examined together. Such plot allows a screening of all parameters and takes into account their possible interactions

performed a multivariate study using the RSM method (Bi et al. 2013). The correlation of several parameters makes it possible to highlight the effects of the solvents and/or the technologies used. The response surface methodology (RSM) is a practical tool for optimizing complex extraction procedures and can evaluate multiple parameters and their interactions (Yang et al. 2008). The RSM is always used in combination with the Box-Behnken design (BBD). Compared to the central composite design (CCD), the BBD avoids the axial points outside the experiment level and requires fewer runs (Marcet et al. 2018). In the case of studies with eutectic solvents, the number of parameters is even greater than in the case of conventional studies. Indeed, it is necessary to consider not only the type of solvent but also the stoichiometries of the components or the proportion of water involved. This makes the studies based on the RSM methods particularly relevant.

The difficulty in choosing the parameters of extraction using eutectic solvents is due to the range of parameters to be controlled (plant-to-solvent ratio, extraction temperature, extraction duration, eutectic mixture used, percentage of water, technology extraction, etc.), often addressed in an empirical manner. Studying parameters screening by using the response surface method allows a rationalization of the impact of the different factors on extraction rates or even on the type of metabolites extracted. This method allows a multifactorial study that highlights the synergies of the parameters studied.

7.4.5 Polyphenol Recovery

When the objectives are to obtain polyphenol metabolites free from solvents, especially deep eutectic solvent herein, recovery of polyphenols is necessary. Several methods for separating eutectic solvents from extracted polyphenols have already been described in the literature. Most of these methods are based on what had already been developed with ionic liquids. The elimination of the solvent was generally regarded as a brake for the development of ionic liquids and now eutectics in industrial extraction processes. Indeed, these solvents have a low vapor pressure that makes them safe but almost impossible to evaporate with traditional protocols. Elimination of eutectic solvents often leads to a significant decrease in the overall efficiency of the extraction protocol.

The use of microporous resin type AB-8 was found to be efficient for the separation of the eutectic solvent from the extracted polyphenols with a recovery rate of almost 72% (Cui et al. 2018). The major drawback of this method remains the large amount of solvent necessary to separate the compounds. Indeed, for 24 mL of extract, 270 mL of deionized water and 270 mL of 95% ethanol were necessary (Cui et al. 2018). Similarly, C18 type SPE columns were used to separate the eutectic solvents from the extracted flavonoids (Mansur et al. 2019). An excellent recovery rate of 97% was obtained. But here again, a large amount of solvents was required: for 0.5 mL of extract, 6 mL of acidified water and 8 mL of methanol were employed.

However, for certain applications (e.g., cosmetics or food supplements), it is possible to retain the solvent with the extracted polyphenols and use this mixture directly (Jeong et al. 2017; Choi and Verpoorte 2019). Thus, Jeong and coworkers noted that eutectic solvents based on betaine and glycerol did not give better extraction results than traditional organic solvents. It has been however proposed to use the extracts without separating the metabolites from the eutectic solvents. The eutectic solvent would be an integral part of the formulation. As a result, the choice of the components of the eutectics should be compatible with cosmetic regulations. Taken together, the results of this study demonstrate that the mixture composed of betaine, glycerol, and glucose can be applied on a large scale as a multifunctional and cheap medium, i.e., for efficient extraction and stable storage of catechins, being at the same time an active ingredient. For example, green tea extracts prepared in this way could be readily used in cosmetic products or in pharmaceutical formulations for skin (Jeong et al. 2017).

We can mention now that beyond the separation from the deep eutectic solvent extracting media, the goal of recovery is to obtain metabolites from complex matrices with different degrees of purity. In some cases, such as in the pharmaceutical field, it is necessary to obtain molecules of the highest purity. In other areas such as cosmetics or the food industry, it is sufficient to obtain a mixture of molecules.

7.5 Quantification, Identification and Antioxidant Activity of Polyphenols

7.5.1 Identification and Quantification of Polyphenols

An important point is the identification and quantification of the polyphenolics in the extracts. Different methods have been employed taking into account that solvents cannot be simply evaporated as in the case of traditional solvents. Figure 7.10 gathers all the different methods that have been employed. Two major approaches currently exist, direct characterization or characterization after removal of the eutectic solvent, notably by using liquid/liquid extraction or chromatographic separation systems such as solid phase extraction (SPE) columns.

The amount of total phenolic contents in extracts is generally quantified by colorimetric methods based on calibration curves derived from known amounts of purified standards such as gallic acid or catechin. The most widely employed assay is the Folin-Ciocalteu method (Fig. 7.11), which displays several advantages (Singleton et al. 1999). It is quite easy to handle and rapid and does not require heavy equipment. It covers all the range of phenolics including monophenolics, contrariwise to other colorimetric methods based on iron or aluminum salts that cannot react with monophenols (Singleton et al. 1999). For instance, the determination of total flavonoids is very often conducted by using $AlCl_3$ and expressed as rutin or (+)-catechin equivalent.

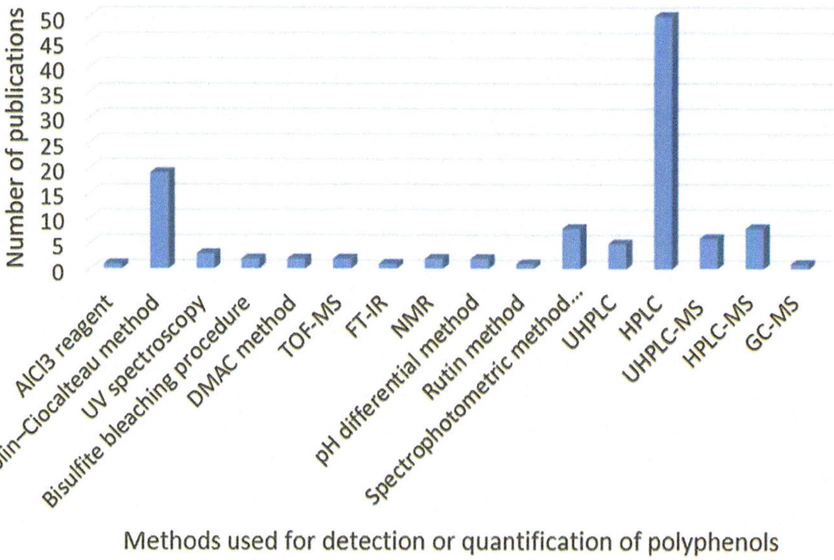

Fig. 7.10 Number of publications reporting the various methods used for the determination and quantification of polyphenolics in deep eutectic solvent extracts. HPLC and Folin-Ciocalteu colorimetric assay are the most frequently used, but the latter remains relative and probably more delicate to interpret compared to the former

Fig. 7.11 Folin-Ciocalteu test for a series of extracts (left) and a calibration using gallic acid at different concentrations (right)

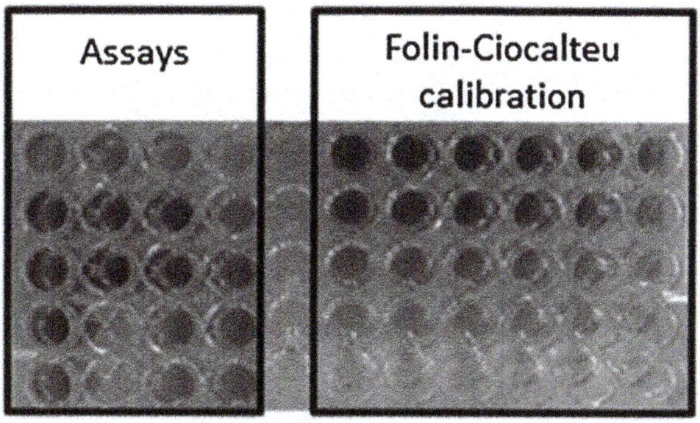

The Folin-Ciocalteu assay has been applied to extraction processes using natural deep eutectic solvent. Most of the works use this method to quantify and to assess the efficacy of polyphenol extraction in those media (Manousaki et al. 2016; Bakirtzi et al. 2016; Patsea et al. 2017; Georgantzi et al. 2017; Cao et al. 2018c; Skulcova et al. 2018; Athanasiadis et al. 2018a, b; Jeong et al. 2018; Yoo et al. 2018; Chanioti and Tzia 2018; Mocan et al. 2019; Pavić et al. 2019; Mamilla et al. 2019; Pal and Jadeja 2019; Obluchinskaya et al. 2019; Rajha et al. 2019; El Kantar et al. 2019). The Folin-Ciocalteu reagent consists in mixture of phosphomolybdic and phosphotungstic acids, which lead to the formation of blue oxides of tungsten (W_8O_{23}) and molybdenum (Mo_8O_{23}) in the presence of reductive species. The reaction occurs under alkaline conditions, with sodium carbonate. The natural deep eutectic solvent extracts are simply diluted in water and/or methanol (1–10%) and sometimes centrifuged prior to dilution. The resulting blue coloration is spectrophotometrically detected in the wavelength range 675–765 nm (Fig. 7.11), and the corresponding absorbance is directly proportional to the concentration of "phenolics" (Georgantzi et al. 2017; Chanioti and Tzia 2018; Wan Mahmood et al. 2019; Obluchinskaya et al. 2019).

The quantification remains relative because it is expressed as standard equivalents, mostly gallic acid, but could be expressed as catechin or even phloroglucinol equivalents (Obluchinskaya et al. 2019). Consequently, Folin-Ciocalteu assay could not be considered as a direct titration of polyphenols in the extracts, even if obtained results have been shown to correlate very often with quantification based on chromatographic methods in conventional media (Singleton et al. 1999). Secondly, Folin-Ciocalteu assay is not specific toward polyphenols. It also reacts with monophenolics (which are not true polyphenols as explained in Sect. 7.2), and many other reductive species may interfere in the assay (Singleton et al. 1999). The interference could be of several natures: inhibiting, additive, or enhancing (Singleton et al. 1999). Especially, the components of natural deep eutectic solvent such as sugars, citric acid, diols, etc., may be potent interferents, sometimes inducing a "positive false." However, only a very few works mentioned blank experiments that were carried out with "pure" natural deep eutectic solvent (Jeong et al. 2018; Chanioti and Tzia 2018; Panić et al. 2019b).

Total flavonoids' content in natural deep eutectic solvent extracts was also reported utilizing $AlCl_3$ spectrophotometric assay, although less often than the Folin-Ciocalteu test (Manousaki et al. 2016; Bakirtzi et al. 2016; Patsea et al. 2017, Georgantzi et al. 2017; Athanasiadis et al. 2018a, b; Jeong et al. 2018; Yoo et al. 2018). The colorimetric assays are very often coupled to chromatographic analyses, while some works are only based on these latter analyses.

Fortunately, (natural) deep eutectic solvent extracts are compatible with the most popular chromatographic separation techniques (Fernández et al. 2018b). After extraction, the deep eutectic solvent phases containing the target polyphenol analytes could be directly introduced into HPLC (Yang et al. 2016; Duan et al. 2016; Chanioti and Tzia 2018; Wan Mahmood et al. 2019; Panić et al. 2019b; Rajha et al. 2019; El Kantar et al. 2019) or LC systems (Bajkacz and Adamek 2017; Athanasiadis et al. 2018a, b, c). In some works, the recovery of polyphenols from (natural) deep eutectic solvent phases has been performed through SPE, SLE, or addition of antisolvent prior to chromatography (García et al. 2016; Fu et al. 2017c; Wang et al. 2017; Cao et al. 2018c; Mamilla et al. 2019; Pal and Jadeja 2019; Tian et al. 2019). Only the chromatographic analyses could be considered as true titration of polyphenols in extracts, allowing a clean identification along with quantification of specific polyphenols.

In works using combination of Folin-Ciocalteu and chromatography, some discrepancies could be noted, although rough trends remain acceptable to a certain point. For instance, the extraction of polyphenols in olive pomace was achieved in a series of deep eutectic solvents made from mixture of choline chloride and either citric acid, lactic acid, maltose, or glycerol by using different assisted extraction techniques (Chanioti and Tzia 2018). Total phenolic content obtained from Folin-Ciocalteu was compared to HPLC analyses. Generally, the two values were hardly correlated (Chanioti and Tzia 2018).

Paradiso et al. have also proposed a direct spectrophotometric analysis of deep eutectic solvent extracts of olive oil (Paradiso et al. 2016a). Deep eutectic solvent based on glucose and lactic acid was used to extract phenolic compounds in extra virgin olive oil. By simply measuring the absorption of the deep eutectic solvent extracts at few wavelengths, a screening of the total phenolic content of the oils could be performed, reducing significantly the use of hazardous solvents and reagents (Paradiso et al. 2016a).

Electrochemical techniques have been also used to quantify the amount of polyphenols, mainly flavonoids, in complex matrices such as wine (Makhotkina and Kilmartin 2010, 2012; Šeruga et al. 2011), tea or coffee (Kilmartin and Hsu 2003; Piljac-Žegarac et al. 2010), beer (Oliveira Neto et al. 2017a, b), or human urine sample (Adam et al. 2007). Following these works, detection of quercetin has been studied in highly diluted natural deep eutectic solvent (concentration of deep eutectic solvent lower than 10% in phosphate buffer) using screen-printed electrodes (Gomez et al. 2016). The authors showed that an increase of deep eutectic solvent concentration had a deleterious effect on the quercetin electrochemical signal. The method was then applied to onion extracts and compared with HPLC, showing good correlation (Gomez et al. 2016).

7.5.2 *Antioxidant Activity*

Along with the total phenolic content, antioxidant activity is also systematically examined in the deep eutectic solvent extracts using classical assays (Galili and Hovav 2014). The reducing power is generally determined by ferric reducing-antioxidant power assay (FRAP), based on the reduction of ferric tripyridyltriazine complex to the intensively blue ferrous tripyridyltriazine complex at low pH. The complex formation is monitored by visible electronic spectrum at *ca.* 620 nm. The ferric reducing-antioxidant power values in deep eutectic solvent extracts are expressed as μmol of ascorbic acid or trolox per g of dry matter (Bakirtzi et al. 2016; Georgantzi et al. 2017; Athanasiadis et al. 2018a, b, c; Jeong et al. 2018; Yoo et al. 2018; Pal and Jadeja 2019). Importantly, ferric reducing-antioxidant power (FRAP) assay cannot detect antioxidant activity of compounds acting by radical quenching. Radical scavenging activity is generally conducted with DPPH assay or alternatively by using ABTS or ORAC assays although less often employed with deep eutectic solvent extracts. DPPH (2,2-diphenyl-1-picrylhydrazyl hydrate) and $ABTS^{\bullet+}$ (2,2'-azino-bis(3-ethyl-benzothiazoline-6-sulfonic) acid) are stable radical molecules, which display intense violet and green-blue color, respectively. The DPPH and $ABTS^{\bullet+}$ assays are both based on the quenching of the colored radicals because of their reduction with the polyphenols. From the decrease of the absorbance, results are calibrated with ascorbic acid or trolox (μmol per g of dry matter). The ORAC, oxygen radical absorbance capacity, assay is based on inhibition of the activity of reactive species by an antioxidant, resulting in loss of emission of fluorescence of phycoerythrin or fluorescein (Galili and Hovav 2014). In contrast to the other assays, the ORAC method combines both inhibition time and degree of inhibition expressed as a single quantity. This assay measures the kinetics of the decrease in fluorescence for each sample compared to a blank by plotting fluorescence emission *versus* time. ORAC values are generally given as trolox equivalents. Widely used until 2012, this assay has been withdrawn as classical antioxidant assay because there is no physiological proof that ORAC values have any true biological significance (Schaich et al. 2015). Other biological studies requiring expensive kits or cell culture could be employed, as well as pulse radiolysis (Quideau et al. 2011) or electrochemical (René et al. 2010) techniques. But the majority of studies have reported antioxidant activity through assays described above due to their relative simplicity, notably DPPH and $ABTS^{•+}$.

As a result, they are similarly widely employed with deep eutectic solvent extraction. As an illustration, we could quote the extraction of microalgae polyphenols in polyol-based deep eutectic solvent (Wan Mahmood et al. 2019). The extracts obtained in conventional solvents (ethyl acetate and water) gave much lower antioxidant activity (DPPH assay) than those provided by deep eutectic solvent (Wan Mahmood et al. 2019). The deep eutectic solvent extracts presenting the best antioxidant activity were also characterized by the largest total phenolic contents as determined by Folin-Ciocalteu (Wan Mahmood et al. 2019). It is then tempting to directly correlate the high content of phenolic compounds to the high antioxidant activity.

However, it is interesting to consider another example: the extraction of phenolic compounds from onion peels in deep eutectic solvent based on choline chloride associated with urea and water (1:2:4), sucrose and water (4:1:8), and D-sorbitol and water (3:1:10). Ferric reducing-antioxidant power (FRAP) was almost five times higher for extract in choline chloride:urea:water than in conventional aqueous methanol or water extract, while the two other deep eutectic solvents gave twice the values. In contrast, onion peel extracts in aqueous methanol or in choline chloride:sorbitol:water were found to have a better antioxidant activity in terms of DPPH assay. The DPPH trend does not fully correlate with the total phenolic content (lowest value for extraction with choline chloride:sorbitol:water).

The nature of the antioxidant activity mechanism (electron transfer or radical scavenging) may be different according to the solvents used for extraction, which remained in the extracts, possibly explaining these different results.

In another example, Jeong et al. have evaluated the total phenolic and flavonoid contents of peppermint leaves (*Mentha piperita*) extracted notably with choline chloride:glucose (5:2) (Jeong et al. 2018). These values were obtained with the classical colorimetric assays. They found a high correlation ($r > 0.9$; $p < 0.05$) of the amount of phenolic and flavonoid compounds with the antioxidant activity, when using the ferric reducing-antioxidant power and ABTS assays, but not when using the DPPH assay (Jeong et al. 2018).

These few examples show that a direct comparison between studies is difficult, namely, if we consider that many combinations are possible for synthesizing deep eutectic solvent, and that their properties stem from this composition variation. Hence, caution is necessary when comparing the profitability between the extraction methods. It is reasonable to expect that enhanced extraction yields of polyphenols by utilizing (natural) deep eutectic solvent will increase the antioxidant activity of the extracts. Like in Folin-Ciocalteu assay, it is worth noting that deep eutectic solvent could directly affect the antioxidant activity when they are still present in the extract itself. Natural deep eutectic solvent alone has been reported to display antioxidant activity in the DPPH assay and may lead to biased results (Skulcova et al. 2018). As for Folin-Ciocalteu, best practice asks for prior evaluation of natural deep eutectic solvent alone for antioxidant activity assays.

Finally, hydrogen bond governs the formation of deep eutectic solvent, and polyphenols are prone to hydrogen bonding as well. A synergic effect of the deep eutectic solvent and polyphenols could be then envisioned. We may anticipate that this phenomenon could definitively affect antioxidant properties beyond the extraction yield. This effect could be expected to be positive as reported by Durand et al. who described that formulation of antioxidants (α-tocopherol, hydroxytyrosol, CR6, ascorbic acid) in natural deep eutectic solvent could greatly improve their activity (Durand et al. 2017). But we cannot neither exclude prooxidant activity since plant polyphenols bearing catechol and/or pyrogallol moieties have been demonstrated to exert prooxidant properties on specific occasions (Quideau et al. 2011).

7.6 Conclusions

Due to their specific structure, and in particular their ability to form hydrogen bonds, deep eutectic solvents are good solvents for the extraction of polyphenols. In addition, due to their low volatility, they can be advantageously combined with advanced extraction technologies like microwave- or ultrasound-assisted extraction, provided that the temperature remains controlled for avoiding degradation of the temperature-sensitive polyphenols.

Despite the large number of possible combinations for preparing deep eutectic solvents with tunable properties, research about deep eutectic solvents is still in its infancy, and only a few structures have been exploited for polyphenol extraction, mainly based on choline chloride. One of the main limitations of the use of deep eutectic solvents, other than choline chloride, is their viscosity at room temperature. However, a few deep eutectic solvents show low viscosity, and some could be specifically designed for lowering viscosity. Thus, plenty of deep eutectic solvents dedicated to efficient polyphenol extraction remain to be discovered.

A consequence of this huge potentiality is the need for **(i)** physicochemical characterization of these new media and **(ii)** rationalization of their impact along with a thorough cross-comparison of literature data. The issue is far from being easy, notably because experimental evidence of deep eutectic solvent formation is still challenging. Indeed, some mixtures are not necessarily deep eutectic solvent, i.e., presenting an anomalous fusion temperature lower than that resulting from the ideal mixture of the components. They also include a significant amount of water that is intrinsically part of the mixture. In the same context, we have also observed esterification reaction between components of deep eutectic solvents formed by the combination of acids and alcohols. Comparison of extraction efficiencies in each component of deep eutectic solvents beyond deep eutectic solvent itself should be equally important to achieve (providing that the component would be a liquid) or alternatively by using different mixture compositions. Such an evaluation has not been performed in any of publications in the field. It will be of great interest to highlight any added value of having a deep eutectic solvent rather than a simple solvent mixture. Another key point is related to the methodologies for quantifying polyphenols in extracts. While most of chromatography techniques are compatible with deep eutectic solvents, care has to be taken with colorimetric protocols.

Although a meta-analysis of published works remains difficult to achieve in order to assess the benefit of deep eutectic solvents for polyphenol extraction, several points are worth mentioning, making them highly promising media. Regarding the performance of extraction, deep eutectic solvents are not necessarily better than conventional solvents or ionic liquids. However, many authors have highlighted that deep eutectic solvents allow a better stability and a better conservation of the extracted polyphenols. As a result, recovery extraction could be bypassed while systematically required when ionic liquids are used as extractant media. Many authors have shown that eutectic solvents could retain the antioxidant properties of polyphenols longer than conventional organic solvents. Interestingly, due to the harmlessness of natural deep eutectic solvent, natural deep eutectic solvent extracts can be anticipated to be used directly for many applications of polyphenol extracts, namely, cosmetics. In that case, natural deep eutectic solvent could even be designed as a part of product formulation.

To summarize, deep eutectic solvents may offer broader industrial perspectives than classical solvents because they are globally cheaper, eco-friendly, and more socially acceptable at the moment.

References

Abbott AP, Capper G, Davies DL et al (2001) Preparation of novel, moisture-stable, Lewis-acidic ionic liquids containing quaternary ammonium salts with functional side chains. Chem Commun:2010–2011. https://doi.org/10.1039/b106357j

Abbott AP, Capper G, Davies DL et al (2003) Novel solvent properties of choline chloride/urea mixtures. Chem Commun:70–71. https://doi.org/10.1039/b210714g

Abbott AP, Boothby D, Capper G et al (2004) Deep eutectic solvents formed between choline chloride and carboxylic acids: versatile alternatives to ionic liquids. J Am Chem Soc 126:9142–9147. https://doi.org/10.1021/ja048266j

Abbott AP, Cullis PM, Gibson MJ et al (2007) Extraction of glycerol from biodiesel into a eutectic based ionic liquid. Green Chem 9:868. https://doi.org/10.1039/b702833d

Abranches DO, Martins MAR, Silva LP et al (2019) Phenolic hydrogen bond donors in the formation of non-ionic deep eutectic solvents: the quest for type V DES. Chem Commun 55:10253–10256. https://doi.org/10.1039/C9CC04846D

Achat S, Rakotomanomana N, Madani K, Dangles O (2016) Antioxidant activity of olive phenols and other dietary phenols in model gastric conditions: Scavenging of the free radical DPPH and inhibition of the haem-induced peroxidation of linoleic acid. Food Chem 213:135–142. https://doi.org/10.1016/j.foodchem.2016.06.076

Adam V, Mikelova R, Hubalek J et al (2007) Utilizing of Square Wave Voltammetry to detect flavonoids in the presence of human urine. Sensors 7:2402–2418. https://doi.org/10.3390/s7102402

Ali MC, Chen J, Zhang H et al (2019) Effective extraction of flavonoids from Lycium barbarum L. fruits by deep eutectic solvents-based ultrasound-assisted extraction. Talanta 203:16–22. https://doi.org/10.1016/j.talanta.2019.05.012

Aroso IM, Paiva A, Reis RL, Duarte ARC (2017) Natural deep eutectic solvents from choline chloride and betaine – physicochemical properties. J Mol Liq 241:654–661. https://doi.org/10.1016/j.molliq.2017.06.051

Athanasiadis V, Grigorakis S, Lalas S, Makris DP (2018a) Methyl β-cyclodextrin as a booster for the extraction for Olea europaea leaf polyphenols with a bio-based deep eutectic solvent. Biomass Convers Biorefinery 8:345–355. https://doi.org/10.1007/s13399-017-0283-5

Athanasiadis V, Grigorakis S, Lalas S, Makris DP (2018b) Highly efficient extraction of antioxidant polyphenols from olea europaea leaves using an eco-friendly glycerol/glycine deep eutectic solvent. Waste Biomass Valorization 9:1985–1992. https://doi.org/10.1007/s12649-017-9997-7

Athanasiadis V, Grigorakis S, Lalas S, Makris DP (2018c) Stability effects of methyl β-cyclodextrin on Olea europaea leaf extracts in a natural deep eutectic solvent. Eur Food Res Technol 244:1783–1792. https://doi.org/10.1007/s00217-018-3090-8

Azwanida N (2015) A review on the extraction methods use in medicinal plants, principle, strength and limitation. Med Aromat Plants 04. https://doi.org/10.4172/2167-0412.1000196

Bajkacz S, Adamek J (2017) Evaluation of new natural deep eutectic solvents for the extraction of isoflavones from soy products. Talanta 168:329–335. https://doi.org/10.1016/j.talanta.2017.02.065

Bakirtzi C, Triantafyllidou K, Makris DP (2016) Novel lactic acid-based natural deep eutectic solvents: Efficiency in the ultrasound-assisted extraction of antioxidant polyphenols from common native Greek medicinal plants. J Appl Res Med Aromat Plants 3:120–127. https://doi.org/10.1016/j.jarmap.2016.03.003

Belščak-Cvitanović A, Durgo K, Huđek A et al (2018) 1 – overview of polyphenols and their properties. In: Galanakis CM (ed) Polyphenols: properties, recovery, and applications. Woodhead Publishing, Duxford, pp 3–44

Benvenutti L, Zielinski AAF, Ferreira SRS (2019) Which is the best food emerging solvent: IL, DES or NADES? Trends Food Sci Technol 90:133–146. https://doi.org/10.1016/j.tifs.2019.06.003

Bi W, Tian M, Row KH (2013) Evaluation of alcohol-based deep eutectic solvent in extraction and determination of flavonoids with response surface methodology optimization. J Chromatogr A 1285:22–30. https://doi.org/10.1016/j.chroma.2013.02.041

Bohn T (2014) Dietary factors affecting polyphenol bioavailability. Nutr Rev 72:429–452. https://doi.org/10.1111/nure.12114

Bosiljkov T, Dujmić F, Cvjetko Bubalo M et al (2017) Natural deep eutectic solvents and ultrasound-assisted extraction: green approaches for extraction of wine lees anthocyanins. Food Bioprod Process 102:195–203. https://doi.org/10.1016/j.fbp.2016.12.005

Boulogne I, Petit P, Ozier-Lafontaine H et al (2012) Insecticidal and antifungal chemicals produced by plants: a review. Environ Chem Lett 10:325–347. https://doi.org/10.1007/s10311-012-0359-1

Burghoff B, Goetheer ELV, de Haan AB (2008) COSMO-RS-based extractant screening for phenol extraction as model system. Ind Eng Chem Res 47:4263–4269. https://doi.org/10.1021/ie7017405

Cao J, Chen L, Li M et al (2018a) Efficient extraction of proanthocyanidin from Ginkgo biloba leaves employing rationally designed deep eutectic solvent-water mixture and evaluation of the antioxidant activity. J Pharm Biomed Anal 158:317–326. https://doi.org/10.1016/j.jpba.2018.06.007

Cao J, Chen L, Li M et al (2018b) Two-phase systems developed with hydrophilic and hydrophobic deep eutectic solvents for simultaneously extracting various bioactive compounds with different polarities. Green Chem 20:1879–1886. https://doi.org/10.1039/C7GC03820H

Cao Q, Li J, Xia Y et al (2018c) Green extraction of six phenolic compounds from rattan (Calamoideae faberii) with Deep Eutectic Solvent by homogenate-assisted vacuum-cavitation method. Molecules 24:113

Casazza AA, Aliakbarian B, Mantegna S et al (2010) Extraction of phenolics from Vitis vinifera wastes using non-conventional techniques. J Food Eng 100:50–55. https://doi.org/10.1016/j.jfoodeng.2010.03.026

Chanioti S, Tzia C (2018) Extraction of phenolic compounds from olive pomace by using natural deep eutectic solvents and innovative extraction techniques. Innov Food Sci Emerg Technol 48:228–239. https://doi.org/10.1016/j.ifset.2018.07.001

Chen J, Liu M, Wang Q et al (2016) Deep Eutectic Solvent-based microwave-assisted method for extraction of hydrophilic and hydrophobic Components from Radix Salviae miltiorrhizae. Molecules 21:1383. https://doi.org/10.3390/molecules21101383

Cheynier V (2005) Polyphenols in foods are more complex than often thought. Am J Clin Nutr 81:223S–229S. https://doi.org/10.1093/ajcn/81.1.223S

Choi YH, Verpoorte R (2019) Green solvents for the extraction of bioactive compounds from natural products using ionic liquids and deep eutectic solvents. Curr Opin Food Sci 26:87–93. https://doi.org/10.1016/j.cofs.2019.04.003

Choi YH, van Spronsen J, Dai Y et al (2011) Are natural deep eutectic solvents the missing link in understanding cellular metabolism and physiology? Plant Physiol 156:1701–1705. https://doi.org/10.1104/pp.111.178426

Crampon C, Boutin O, Badens E (2011) Supercritical Carbon Dioxide extraction of molecules of interest from microalgae and seaweeds. Ind Eng Chem Res 50:8941–8953. https://doi.org/10.1021/ie102297d

Creis (n.d.) EAG Erwan; Potin, Philippe Ubiquitous phlorotannins prospects and perspectives. Blue Biotechnol:67–116. https://doi.org/10.1002/9783527801718.ch3

Cren-Olivé C, Hapiot P, Pinson J, Rolando C (2002) Free radical chemistry of Flavan-3-ols: determination of thermodynamic parameters and of kinetic reactivity from short (ns) to long (ms) time scale. J Am Chem Soc 124:14027–14038. https://doi.org/10.1021/ja0262434

Cui Q, Peng X, Yao X-H et al (2015) Deep eutectic solvent-based microwave-assisted extraction of genistin, genistein and apigenin from pigeon pea roots. Sep Purif Technol 150:63–72. https://doi.org/10.1016/j.seppur.2015.06.026

Cui Q, Liu J-Z, Wang L-T et al (2018) Sustainable deep eutectic solvents preparation and their efficiency in extraction and enrichment of main bioactive flavonoids from sea buckthorn leaves. J Clean Prod 184:826–835. https://doi.org/10.1016/j.jclepro.2018.02.295

Cunha SC, Fernandes JO (2018) Extraction techniques with deep eutectic solvents. TrAC Trends Anal Chem 105:225–239. https://doi.org/10.1016/j.trac.2018.05.001

Cvjetko Bubalo M, Ćurko N, Tomašević M et al (2016) Green extraction of grape skin phenolics by using deep eutectic solvents. Food Chem 200:159–166. https://doi.org/10.1016/j.foodchem.2016.01.040

Cvjetko Bubalo M, Vidović S, Radojčić Redovniković I, Jokić S (2018) New perspective in extraction of plant biologically active compounds by green solvents. Food Bioprod Process 109:52–73. https://doi.org/10.1016/j.fbp.2018.03.001

Dahmoune F, Nayak B, Moussi K et al (2015) Optimization of microwave-assisted extraction of polyphenols from Myrtus communis L. leaves. Food Chem 166:585–595. https://doi.org/10.1016/j.foodchem.2014.06.066

Dai F, Chen W-F, Zhou B (2008) Antioxidant synergism of green tea polyphenols with α-tocopherol and l-ascorbic acid in SDS micelles. Biochimie 90:1499–1505. https://doi.org/10.1016/j.biochi.2008.05.007

Dai Y, van Spronsen J, Witkamp G-J et al (2013a) Ionic Liquids and Deep Eutectic Solvents in natural products research: mixtures of solids as extraction solvents. J Nat Prod 76:2162–2173. https://doi.org/10.1021/np400051w

Dai Y, van Spronsen J, Witkamp G-J et al (2013b) Natural deep eutectic solvents as new potential media for green technology. Anal Chim Acta 766:61–68. https://doi.org/10.1016/j.aca.2012.12.019

Dai Y, Witkamp G-J, Verpoorte R, Choi YH (2013c) Natural Deep Eutectic Solvents as a new extraction media for phenolic metabolites in Carthamus tinctorius L. Anal Chem 85:6272–6278. https://doi.org/10.1021/ac400432p

Dai Y, Witkamp G-J, Verpoorte R, Choi YH (2015) Tailoring properties of natural deep eutectic solvents with water to facilitate their applications. Food Chem 187:14–19. https://doi.org/10.1016/j.foodchem.2015.03.123

Dai Y, Rozema E, Verpoorte R, Choi YH (2016) Application of natural deep eutectic solvents to the extraction of anthocyanins from Catharanthus roseus with high extractability and stability replacing conventional organic solvents. J Chromatogr A 1434:50–56. https://doi.org/10.1016/j.chroma.2016.01.037

de Heer MI, Korth H-G, Mulder P (1999) Poly methoxy phenols in solution: O–H bond dissociation enthalpies, structures, and hydrogen bonding. J Org Chem 64:6969–6975. https://doi.org/10.1021/jo9901485

Deng J, Yang H, Capanoglu E et al (2018) Technological aspects and stability of polyphenols. In: Polyphenols: properties, recovery, and applications. Elsevier, Oxford, pp 295–323

Du Y, Guo H, Lou H (2007) Grape seed polyphenols protect cardiac cells from apoptosis via induction of endogenous antioxidant enzymes. J Agric Food Chem 55:1695–1701. https://doi.org/10.1021/jf063071b

Duan L, Dou L-L, Guo L et al (2016) Comprehensive evaluation of Deep Eutectic Solvents in extraction of bioactive natural products. ACS Sustain Chem Eng 4:2405–2411. https://doi.org/10.1021/acssuschemeng.6b00091

Duan L, Zhang W-H, Zhang Z-H et al (2019) Evaluation of natural deep eutectic solvents for the extraction of bioactive flavone C-glycosides from Flos Trollii. Microchem J 145:180–186. https://doi.org/10.1016/j.microc.2018.10.031

Durand E, Lecomte J, Upasani R et al (2017) Evaluation of the ROS inhibiting activity and mitochondrial targeting of phenolic compounds in fibroblast cells Model system and enhancement of efficiency by Natural Deep Eutectic Solvent (NADES) formulation. Pharm Res 34:1134–1146. https://doi.org/10.1007/s11095-017-2124-4

Dwamena AK (2019) Recent advances in hydrophobic deep eutectic solvents for extraction. Separations 6:9. https://doi.org/10.3390/separations6010009

El Kantar S, Rajha HN, Boussetta N et al (2019) Green extraction of polyphenols from grapefruit peels using high voltage electrical discharges, deep eutectic solvents and aqueous glycerol. Food Chem 295:165–171. https://doi.org/10.1016/j.foodchem.2019.05.111

Faggian M, Sut S, Perissutti B et al (2016) Natural Deep Eutectic Solvents (NADES) as a tool for bioavailability improvement: pharmacokinetics of Rutin dissolved in Proline/Glycine after oral administration in rats: possible application in nutraceuticals. Molecules 21:1531. https://doi.org/10.3390/molecules21111531

Feduraev P, Chupakhina G, Maslennikov P et al (2019) Variation in phenolic compounds content and antioxidant activity of different plant organs from Rumex crispus L. and Rumex obtusifolius L. at different growth stages. Antioxidants 8:237. https://doi.org/10.3390/antiox8070237

Fernández M, de LÁ, Espino M, Gomez FJV, Silva MF (2018a) Novel approaches mediated by tailor-made green solvents for the extraction of phenolic compounds from agro-food industrial by-products. Food Chem 239:671–678. https://doi.org/10.1016/j.foodchem.2017.06.150

Fernández M, de los Á, Boiteux J, Espino M et al (2018b) Natural deep eutectic solvents-mediated extractions: the way forward for sustainable analytical developments. Anal Chim Acta 1038:1–10. https://doi.org/10.1016/j.aca.2018.07.059

Ferrone V, Genovese S, Carlucci M et al (2018) A green deep eutectic solvent dispersive liquid-liquid micro-extraction (DES-DLLME) for the UHPLC-PDA determination of oxyprenylated phenylpropanoids in olive, soy, peanuts, corn, and sunflower oil. Food Chem 245:578–585. https://doi.org/10.1016/j.foodchem.2017.10.135

Florindo C, Monteiro NV, Ribeiro BD et al (2019) Hydrophobic deep eutectic solvents for purification of water contaminated with Bisphenol-A. J Mol Liq:111841. https://doi.org/10.1016/j.molliq.2019.111841

Fraige K, Arrua RD, Sutton AT et al (2019) Using natural deep eutectic solvents for the extraction of metabolites in *Byrsonima intermedia* leaves. J Sep Sci 42:591–597. https://doi.org/10.1002/jssc.201800905

Fu N, Li L, Liu X et al (2017a) Specific recognition of polyphenols by molecularly imprinted polymers based on a ternary deep eutectic solvent. J Chromatogr A 1530:23–34. https://doi.org/10.1016/j.chroma.2017.11.011

Fu N, Liu X, Li L et al (2017b) Ternary choline chloride/caffeic acid/ethylene glycol deep eutectic solvent as both a monomer and template in a molecularly imprinted polymer. J Sep Sci 40:2286–2291. https://doi.org/10.1002/jssc.201700146

Fu N, Lv R, Guo Z et al (2017c) Environmentally friendly and non-polluting solvent pretreatment of palm samples for polyphenol analysis using choline chloride deep eutectic solvents. J Chromatogr A 1492:1–11. https://doi.org/10.1016/j.chroma.2017.02.036

Fukumoto K, Yoshizawa M, Ohno H (2005) Room temperature Ionic Liquids from 20 natural amino acids. J Am Chem Soc 127:2398–2399. https://doi.org/10.1021/ja043451i

Galili S, Hovav R (2014) Chapter 16 – determination of polyphenols, flavonoids, and antioxidant capacity in dry seeds. In: Watson RR (ed) Polyphenols in plants. Academic Press, San Diego, pp 305–323

Gao Y, Arritt SW, Twamley B, Shreeve JM (2005) Guanidinium-based ionic liquids. Inorg Chem 44:1704–1712. https://doi.org/10.1021/ic048513k

García A, Rodríguez-Juan E, Rodríguez-Gutiérrez G et al (2016) Extraction of phenolic compounds from virgin olive oil by deep eutectic solvents (DESs). Food Chem 197:554–561. https://doi.org/10.1016/j.foodchem.2015.10.131

Georgantzi C, Lioliou A-E, Paterakis N, Makris D (2017) Combination of Lactic Acid-Based Deep Eutectic Solvents (DES) with β-Cyclodextrin: performance screening using ultrasound-assisted extraction of polyphenols from selected native Greek medicinal plants. Agronomy 7:54. https://doi.org/10.3390/agronomy7030054

Giftson JS, Jayanthi S, Nalini N (2010) Chemopreventive efficacy of gallic acid, an antioxidant and anticarcinogenic polyphenol, against 1,2-dimethyl hydrazine induced rat colon carcinogenesis. Invest New Drugs 28:251–259. https://doi.org/10.1007/s10637-009-9241-9

Goiris K, Muylaert K, Fraeye I et al (2012) Antioxidant potential of microalgae in relation to their phenolic and carotenoid content. J Appl Phycol 24:1477–1486. https://doi.org/10.1007/s10811-012-9804-6

Gomez FJV, Espino M, de los Angeles Fernandez M et al (2016) Enhanced electrochemical detection of quercetin by natural deep eutectic solvents. Anal Chim Acta 936:91–96. https://doi.org/10.1016/j.aca.2016.07.022

Goula AM, Ververi M, Adamopoulou A, Kaderides K (2017) Green ultrasound-assisted extraction of carotenoids from pomegranate wastes using vegetable oils. Ultrason Sonochem 34:821–830. https://doi.org/10.1016/j.ultsonch.2016.07.022

Guo N, Ping-Kou, Jiang Y-W et al (2019) Natural deep eutectic solvents couple with integrative extraction technique as an effective approach for mulberry anthocyanin extraction. Food Chem 296:78–85. https://doi.org/10.1016/j.foodchem.2019.05.196

Gutiérrez MC, Ferrer ML, Mateo CR, del Monte F (2009) Freeze-drying of aqueous solutions of deep eutectic solvents: a suitable approach to deep eutectic suspensions of self-assembled structures. Langmuir 25:5509–5515. https://doi.org/10.1021/la900552b

Hajimahmoodi M, Faramarzi MA, Mohammadi N et al (2010) Evaluation of antioxidant properties and total phenolic contents of some strains of microalgae. J Appl Phycol 22:43–50. https://doi.org/10.1007/s10811-009-9424-y

Hamany Djande CY, Piater LA, Steenkamp PA et al (2018) Differential extraction of phytochemicals from the multipurpose tree, Moringa oleifera, using green extraction solvents. South Afr J Bot 115:81–89. https://doi.org/10.1016/j.sajb.2018.01.009

Haslam E, Cai Y (1994) Plant polyphenols (vegetable tannins): gallic acid metabolism. Nat Prod Rep 11:41–66. https://doi.org/10.1039/NP9941100041

Hayyan A, Mjalli FS, AlNashef IM et al (2013) Glucose-based deep eutectic solvents: Physical properties. J Mol Liq 178:137–141. https://doi.org/10.1016/j.molliq.2012.11.025

Hollman PCH (2001) Evidence for health benefits of plant phenols: local or systemic effects? J Sci Food Agric 81:842–852. https://doi.org/10.1002/jsfa.900

Huang Y, Feng F, Jiang J et al (2017) Green and efficient extraction of rutin from tartary buckwheat hull by using natural deep eutectic solvents. Food Chem 221:1400–1405. https://doi.org/10.1016/j.foodchem.2016.11.013

Huang H, Xu Q, Belwal T et al (2019a) Ultrasonic impact on viscosity and extraction efficiency of polyethylene glycol: a greener approach for anthocyanins recovery from purple sweet potato. Food Chem 283:59–67. https://doi.org/10.1016/j.foodchem.2019.01.017

Huang J, Guo X, Xu T et al (2019b) Ionic deep eutectic solvents for the extraction and separation of natural products. J Chromatogr A 1598:1–19. https://doi.org/10.1016/j.chroma.2019.03.046

Ivanović M, Alañón ME, Arráez-Román D, Segura-Carretero A (2018) Enhanced and green extraction of bioactive compounds from Lippia citriodora by tailor-made natural deep eutectic solvents. Food Res Int 111:67–76. https://doi.org/10.1016/j.foodres.2018.05.014

Jancheva M, Grigorakis S, Loupassaki S, Makris DP (2017) Optimised extraction of antioxidant polyphenols from Satureja thymbra using newly designed glycerol-based natural low-transition temperature mixtures (LTTMs). J Appl Res Med Aromat Plants 6:31–40. https://doi.org/10.1016/j.jarmap.2017.01.002

Jeliński T, Cysewski P (2018) Application of a computational model of natural deep eutectic solvents utilizing the COSMO-RS approach for screening of solvents with high solubility of rutin. J Mol Model 24:180. https://doi.org/10.1007/s00894-018-3700-1

Jeong KM, Zhao J, Jin Y et al (2015) Highly efficient extraction of anthocyanins from grape skin using deep eutectic solvents as green and tunable media. Arch Pharm Res 38:2143–2152. https://doi.org/10.1007/s12272-015-0678-4

Jeong KM, Ko J, Zhao J et al (2017) Multi-functioning deep eutectic solvents as extraction and storage media for bioactive natural products that are readily applicable to cosmetic products. J Clean Prod 151:87–95. https://doi.org/10.1016/j.jclepro.2017.03.038

Jeong KM, Jin Y, Yoo DE et al (2018) One-step sample preparation for convenient examination of volatile monoterpenes and phenolic compounds in peppermint leaves using deep eutectic solvents. Food Chem 251:69–76. https://doi.org/10.1016/j.foodchem.2018.01.079

Joana Gil-Chávez G, Villa JA, Fernando Ayala-Zavala J et al (2013) Technologies for extraction and production of bioactive compounds to be used as nutraceuticals and food ingredients: an overview. Compr Rev Food Sci Food Saf 12:5–23. https://doi.org/10.1111/1541-4337.12005

Kadam SU, Tiwari BK, O'Donnell CP (2013) Application of novel extraction technologies for bioactives from Marine Algae. J Agric Food Chem 61:4667–4675. https://doi.org/10.1021/jf400819p

Kanberoglu GS, Yilmaz E, Soylak M (2019) Application of deep eutectic solvent in ultrasound-assisted emulsification microextraction of quercetin from some fruits and vegetables. J Mol Liq 279:571–577. https://doi.org/10.1016/j.molliq.2019.01.130

Karageorgou I, Grigorakis S, Lalas S, Makris DP (2017) Enhanced extraction of antioxidant polyphenols from Moringa oleifera Lam. leaves using a biomolecule-based low-transition temperature mixture. Eur Food Res Technol 243:1839–1848. https://doi.org/10.1007/s00217-017-2887-1

Khezeli T, Daneshfar A, Sahraei R (2016) A green ultrasonic-assisted liquid–liquid microextraction based on deep eutectic solvent for the HPLC-UV determination of ferulic, caffeic and cinnamic acid from olive, almond, sesame and cinnamon oil. Talanta 150:577–585. https://doi.org/10.1016/j.talanta.2015.12.077

Kilmartin PA, Hsu CF (2003) Characterisation of polyphenols in green, oolong, and black teas, and in coffee, using cyclic voltammetry. Food Chem 82:501–512. https://doi.org/10.1016/S0308-8146(03)00066-9

Klejdus B, Lojková L, Plaza M et al (2010) Hyphenated technique for the extraction and determination of isoflavones in algae: Ultrasound-assisted supercritical fluid extraction followed by fast chromatography with tandem mass spectrometry. J Chromatogr A 1217:7956–7965. https://doi.org/10.1016/j.chroma.2010.07.020

Komaty S, Letertre M, Dang HD et al (2016) Sample preparation for an optimized extraction of localized metabolites in lichens: application to Pseudevernia furfuracea. Talanta 150:525–530. https://doi.org/10.1016/j.talanta.2015.12.081

Kováčik J, Klejdus B, Bačkor M (2010) Physiological responses of Scenedesmus quadricauda (Chlorophyceae) to UV-A and UV-C Light. Photochem Photobiol 86:612–616. https://doi.org/10.1111/j.1751-1097.2010.00708.x

Lapkin AA, Plucinski PK, Cutler M (2006) Comparative assessment of technologies for extraction of Artemisinin. J Nat Prod 69:1653–1664. https://doi.org/10.1021/np060375j

Larrauri JA, Rupérez P, Saura-Calixto F (1997) Effect of drying temperature on the stability of polyphenols and antioxidant activity of Red Grape Pomace Peels. J Agric Food Chem 45:1390–1393. https://doi.org/10.1021/jf960282f

Li AS-H, Bandy B, Tsang S-S, Davison AJ (2000) DNA-breaking versus DNA-protecting activity of four phenolic compounds in vitro. Free Radic Res 33:551–566. https://doi.org/10.1080/10715760000301091

Li Y, Skouroumounis GK, Elsey GM, Taylor DK (2011) Microwave-assistance provides very rapid and efficient extraction of grape seed polyphenols. Food Chem 129:570–576. https://doi.org/10.1016/j.foodchem.2011.04.068

Li J, Han Z, Zou Y, Yu B (2015) Efficient extraction of major catechins in Camellia sinensis leaves using green choline chloride-based deep eutectic solvents. RSC Adv 5:93937–93944. https://doi.org/10.1039/C5RA15830C

Li L, Liu J-Z, Luo M et al (2016) Efficient extraction and preparative separation of four main isoflavonoids from Dalbergia odorifera T. Chen leaves by deep eutectic solvents-based negative pressure cavitation extraction followed by macroporous resin column chromatography. J Chromatogr B 1033–1034:40–48. https://doi.org/10.1016/j.jchromb.2016.08.005

Liu Y, Friesen JB, McAlpine JB et al (2018) Natural deep eutectic solvents: properties, applications, and perspectives. J Nat Prod 81:679–690. https://doi.org/10.1021/acs.jnatprod.7b00945

Ma Y, Liu M, Tan T et al (2018) Deep eutectic solvents used as extraction solvent for the determination of flavonoids from Camellia oleifera flowers by high-performance liquid chromatography. Phytochem Anal 29:639–648. https://doi.org/10.1002/pca.2777

Makhotkina O, Kilmartin P (2010) The use of cyclic voltammetry for wine analysis: determination of polyphenols and free sulfur dioxide. Anal Chim Acta 668:155–165. https://doi.org/10.1016/j.aca.2010.03.064

Makhotkina O, Kilmartin PA (2012) The phenolic composition of Sauvignon blanc juice profiled by cyclic voltammetry. Electrochimica Acta 83:188–195. https://doi.org/10.1016/j.electacta.2012.07.101

Mamilla JLK, Novak U, Grilc M, Likozar B (2019) Natural deep eutectic solvents (DES) for fractionation of waste lignocellulosic biomass and its cascade conversion to value-added bio-based chemicals. Biomass Bioenergy 120:417–425. https://doi.org/10.1016/j.biombioe.2018.12.002

Manach C, Scalbert A, Morand C et al (2004) Polyphenols: food sources and bioavailability. Am J Clin Nutr 79:727–747. https://doi.org/10.1093/ajcn/79.5.727

Manach C, Williamson G, Morand C et al (2005) Bioavailability and bioefficacy of polyphenols in humans. I. Review of 97 bioavailability studies. Am J Clin Nutr 81:230S–242S. https://doi.org/10.1093/ajcn/81.1.230S

Manousaki A, Jancheva M, Grigorakis S, Makris D (2016) Extraction of antioxidant phenolics from agri-food waste biomass using a newly designed glycerol-based natural low-transition temperature mixture: a comparison with conventional eco-friendly solvents. Recycling 1:194–204. https://doi.org/10.3390/recycling1010194

Mansur AR, Song N-E, Jang HW et al (2019) Optimizing the ultrasound-assisted deep eutectic solvent extraction of flavonoids in common buckwheat sprouts. Food Chem 293:438–445. https://doi.org/10.1016/j.foodchem.2019.05.003

Marcet I, Salvadores M, Rendueles M, Díaz M (2018) The effect of ultrasound on the alkali extraction of proteins from eggshell membranes: the effect of ultrasound on the alkali extraction of proteins from eggshell membranes. J Sci Food Agric 98:1765–1772. https://doi.org/10.1002/jsfa.8651

Marcus Y (2019) Deep eutectic solvents. Springer International Publishing, Cham

Martins MAR, Pinho SP, Coutinho JAP (2019) Insights into the nature of eutectic and deep eutectic mixtures. J Solut Chem 48:962–982. https://doi.org/10.1007/s10953-018-0793-1

Meng Z, Zhao J, Duan H et al (2018) Green and efficient extraction of four bioactive flavonoids from Pollen Typhae by ultrasound-assisted deep eutectic solvents extraction. J Pharm Biomed Anal 161:246–253. https://doi.org/10.1016/j.jpba.2018.08.048

Michalak I, Chojnacka K (2015) Algae as production systems of bioactive compounds. Eng Life Sci 15:160–176. https://doi.org/10.1002/elsc.201400191

Migliorati V, Sessa F, D'Angelo P (2019) Deep eutectic solvents: A structural point of view on the role of the cation. Chem Phys Lett X 2:100001. https://doi.org/10.1016/j.cpletx.2018.100001

Mocan A, Diuzheva A, Bădărău S et al (2019) Liquid phase and microwave-assisted extractions for multicomponent phenolic pattern determination of five romanian galium species coupled with bioassays. Molecules 24:1226

Molnar M, Jakovljević M, Jokić S (2018) Optimization of the process conditions for the extraction of Rutin from *Ruta graveolens* L. by choline chloride based Deep Eutectic Solvents. Solvent Extr Res Dev Jpn 25:109–116. https://doi.org/10.15261/serdj.25.109

Mouratoglou E, Malliou V, Makris DP (2016) Novel glycerol-based natural eutectic mixtures and their efficiency in the ultrasound-assisted extraction of antioxidant polyphenols from agri-food waste biomass. Waste Biomass Valorization 7:1377–1387. https://doi.org/10.1007/s12649-016-9539-8

Naczk M, Shahidi F (2004) Extraction and analysis of phenolics in food. J Chromatogr A 1054:95–111. https://doi.org/10.1016/j.chroma.2004.08.059

Naikoo MI, Dar MI, Raghib F et al (2019) Chapter 9 – role and regulation of plants phenolics in abiotic stress tolerance: an overview. In: MIR K, Reddy PS, Ferrante A, Khan NA (eds) Plant signaling molecules. Woodhead Publishing, Duxford, pp 157–168

Nam MW, Zhao J, Lee MS et al (2015) Enhanced extraction of bioactive natural products using tailor-made deep eutectic solvents: application to flavonoid extraction from Flos sophorae. Green Chem 17:1718–1727. https://doi.org/10.1039/C4GC01556H

Neudörffer A, Desvergne J-P, Bonnefont-Rousselot D et al (2006) Protective effects of 4-hydroxycinnamic ethyl ester derivatives and related dehydrodimers against oxidation of LDL: radical scavengers or metal chelators? J Agric Food Chem 54:1898–1905. https://doi.org/10.1021/jf052923p

Obluchinskaya ED, Daurtseva AV, Pozharitskaya ON et al (2019) Natural Deep Eutectic Solvents as alternatives for extracting phlorotannins from brown algae. Pharm Chem J 53:243–247. https://doi.org/10.1007/s11094-019-01987-0

Oliveira Neto JR, de Oliveira TS, Ghedini PC et al (2017a) Antioxidant and vasodilatory activity of commercial beers. J Funct Foods 34:130–138. https://doi.org/10.1016/j.jff.2017.04.019

Oliveira Neto JR, Lopes de Macêdo IY, Lopes de Oliveira NR et al (2017b) Antioxidant capacity and total phenol content in hop and malt commercial samples. Electroanalysis 29:2788–2792. https://doi.org/10.1002/elan.201700492

Ozturk B, Parkinson C, Gonzalez-Miquel M (2018) Extraction of polyphenolic antioxidants from orange peel waste using deep eutectic solvents. Sep Purif Technol 206:1–13. https://doi.org/10.1016/j.seppur.2018.05.052

Paiva A, Craveiro R, Aroso I et al (2014) Natural deep eutectic solvents – solvents for the 21st century. ACS Sustain Chem Eng 2:1063–1071. https://doi.org/10.1021/sc500096j

Pal CBT, Jadeja GC (2019) Deep eutectic solvent-based extraction of polyphenolic antioxidants from onion (Allium cepa L.) peel. J Sci Food Agric 99:1969–1979. https://doi.org/10.1002/jsfa.9395

Panić M, Gunjević V, Cravotto G, Radojčić Redovniković I (2019a) Enabling technologies for the extraction of grape-pomace anthocyanins using natural deep eutectic solvents in up-to-half-litre batches extraction of grape-pomace anthocyanins using NADES. Food Chem 300:125185. https://doi.org/10.1016/j.foodchem.2019.125185

Panić M, Radić Stojković M, Kraljić K et al (2019b) Ready-to-use green polyphenolic extracts from food by-products. Food Chem 283:628–636. https://doi.org/10.1016/j.foodchem.2019.01.061

Paradiso VM, Clemente A, Summo C et al (2016a) Towards green analysis of virgin olive oil phenolic compounds: extraction by a natural deep eutectic solvent and direct spectrophotometric detection. Food Chem 212:43–47. https://doi.org/10.1016/j.foodchem.2016.05.082

Paradiso VM, Clemente A, Summo C et al (2016b) Extraction of phenolic compounds from extra virgin olive oil by a natural deep eutectic solvent: data on UV absorption of the extracts. Data Brief 8:553–556. https://doi.org/10.1016/j.dib.2016.05.076

Passos H, Tavares DJP, Ferreira AM et al (2016) Are aqueous biphasic systems composed of Deep Eutectic Solvents ternary or quaternary Systems? ACS Sustain Chem Eng 4:2881–2886. https://doi.org/10.1021/acssuschemeng.6b00485

Patsea M, Stefou I, Grigorakis S, Makris DP (2017) Screening of natural sodium acetate-based low-transition temperature mixtures (LTTMs) for enhanced extraction of antioxidants and pigments from red vinification solid wastes. Environ Process 4:123–135. https://doi.org/10.1007/s40710-016-0205-8

Pavić V, Flačer D, Jakovljević M et al (2019) Assessment of total phenolic content, In Vitro antioxidant and antibacterial activity of Ruta graveolens L. extracts obtained by choline chloride based natural deep eutectic solvents. Plants 8:69. https://doi.org/10.3390/plants8030069

Pena-Pereira F, Namieśnik J (2014) Ionic liquids and deep eutectic mixtures: sustainable solvents for extraction processes. ChemSusChem 7:1784–1800. https://doi.org/10.1002/cssc.201301192

Peng X, Duan M-H, Yao X-H et al (2016) Green extraction of five target phenolic acids from Lonicerae japonicae Flos with deep eutectic solvent. Sep Purif Technol 157:249–257. https://doi.org/10.1016/j.seppur.2015.10.065

Peng F, Xu P, Zhao B-Y et al (2018) The application of deep eutectic solvent on the extraction and in vitro antioxidant activity of rutin from Sophora japonica bud. J Food Sci Technol 55:2326–2333. https://doi.org/10.1007/s13197-018-3151-9

Perron NR, Brumaghim JL (2009) A review of the antioxidant mechanisms of polyphenol compounds related to iron binding. Cell Biochem Biophys 53:75–100. https://doi.org/10.1007/s12013-009-9043-x

Pietta P-G (2000) Flavonoids as antioxidants. J Nat Prod 63:1035–1042. https://doi.org/10.1021/np9904509

Piljac-Žegarac J, Valek L, Stipčević T, Martinez S (2010) Electrochemical determination of antioxidant capacity of fruit tea infusions. Food Chem 121:820–825. https://doi.org/10.1016/j.foodchem.2009.12.090

Plechkova NV, Seddon KR (2008) Applications of ionic liquids in the chemical industry. Chem Soc Rev 37:123–150. https://doi.org/10.1039/B006677J

Qi X-L, Peng X, Huang Y-Y et al (2015) Green and efficient extraction of bioactive flavonoids from Equisetum palustre L. by deep eutectic solvents-based negative pressure cavitation method combined with macroporous resin enrichment. Ind Crops Prod 70:142–148. https://doi.org/10.1016/j.indcrop.2015.03.026

Quideau S, Deffieux D, Douat-Casassus C, Pouységu L (2011) Plant polyphenols: chemical properties, biological activities, and synthesis. Angew Chem Int Ed 50:586–621. https://doi.org/10.1002/anie.201000044

Radošević K, Ćurko N, Gaurina Srček V et al (2016) Natural deep eutectic solvents as beneficial extractants for enhancement of plant extracts bioactivity. LWT – Food Sci Technol 73:45–51. https://doi.org/10.1016/j.lwt.2016.05.037

Rajan M, Prabhavathy A, Ramesh U (2015) Natural Deep Eutectic Solvent extraction media for Zingiber officinale Roscoe: the study of chemical compositions, antioxidants and antimicrobial activities. Nat Prod J 5:3–13. https://doi.org/10.2174/2210315505011150414094719

Rajha HN, Mhanna T, El Kantar S et al (2019) Innovative process of polyphenol recovery from pomegranate peels by combining green deep eutectic solvents and a new infrared technology. LWT 111:138–146. https://doi.org/10.1016/j.lwt.2019.05.004

René A, Abasq M-L, Hauchard D, Hapiot P (2010) How do phenolic compounds react toward superoxide ion? a simple electrochemical method for evaluating antioxidant capacity. Anal Chem 82:8703–8710. https://doi.org/10.1021/ac101854w

Roche M, Dufour C, Mora N, Dangles O (2005) Antioxidant activity of olive phenols: mechanistic investigation and characterization of oxidation products by mass spectrometry. Org Biomol Chem 3:423–430. https://doi.org/10.1039/B416101G

Romero A, Santos A, Tojo J, Rodríguez A (2008) Toxicity and biodegradability of imidazolium ionic liquids. J Hazard Mater 151:268–273. https://doi.org/10.1016/j.jhazmat.2007.10.079

Ruesgas-Ramón M, Figueroa-Espinoza MC, Durand E (2017) Application of Deep Eutectic Solvents (DES) for phenolic compounds extraction: overview, challenges, and opportunities. J Agric Food Chem 65:3591–3601. https://doi.org/10.1021/acs.jafc.7b01054

Saha SK, Dey S, Chakraborty R (2019) Effect of choline chloride-oxalic acid based deep eutectic solvent on the ultrasonic assisted extraction of polyphenols from Aegle marmelos. J Mol Liq 287:110956. https://doi.org/10.1016/j.molliq.2019.110956

Sang J, Li B, Huang Y et al (2018) Deep eutectic solvent-based extraction coupled with green two-dimensional HPLC-DAD-ESI-MS/MS for the determination of anthocyanins from Lycium ruthenicum Murr. fruit. Anal Methods 10:1247–1257. https://doi.org/10.1039/C8AY00101D

Santoso J, Yoshie Y, Suzuki T (2004) Polyphenolic compounds from seaweeds: distribution and their antioxidative effect. In: Sakaguchi M (ed) Developments in food science. Elsevier, Amsterdam, pp 169–177

Schaich KM, Tian X, Xie J (2015) Hurdles and pitfalls in measuring antioxidant efficacy: a critical evaluation of ABTS, DPPH, and ORAC assays. J Funct Foods 14:111–125. https://doi.org/10.1016/j.jff.2015.01.043

Šeruga M, Novak I, Jakobek L (2011) Determination of polyphenols content and antioxidant activity of some red wines by differential pulse voltammetry, HPLC and spectrophotometric methods. Food Chem 124:1208–1216. https://doi.org/10.1016/j.foodchem.2010.07.047

Shang X, Tan J-N, Du Y et al (2018) Environmentally-friendly extraction of flavonoids from Cyclocarya paliurus (Batal.) Iljinskaja Leaves with Deep Eutectic Solvents and evaluation of their antioxidant activities. Molecules 23:2110. https://doi.org/10.3390/molecules23092110

Shi X, Ye J, Leonard S et al (2000) Antioxidant properties of (-)-epicatechin-3-gallate and its inhibition of Cr (VI)-induced DNA damage and Cr (IV)- or TPA-stimulated NF-κB activation. Mol Cell Biochem 206:125–132. https://doi.org/10.1023/a:1007012403691

Shoubaky GAE, Abdel-Daim MM, Mansour MH, Salem EA (2016) Isolation and identification of a flavone apigenin from marine red alga Acanthophora spicifera with antinociceptive and anti-Inflammatory activities. J Exp Neurosci 10:JEN.S25096. https://doi.org/10.4137/JEN.S25096

Silva LP, Fernandez L, Conceição JHF et al (2018) Design and characterization of sugar-based Deep Eutectic Solvents using conductor-like screening model for real solvents. ACS Sustain Chem Eng 6:10724–10734. https://doi.org/10.1021/acssuschemeng.8b02042

Singh BS, Lobo HR, Pinjari DV et al (2013) Ultrasound and deep eutectic solvent (DES): A novel blend of techniques for rapid and energy efficient synthesis of oxazoles. Ultrason Sonochem 20:287–293. https://doi.org/10.1016/j.ultsonch.2012.06.003

Singleton VL, Orthofer R, Lamuela-Raventós RM (1999) Analysis of total phenols and other oxidation substrates and antioxidants by means of Folin-Ciocalteu reagent. In: Methods in enzymology. Academic Press, Amsterdam, pp 152–178

Skulcova A, Haščičová Z, Hrdlicka L, Sima J, Jablonsky M (2018) Green Solvents based on choline chloride extraction of Spruce Bark (Picea Abies). Cellul Chem Technol 52:171–179

Swain T, Bate-Smith EC (1962) Flavonoid compounds. In: Comparative biochemistry. Academic Press, New York

Tanase C, Coşarcă S, Muntean D-L (2019) A critical review of phenolic compounds extracted from the bark of woody vascular plants and their potential biological activity. Molecules 24:1182. https://doi.org/10.3390/molecules24061182

Tang B, Bi W, Tian M, Row KH (2012) Application of ionic liquid for extraction and separation of bioactive compounds from plants. J Chromatogr B 904:1–21. https://doi.org/10.1016/j.jchromb.2012.07.020

Tang B, Park HE, Row KH (2015) Simultaneous extraction of flavonoids from Chamaecyparis obtusa using Deep Eutectic Solvents as additives of conventional extractions solvents. J Chromatogr Sci 53:836–840. https://doi.org/10.1093/chromsci/bmu108

Tang W, Li G, Chen B et al (2017) Evaluating ternary deep eutectic solvents as novel media for extraction of flavonoids from *Ginkgo biloba*. Sep Sci Technol 52:91–99. https://doi.org/10.1080/01496395.2016.1247864

Tanna B, Mishra A (2018) Metabolites unravel nutraceutical potential of edible seaweeds: an emerging source of functional food. Compr Rev Food Sci Food Saf 17:1613–1624. https://doi.org/10.1111/1541-4337.12396

Tian H, Wang J, Li Y et al (2019) Recovery of natural products from Deep Eutectic Solvents by mimicking denaturation. ACS Sustain Chem Eng 7:9976–9983. https://doi.org/10.1021/acssuschemeng.9b01012

Tiwari BK (2015) Ultrasound: a clean, green extraction technology. TrAC Trends Anal Chem 71:100–109. https://doi.org/10.1016/j.trac.2015.04.013

Treutter D (2006) Significance of flavonoids in plant resistance: a review. Environ Chem Lett 4:147–157. https://doi.org/10.1007/s10311-006-0068-8

Tsao R (2010) Chemistry and biochemistry of dietary polyphenols. Nutrients 2:1231–1246. https://doi.org/10.3390/nu2121231

van Osch DJGP, Parmentier D, Dietz CHJT et al (2016) Removal of alkali and transition metal ions from water with hydrophobic deep eutectic solvents. Chem Commun 52:11987–11990. https://doi.org/10.1039/C6CC06105B

Vanda H, Dai Y, Wilson EG et al (2018) Green solvents from ionic liquids and deep eutectic solvents to natural deep eutectic solvents. Comptes Rendus Chim 21:628–638. https://doi.org/10.1016/j.crci.2018.04.002

Vaya J, Mahmood S, Goldblum A et al (2003) Inhibition of LDL oxidation by flavonoids in relation to their structure and calculated enthalpy. Phytochemistry 62:89–99. https://doi.org/10.1016/S0031-9422(02)00445-4

Velmurugan BK, Rathinasamy B, Lohanathan BP et al (2018) Neuroprotective role of phytochemicals. Molecules 23:2485

Ventura SPM, Silva FA, Quental MV et al (2017) Ionic-liquid-mediated extraction and separation processes for bioactive compounds: past, present, and future trends. Chem Rev 117:6984–7052. https://doi.org/10.1021/acs.chemrev.6b00550

Vieira V, Prieto MA, Barros L et al (2018) Enhanced extraction of phenolic compounds using choline chloride based deep eutectic solvents from Juglans regia L. Ind Crops Prod 115:261–271. https://doi.org/10.1016/j.indcrop.2018.02.029

Volf I, Ignat I, Neamtu M, Popa V (2014) Thermal stability, antioxidant activity, and photo-oxidation of natural polyphenols. Chem Pap 68. https://doi.org/10.2478/s11696-013-0417-6

Wan Mahmood WMA, Lorwirachsutee A, Theodoropoulos C, Gonzalez-Miquel M (2019) Polyol-based Deep Eutectic Solvents for extraction of natural polyphenolic antioxidants from *Chlorella vulgaris*. ACS Sustain Chem Eng 7:5018–5026. https://doi.org/10.1021/acssuschemeng.8b05642

Wang T, Jonsdottir R, Olafsdottir G, Kristinsson HG (2014) Antioxidant properties of marine macroalgae. In: Kristinsson HG (ed) Antioxidants and functional components in aquatic foods. John Wiley & Sons Ltd, Chichester, pp 283–317

Wang T, Jiao J, Gai Q-Y et al (2017) Enhanced and green extraction polyphenols and furanocoumarins from Fig (Ficus carica L.) leaves using deep eutectic solvents. J Pharm Biomed Anal 145:339–345. https://doi.org/10.1016/j.jpba.2017.07.002

Wang H, Ma X, Cheng Q et al (2018a) Deep Eutectic Solvent-based microwave-assisted extraction of Baicalin from *Scutellaria baicalensis* Georgi. J Chem 2018:1–10. https://doi.org/10.1155/2018/9579872

Wang H, Ma X, Cheng Q et al (2018b) Deep Eutectic Solvent-based ultrahigh pressure extraction of Baicalin from Scutellaria baicalensis Georgi. Molecules 23:3233. https://doi.org/10.3390/molecules23123233

Wang G, Cui Q, Yin L-J et al (2019) Efficient extraction of flavonoids from Flos Sophorae Immaturus by tailored and sustainable deep eutectic solvent as green extraction media. J Pharm Biomed Anal 170:285–294. https://doi.org/10.1016/j.jpba.2018.12.032

Wei Z, Qi X, Li T et al (2015a) Application of natural deep eutectic solvents for extraction and determination of phenolics in Cajanus cajan leaves by ultra performance liquid chromatography. Sep Purif Technol 149:237–244. https://doi.org/10.1016/j.seppur.2015.05.015

Wei Z-F, Wang X-Q, Peng X et al (2015b) Fast and green extraction and separation of main bioactive flavonoids from Radix Scutellariae. Ind Crops Prod 63:175–181. https://doi.org/10.1016/j.indcrop.2014.10.013

Williams RJ, Spencer JPE, Rice-Evans C (2004) Flavonoids: antioxidants or signalling molecules? Free Radic Biol Med 36:838–849. https://doi.org/10.1016/j.freeradbiomed.2004.01.001

Wright JS, Johnson ER, DiLabio GA (2001) Predicting the activity of phenolic antioxidants: theoretical method, analysis of substituent effects, and application to major families of antioxidants. J Am Chem Soc 123:1173–1183. https://doi.org/10.1021/ja002455u

Xia B, Yan D, Bai Y et al (2015) Determination of phenolic acids in Prunella vulgaris L.: a safe and green extraction method using alcohol-based deep eutectic solvents. Anal Methods 7:9354–9364. https://doi.org/10.1039/C5AY02035B

Xu M, Ran L, Chen N et al (2019) Polarity-dependent extraction of flavonoids from citrus peel waste using a tailor-made deep eutectic solvent. Food Chem 297:124970. https://doi.org/10.1016/j.foodchem.2019.124970

Yan M-M, Liu W, Fu Y-J et al (2010) Optimisation of the microwave-assisted extraction process for four main astragalosides in Radix Astragali. Food Chem 119:1663–1670. https://doi.org/10.1016/j.foodchem.2009.09.021

Yang B, Zhao M, Jiang Y (2008) Optimization of tyrosinase inhibition activity of ultrasonic-extracted polysaccharides from longan fruit pericarp. Food Chem 110:294–300. https://doi.org/10.1016/j.foodchem.2008.01.067

Yang F-X, Xu P, Yang J-G et al (2016) Efficient separation and purification of anthocyanins from saskatoon berry by using low transition temperature mixtures. RSC Adv 6:104582–104590. https://doi.org/10.1039/C6RA22912C

Yang D, Wang Y, Peng J et al (2019) A green deep eutectic solvents microextraction coupled with acid-base induction for extraction of trace phenolic compounds in large volume water samples. Ecotoxicol Environ Saf 178:130–136. https://doi.org/10.1016/j.ecoenv.2019.04.021

Yao X-H, Zhang D-Y, Duan M-H et al (2015) Preparation and determination of phenolic compounds from Pyrola incarnata Fisch. with a green polyols based-deep eutectic solvent. Sep Purif Technol 149:116–123. https://doi.org/10.1016/j.seppur.2015.03.037

Yılmaz GG, Pinchetti JLG, Cifuentes A et al (2019) Comparison of extraction techniques and surfactants for the isolation of total polyphenols and phlorotannins from the brown algae Lobophora variegata. Anal Lett 52:2724–2740. https://doi.org/10.1080/00032719.2019.1597878

Yoo DE, Jeong KM, Han SY et al (2018) Deep eutectic solvent-based valorization of spent coffee grounds. Food Chem 255:357–364. https://doi.org/10.1016/j.foodchem.2018.02.096

Yoshie Y, Wang W, Petillo D, Suzuki T (2000) Distribution of catechins in Japanese seaweeds. Fish Sci 66:998–1000. https://doi.org/10.1046/j.1444-2906.2000.00160.x

Zainal-Abidin MH, Hayyan M, Hayyan A, Jayakumar NS (2017) New horizons in the extraction of bioactive compounds using deep eutectic solvents: A review. Anal Chim Acta 979:1–23. https://doi.org/10.1016/j.aca.2017.05.012

Zakaria SM, Kamal SMM, Harun MR et al (2017) Subcritical water technology for extraction of phenolic compounds from Chlorella sp. microalgae and assessment on its antioxidant activity. Molecules 22:1105. https://doi.org/10.3390/molecules22071105

Zhang H, Tang B, Row K (2014) Extraction of catechin compounds from green tea with a new green solvent. Chem Res Chin Univ 30:37–41. https://doi.org/10.1007/s40242-014-3339-0

Zhao B-Y, Xu P, Yang F-X et al (2015) Biocompatible Deep Eutectic Solvents based on choline chloride: characterization and application to the extraction of Rutin from Sophora japonica. ACS Sustain Chem Eng 3:2746–2755. https://doi.org/10.1021/acssuschemeng.5b00619

Zhou B, Wu L-M, Yang L, Liu Z-L (2005) Evidence for α-tocopherol regeneration reaction of green tea polyphenols in SDS micelles. Free Radic Biol Med 38:78–84. https://doi.org/10.1016/j.freeradbiomed.2004.09.023

Zhou P, Wang X, Liu P et al (2018) Enhanced phenolic compounds extraction from Morus alba L. leaves by deep eutectic solvents combined with ultrasonic-assisted extraction. Ind Crops Prod 120:147–154. https://doi.org/10.1016/j.indcrop.2018.04.071

Zhuang B, Dou L-L, Li P, Liu E-H (2017) Deep eutectic solvents as green media for extraction of flavonoid glycosides and aglycones from Platycladi Cacumen. J Pharm Biomed Anal 134:214–219. https://doi.org/10.1016/j.jpba.2016.11.049

Zubeir LF, van Osch DJGP, Rocha MAA et al (2018) Carbon dioxide solubilities in decanoic acid-based hydrophobic Deep Eutectic Solvents. J Chem Eng Data 63:913–919. https://doi.org/10.1021/acs.jced.7b00534

Index

S. Fourmentin et al. (eds.), *Deep Eutectic Solvents for Medicine, Gas Solubilization and Extraction of Natural Substances*, Environmental Chemistry for a Sustainable World 56, https://doi.org/10.1007/978-3-030-53069-3